THE LOGIC OF THE ARTICLES IN
TRADITIONAL PHILOSOPHY

SYNTHESE HISTORICAL LIBRARY

TEXTS AND STUDIES IN THE HISTORY OF

LOGIC AND PHILOSOPHY

VOLUME 10

E. M. BARTH

THE LOGIC
OF THE ARTICLES
IN TRADITIONAL
PHILOSOPHY

A Contribution to the Study of Conceptual Structures

D. REIDEL PUBLISHING COMPANY

DORDRECHT-HOLLAND / BOSTON-U.S.A.

280255

DE LOGICA VAN DE LIDWOORDEN IN DE
TRADITIONELE FILOSOFIE
First published by Universitaire Pers, Leiden 1971
Translated from the Dutch by E. M. Barth and T. C. Potts

Library of Congress Catalog Card Number 73-94452

ISBN 90 277 0350 7

Published by D. Reidel Publishing Company,
P.O. Box 17, Dordrecht, Holland

Sold and distributed in the U.S.A., Canada and Mexico
by D. Reidel Publishing Company, Inc.
306 Dartmouth Street, Boston,
Mass. 02116, U.S.A.

Printed in The Netherlands by D. Reidel, Dordrecht

To the memory of J. B. Barth

To H. A. J. F. Misset

TABLE OF CONTENTS

PART 2. HISTORICAL SURVEY

PART 3. DESCENT

PART 4. ASCENT

PREFACE

When the original Dutch version of this book was presented in 1971 to the University of Leiden as a thesis for the Doctorate in philosophy, I was prevented by the academic *mores* of that university from expressing my sincere thanks to three members of the Philosophical Faculty for their support of and interest in my pursuits. I take the liberty of doing so now, two and a half years later.

First and foremost I want to thank Professor G. Nuchelmans warmly for his expert guidance of my research. A number of my most important sources were brought to my attention by him. During the whole process of composing this book his criticism and encouragement were carried out in a truly academic spirit. He thereby provided working conditions that are a *sine qua non* for every author who is attempting to approach controversial matters in a scientific manner, conditions which, however, were not easily available at that time.

In a later phase I also came into contact with Professors L. M. de Rijk and J. B. Ubbink, with both of whom I had highly stimulating discussions and exchanges of ideas.

The present edition contains some entirely new sections, *viz.* I-9, IV-29, V-9, V-20, VII-14 (iii), (iv), VII-17 (i), VIII-22, IX-17, IX-19, X-9 and XI-8. Section X-9 was inspired by a remark made by Professor A. Heyting about the desirability of formalizing philosophical arguments and hence the logics upon which they rest. Drs. E. C. W. Krabbe convinced me that SiP-SoP syntax is a quite distinct variant of traditional syntax. Section XI-8 is reprinted from my contribution to *Semantics and Communication* (ed. by C. H. Heidrich) with the permission of the publishers. The rest of the book is substantially the same, but the material has undergone some revision and a certain amount of re-arrangement and further sub-division in the course of translation.

Unless otherwise stated, all translations of quotations from works written in languages other than English are my own, partly because the authorized translations into English of the works I have discussed were not readily available or were non-existent. For the convenience of English-speaking readers I have, nevertheless, included the titles of translations in the bibliography.

It was most fortunate for me that Dr. T. C. Potts of the University of Leeds offered to work through the whole translation of this book in order to check my own effort at writing English. It is hard indeed to say what I would have done without this generous offer. I have also benefited from several talks with him and from many a valuable suggestion and critical remark. I am also grateful to my students B. J. Leideritz, J. B. M. van Rijen, J. S. M. Savenije and H. Visscher for critical remarks; to F. H. H. Schaeffer who compiled the name index and the subject index in this edition; to F. J. Veltman for adapting the bibliography to the present edition; and to W. van Wely for assistance in proof-reading.

Utrecht, May 1974 E. M. BARTH

PREFACE TO THE ORIGINAL EDITION

Two persons whose names occur but sporadically in this book should be mentioned here, for my indebtedness to them is great: Professor Arne Næss and Professor Leonard Linsky. Professor Linsky introduced me to the problems connected with the logical analysis of the use of articles when he was a visiting professor at the University of Amsterdam during the academic year 1960–61. My intellectual debts to my first philosophy teacher, Professor Arne Næss, will be recognized by any reader who is acquainted with his efforts at promoting logical attitudes in discussions and who knows how he has always stressed the importance of seeking optimally clear formulations for the benefit of human communication in general.

It is not without a certain diffidence that I put before the academic world a book in which use is made of thoughts and conclusions set forth by my second philosophy teacher, E. W. Beth, whose untimely death is regretted by all his students. The profusion of ideas, theses, discoveries and suggestions in his works provide an inexhaustible source of material for further investigations for all of us.

I have drawn considerable inspiration from numerous talks with students from various faculties of the University of Amsterdam who, like myself, are convinced of the importance of contemporary logic as a social instrument and of the necessity of extending philosophical research with this object in mind. My thanks are due to all of them.

In the process of making the manuscript ready for press I have been assisted in many ways, and in an expert manner, by Mr. E. C. W. Krabbe. His suggestions and critical remarks have been of great value. The English translation of the last chapter, included in this book under the title *Summary and Conclusion*, was made by Drs. I. van Dam van Isselt.

I thank The Netherlands Organization for the Advancement of Pure Research (Z.W.O.) for financing the publication of this book.

Utrecht, September 1971   E. M. BARTH

ON THE USE OF SYMBOLS AND GRAPHICAL TYPES

(i) The italicized variables (without inverted commas) in this book are variables for elements of a language (syntactic variables). Propositional functions containing syntactic variables are italicized in their entirety. Quine's corners are not used. Instead of bold-faced letters as variables for the designata of the values of the italicized variables, I have used ordinary lower case letters.

Italics are furthermore employed in the titles of works discussed or referred to, and also in order to emphasize or draw special attention to an expression or statement.

A fifth and last employment of italics will be mentioned *sub* (ii) below.

(ii) Double inverted commas are used, in the first place, to point out that we are discussing the sequence of tokens between these commas and not one of its meanings. Secondly, an expression is sometimes put between double inverted commas in order to draw attention to a dubious or idiosyncratic use of language. Finally, double inverted commas are used as quotation marks. A quotation is, by definition, accompanied by a reference to the year of publication of that edition of the quoted work which I have consulted and to the page or pages on which the quoted passage is found. If in connection with an expression in a language other than English these references are not given, then this expression is (according to my prescriptive definition of "quotation") not quoted but used, and in this case the non-English expression is put in italics, unless Greek letters are used.

If a quoted passage is placed on, or starting with, a separate line on the page, then the double inverted commas as quotation marks are omitted.

In what follows, the above conventions are already employed.

(iii) Concepts and judgments are indicated by means of token sequences which are put between single inverted commas. As variables for concepts subjective or objective, and related entities (among which we shall in this book count the "objects" (*Gegenstände*) of idealistic logics), the symbols " '*S*' ", " '*P*' ", etc. are chosen, and not " 'S' ", " 'P' ", etc. This is done both in order to indicate that in the opinion of most traditional logicians there clearly belongs to every term (a term here understood as an expression) a concept (or a *Gegenstand*, or an idea, or some other non-linguistic logically relevant entity), which may be acquired by the application of some philosophical operation or other to the term in question; and also in order to point out clearly that the present study quite consciously focusses on the use of language (languages).

The philosophical operation in question is indicated by single inverted commas. Summing up (cp. II/4): for every general term M of most traditional logics it holds that the token sequence *the M* in the sense (or: in a sense) ιM refers to (names) an entity 'M' of traditional logic. It is therefore ungrammatical according to the above stipulations to write

$$\iota M = {}^{\prime}M^{\prime}.$$

An expression ιM designates or names, whatever M may be, 'M' (or still better: it designates some entity 'ιM'). It is often open to doubt which of the following symbols should be used in a certain context:

$$M, \iota M, {}^{\prime}M^{\prime}, {}^{\prime}\iota M^{\prime}.$$

The traditional theories were not worked out in such detail as to allow for a rational choice between these symbols. Where the opinions of the present author are expressed, M will not be identified with 'M'.

A distinction between lower-case Greek letters, as variables ("ι", "κ",...) or names ("λ",...) for logical degrees in traditional logics, and italicized Greek letters, as variables or names for names of logical degrees, is also carried out in the present translation of the original Dutch edition of this book.

(iv) The word "proposition" is used in the sense of "interpreted sentence" or "meaningful sentence". We may say, with Whately, that a proposition is a "judgment expressed in words" (1859/41).

(v) The use of types and inverted commas in quotations does not conform to the above stipulations but copies exactly the original text quoted. Letter spacing for emphasis in quotations from German writers has been retained to distinguish the emphasis of the original writer from that of the present author.

Fremad dog, I usle rader!
Hær af Ord!

HENDRIK WERGELAND

PART 1

THE PROBLEM

CHAPTER I

INTRODUCTION: PROBLEMS AND SOURCES

1. THE DIVIDED WORLD OF PHILOSOPHY

In our day, almost all professional philosophers who explicitly relate their thinking to some known logical theory have opted for modern logic, a science developed since the second half of the last century, in spite of all the questions which it leaves unsettled. Many logical views which were once commonplace have turned out to be untenable, and are now as a rule avoided. The so-called traditional logic, flourishing among philosophers of the nineteenth century, has practically disappeared from the scene of theoretical philosophy.

This traditional logic was never so systematically formulated as logical theories are nowadays. With very few exceptions its defenders are all dead. An idealistic version of this logic became influential and lingered on for some time in England through the works of Bradley and Bosanquet, but petered out after their deaths in the nineteen-twenties. In Germany, too, the last important works on traditional logic appeared around 1920. The range and the power of modern logic were not so easily estimated then as now, forty or fifty years later.

Roughly speaking, the philosophical world divides into two halves. One part consists of professional philosophers, some of whom have a strong inclination towards modern logic, while others are more sceptical about the scope of present theories, but all of whom have a positive attitude towards its development. On the other side we find professional philosophers who maintain that the development of so-called modern logic is philosophically completely irrelevant or even downright harmful.[1]

At present, philosophers of this latter and seemingly quite heterogeneous group keep aloof from any kind of systematic logical studies and seem to think that they can proceed in a logical vacuum. Sartre's attack on "the law of identity" (1957/33) is seen as an attack on logical thought in general wherever human conditions and problems of life are concerned. Even so, no philosopher has ever held that everything follows from anything. No-one has ever asserted that there is no limit to what may reasonably be called a consequence or conclusion from given points of view. And that is to say that no one is wholly without a logic.

As anybody may verify, both in German, French and other continental philosophical literature, as well as in English philosophical literature written in the continental philosophical tradition, very frequent use is made of the articles – "der", "die", "das" in German, "le" and "la" in French, "de" and "het" in Dutch, "-en", "-a", "-et" in Scandinavian languages and so on – that is, of the definite articles as grammarians call them. Sentences which in German would begin with a definite article are frequently translated into English by indefinite articles, and often also occur without any preceding article or other operator. In any case, the choice between a definite and an indefinite prenex article in English philosophical literature can seldom be said to be philosophically significant. Now and then a philosopher dwells upon this kind of philosophical language, like Bertrand Russell when he said in a critical discussion of Hegel's political philosophy: "The habit of speaking of "*the* State", as if there were only one, is misleading so long as there is no world State" (Russell 1947/768). Russell's view, implicit in this criticism, about what are to count as logically acceptable uses of the definite article is echoed throughout modern logic, technical variants notwithstanding: for example, in the works of Frege, and later in those of Russell, such as in his well-known *theory of descriptions.* But there is also another usage of definite articles in philosophical tradition, very appropriately called "Meinong's beard" by Angelelli, by analogy with "Ockham's razor" (Angelelli 1967/155, 183f.). Meinong's beard accommodates entities which can be referred to with the aid of terms beginning with definite articles in German, French, and many other languages, and even sometimes in English, such as 'die Sprache", "the State", "l'homme", "de hond", or "hunden". Many philosophers have worried about what one ought to understand by such expressions, but, alas, without arriving at any clear answer.

There is no analogy in modern logic to the traditional usage of articles which Angelelli refers to, but does not describe. It is certainly no accident that sentences with definite articles, or, as is often the case in English, indefinite articles, and general sentences with no prenex articles or other prenex operators at all, were found earlier much more frequently in scientific literature, too, than today. The development on this point which has taken place in the social sciences can be illustrated by the following quotation: "The emphasis has shifted from the attempt to discover the characteristics of *the* leader, to an understanding of the leader-follower relationship" (Klineberg 1954/466). Due to the development of a logic of relations by De Morgan and Peirce, from 1860 onwards, in combination with Frege's revision of the logic of "all" and "some", the language form exemplified by "the state" and "die Sprache" gradually disappeared from theoretical logic. The development in the social sciences to which Klineberg refers is not necessarily directly connected with that theoretical logic, but it could be. In any case the similarity between these developments is an interesting symptom of a generally felt need.[2]

2. THE *methodenstreit* IN THE HUMANITIES AND THE SOCIAL SCIENCES

During the last few years, the debate on problems of scientific method has once more flared up. Contrary to what happened in the nineteenth century, the *Methodenstreit* is this time restricted to the humanities and the social sciences, and – to a lesser degree – to economics, while physics, geology, botany, medicine, and mathematics nowadays remain undisturbed. The onlooker, even when he is able to disregard the hue and cry and to ignore what is purely polemical, is struck by the extreme confusion of the discussion. This is often explained as resulting from the great complexity of the problems in question and from their being at the same time so fundamental as to defy clearer formulation. Methodology, epistemology, even ontology, politics, and questions of values and their relationship to scientific thought and behaviour – all this is explicitly held to be at stake by the participants. When the subject of logic is mentioned, this is done in terms of a conflict between "formal logic", which is assumed to stand for one well-defined science, constant from Aristotle to our time except for a few changes of detail that presumably are of little or no philosophical significance, and "dialectical logic", which is likewise assumed to stand for a well-defined science, or rather, for a well-defined and completed doctrine. The actual discussion in terms of "formal" and "dialectical" logic is, from a scientific point of view, without the slightest interest; the intellectual level is the lowest thinkable since the days of Lysenko. I believe that this discussion is entirely misconceived

and that it is quite infertile to carry it on in terms of "form al" and "dialectical logic", or "dialectics".

Nevertheless, the study of logic turns out, somewhat to my surprise,
to be a key which gives access not only to two philosophical worlds, but
even to two different cultures (in C. P. Snow's sense; cp. Snow 1959).
This may sound exaggerated, but it is not entirely indefensible. It is
defended in the latest work by Veatch, himself a protagonist of a traditional logic: "Might it not be that what is decisive... is that there are two
logics operative here and not just one? Moreover, if this is so, might it
not account for the fact of our having two cultures, in Lord Snow's sense,
and not just one – two radically different modes of learning and knowledge, and therewith two radically different types of cultured human per-
sonalities?" (Veatch 1969/11). Veatch is here writing as if it were settled in
advance that both logics can be used for something or other. Only in that
case can one speak, as Veatch does, of two kinds of knowledge.

But although I disagree with Veatch about which logic is the best one,
I am strongly inclined to think that he is right in saying that basically
there are *two*, and not one, or three, or still more, theoretical logics in
current philosophies. But these logics cannot be characterized by calling
them "formal" and "dialectical", respectively. To do so is completely
misleading, both in respect of what has happened in the history of the
theory of thought, and as to what are the really important systematic issues.

When Bocheński summed up his impressions of the Tenth International Congress for Philosophy which was held in 1948, he did so in terms
of mathematical logic *vs.* phenomenology. Philosophers of quite different
and even opposed persuasions, professed Platonists as well as Aristotelians, nominalists, Kantians, and what not, were able to follow each other's
expositions and even to come to important agreements, provided they
had all adopted the modern or "mathematical" logic. Between all these
philosophers and those who based their thinking upon "the phenomenological method" there seemed to be, on the other hand, an unbridgeable
conceptual gap.[3]

I intend to show that this gap can and should be explained in terms of
modern ("mathematical") *vs.* some variant of traditional *logic*. The situation is in this respect no different from what it was in 1948. Especially on
the European continent and eastwards, traditional logic is still, in philosophical practice, very much alive. With a few exceptions, however, the

theoretical traditional logic has, one might say, "gone underground". Except for a small number of books which almost no self-respecting modern logician condescends to look at, philosophers who were themselves trained in traditional logic, or whose academic masters were so trained, do not discourse at length any longer upon logical themes. When confronted with logical problems of some technicality most of them take refuge in a discourse on the "nominalistic" or "platonistic" traits in the "ontological" bases of the schools. I shall attempt to divert the attention of the reader away from this ontological and epistemological talk and to make him focus instead on the theory of inference, especially in connection with such uses of the articles as I discussed in section 1. I shall do this by means of a detailed study of a number of publications on traditional logic. To start with, I shall introduce the reader to those twentieth-century traditional logicians and writers on logic from whose works I will draw my arguments in the chapters to come. In presenting these authors to my readers, I do not pretend to give a *complete* survey of contemporary traditional logicians – for one thing, no contemporary Spanish, Italian, east-European or Latin-American authors on traditional logic will be mentioned at all – but I do think the authors I shall introduce in the two next sections are sufficiently many and come from sufficiently different backgrounds to make the result of a study of their works on logic significant.

3. RECENT PUBLICATIONS IN TRADITIONAL LOGIC: GERMAN IDEALISM AND "DIALECTICS"

The period during which traditional theoretical logic flourished in Germany ended with the publication of a traditional-logical work by Th. Ziehen (1920). The title of this work, *Lehrbuch der Logik auf positivistischer Grundlage mit Berücksichtigung der Geschichte der Logik,* makes it clear that traditional logic was no more tied to a special philosophical school than modern logic is, although the latter does exclude philosophical schools tied too strongly to out-dated logical theories.

Pfänder. The first edition of the *Logik* of A. Pfänder appeared in 1921. Göldel has characterized Pfänder as a *Systematiker der phänomenologischen Logik.* I quote an important passage from Göldel's book: "The "phenomenological" systematics of formal logic in the conventional sense (*i.e.*, logic as the theory of concept, judgment, inference and of the highest principles), which the creator of "phenomenology", Edmund Husserl, fails to give us, is offered by Husserl's only renowned disciple in matters intrinsic to logic as a science, Alexander Pfänder. Pfänder, who in fact in several ways also has affinities with psychology, undertakes particularly to delimit unambiguously the object of logic. His philosophical position coincides with that of Husserl, *i.e.* like Husserl he does not look for the object of the science of logic in the domain of thought, or in a normalized thought, but he assumes, as Husserl did, an ideal Being of the Logical" ("ein ideelles Sein des Logischen", 1935/263f.).

Lenk calls Pfänder's work the last great exposition of the traditional Aristotelian logic (1970/188). It was re-issued in 1963 by H. Spiegelberg. In his preface to this edition Spiegelberg states: "Pfänder's "Logik" is in no need of an introductory preface even today. The inexhaustible demand for this work bears witness to its value as one of the classical texts of the phenomenological movement".

This would seem to justify a close scrutiny of Pfänder's theory of valid inference.

Von Freytag. In 1955 there appeared in Germany a book called *Logik, ihr System und ihr Verhältnis zur Logistik,* by Bruno Baron von Freytag Löringhoff (1961). The meaning of this title becomes apparent as soon as one discovers that the author calls modern logic by the name of "Logistik" and reserves "Logik" for what he calls "classical" logic.[4] Contemporary logical theories were called "logistic" by their adherents up to the thirties, but since the second world war this term has seldom been used by logicians.[5] Nowadays one generally speaks of traditional logic, sometimes called classical logic, and of modern logic.

The author of that book, von Freytag Löringhoff, endeavours to give an exposition of the logic he advocates and at the same time to carry on a polemic with adherents of contemporary logic. His stress on the Aristotelian system of syllogisms, which amounts to only a small part of the modern science of logic,[6] opens no new perspectives, however, and his insistence on an intensional approach is not accompanied by a fully developed intensional logical theory. Anyone who is interested in carefully elaborated intensional approaches to logic will do better to read the works of Rudolf Carnap, Alonzo Church, Jaakko Hintikka, Henry S.

Leonard, Richard Martin, Richard Montague, Willard Van Orman Quine, to mention only a few. Von Freytag refers to two older works on traditional logic, *viz.* those of Erdmann (1907) and Pfänder *(op. cit.)*. So far as positive suggestions for an alternative system of logic are concerned, the work of von Freytag turns out to be a disappointment. There will always be people who cling to falling stock, but as long as they operate alone they are of little importance. The question then arises whether von Freytag is representative of a larger group of philosophers.

When I started to write this book, no other *post-war* defender of traditional logic apart from von Freytag was known to me who had tried to explain his logical theories in any detail. I therefore sat down to read his book closely.

It turns out that von Freytag Löringhoff gives an important role in his logic to terms of the form *der M, das M, die M* with a substantive or adjective *M* in the singular, and to the concepts that are concealed behind them, but on the other hand he does not make clear in his book what is the logic of such expressions. This is very peculiar. There is a truly striking difference on this point between his attitude and that of Bertrand Russell, who in his *Introduction to Mathematical Philosophy* announces: "... in this chapter we shall consider the word *the* in the singular, and in the next chapter we 'shall consider the word *the* in the plural. It may be thought excessive to devote two chapters to one word, but to the philosophical mathematician it is a word of very great importance; like Browning's Grammarian with the enclitic δε, I would give the doctrine of this word if I were 'dead from the waist down' and not merely in prison" (1963/167). It is not difficult to see that the theory of "the" is of no less importance for other scientific and philosophical disciplines, for instance for the social sciences and for ethics, than for the philosophy of mathematics.

Von Freytag's silence about the logic of the articles contrasts sharply with the attitude of Russell and other modern logicians in this matter. His fellow-countryman Heinrich Scholz also observed that traditional logicians have never stated their account of the logic of the articles; his opinion is succinctly expressed and deserves mention here: "And it is not only freedom from contradiction which was taken seriously for the first time in logistics, but also other things that are of importance for drawing conclusions correctly, and which were completely neglected in the classical logic. I am thinking of the analysis of the *definite article* in expressions of the form "The so-and-so". To provide an accurate logical analysis of these expressions is a vital preparation for the formulation of correct rules of inference. To this day [1936, E.M.B.], classical logic has not even suspected that there is a considerable task here, which falls within the competence of the logician" (1967/68f.).

Albrecht on ekthesis. Von Freytag occasionally refers to a monograph by W. Albrecht, called *Die Logik der Logistik*. Von Freytag assumes that this author as well as Veatch has the same logic ("Logik") in mind as he himself. Albrecht's monograph was published in 1954, that is to say two years before Beth's article *Über Lockes 'Allgemeines Dreieck'*, to be discussed below. Beth probably never knew about Albrecht's essay. This is to be regretted, because Albrecht connects his attacks upon modern logic precisely with that problem of *ekthesis* which according to Beth lies behind the problem of "the general triangle". Since neither Pfänder nor von Freytag mention ekthesis, Albrecht's monograph is a welcome supplementation, upon which we shall draw in Chapter X.

Jacoby. The first edition of von Freytag's *Logik* appeared in 1955. In 1962 he was supported in a fierce polemical work by Günther Jacoby: *Die Ansprüche der Logistiker auf die Logik und ihre Geschichtsschreibung* (Jacoby 1962). This is in every respect an attempt to justify the characterization of logic as a doctrine of identity which we found in the work of von Freytag. Jacoby had already drawn his sword against modern logic in Jena in 1951 (Protokoll/44f.), followed by von Freytag the year after (Symphilosophein/161f.).

Jacoby, his comrade-in-arms, speaks like von Freytag about *Logik* as opposed to *Logistik*. Among the students of *Logik* he reckons the following thinkers: Aristotle, Galen, Boethius, Abelard, Albert the Great, Thomas Aquinas, Duns Scotus, William of Ockham, Arnauld, Nicole, Pascal, Geulinx, Leibniz, Wolff, Kant, Herbert, Bolzano, Hamilton, Mill, Lotze, Sigwart, and Wundt (1962/132). It is noteworthy that neither he nor von Freytag mentions the name of De Morgan although he wrote extensively on the nature of the syllogism.[7] That they do not, is interesting in the light of the fact that De Morgan was one of the pioneers whose work led to the acceptance of relational propositions as elementary (atomic) categorical propositions in predicate logic. Anyone who does not want to take this step will probably be inclined to minimalize the importance of De Morgan's work.

Carnap once said that traditional logic was ailing and anaemic (1929/1), probably with nineteenth-century logic primarily in mind. But Jacoby does not accept this judgment. The *Logik* is not at all anaemic or ailing, he thinks, nor has this been the case during the last hundred years in the works of "J. St. Mill, Lotze, Sigwart, Ueberweg, Erdmann, Ziehen, Pfänder, von Freytag" (*loc. cit.*).

But here we must be on our guard, lest from the words of Jacoby we draw the conclusion that in matters of logic Mill, Lotze and Sigwart held and published views which in broad outline coincide with those of Jacoby and von Freytag. As we shall see later on this is by no means the case (cp. IV-26 and VII-27, 28). Von Freytag's characterization of *Logik* as a doctrine of identity points in directions far away from the logical work of these three students of logic. The situation is therefore quite different from what Jacoby and von Freytag make it appear to be. They have little reason to suggest that Mill, Lotze and Sigwart, together with Aristotle and many others, would have been antagonistic towards the present development in logic if they had been able to take cognizance of it in its full panoply. One could maintain just as plausibly that if these enthusiastic logicians from the past were all of a sudden to be brought alive again, then they would show great interest in modern logical views and activities and would put their talents to the further scientific development of this discipline.

I prefer to leave both of these counterfactual conditionals right there, since it is still a much debated open problem how propositions of this kind should be evaluated.

Bloch. In Jena Jacoby was seconded by the well-known "dialectical" philosopher Ernst Bloch, who stressed the positive relationship between the "good old formal logic, which was conscious of its own limits" and "dialectics". The good, old, formal logic is the Aristotelian, or perhaps rather the later traditional logic. "Thus the traditional ["überlieferte"] formal *logic* with its oldest 'law of thought' [the Law of Non-Contradiction, E.M.B.], which was formulated by Aristotle, is entirely linked up with dialectic" (Protokoll/40). This explicit preference for traditional to modern logic by an adherent of Hegelian or "dialectic" logic is of great interest. Unfortunately Bloch has not, to my knowledge, made any attempt at further clarification of his ideas on logic as von Freytag has done, nor of his ideas on the history of logic, in the manner of Jacoby.

Bolland. In The Netherlands, the only influential traditional logics published in this century were written either by Hegelians or by neo-Thomists. Among the first, G. J. P. J. Bolland has been tremendously influential and his teachings are presently undergoing a revival. *J. Hessing* was one of his pupils. I shall mention their logic-texts only very occasionally. They do not usually offer any particular clarification of the problems with which earlier logicians were confronted; it is more revealing to study Hegel's own logic.

Eley. A good example of how many philosophers of a phenomenological bent try to escape from anything that looks like a technical logical question is offered by Eley's recent resort to "meta-criticism" of "the" formal logic (Eley 1969). Eley points out that "the difference between *name* and *description* is not at stake in this work" (372), and that "a confrontation with Russell's *Theory of Descriptions* and the linguistic school will not be attempted" (28 n.). "The problem of universals cannot be treated in a purely formal manner" (45). "The *formal structure* [*"formale Ausbau"*] of logic was not at issue in this discussion, only its *philosophical* basis" (375).

Eley's book is a plea for a phenomenological "foundation" for *modern* logic, which he does not want to reject. This foundation turns out, however, to consist in a traditional vernacular of subject and predicate (104, 175, 190 – a complete misunderstanding of Frege!), of "the copula as a relation of identity" (373), "identity in difference" (374) and "the contradictory nature of the immediate world "("Lebenswelt"; 284). Throughout his book he refers both to Hegel and to Husserl as the positive sources of his own thought. He is impressed by the dialogical turn given to logic by Lorenzen, but cannot accept Kamlah and Lorenzen's distinction between identity and the relation of predication (373f.). His "transcendental criticism" of modern logic consists in the subordination of modern logic to traditional logic, which assumes the role of its "philosophical basis".

4. Recent Publications: Neo-Thomistic Logic. What-Logic and Relating-Logic

Veatch. Apart from these German authors, the American philosopher Henry B. Veatch advocates a return to a traditional logic in his book *Intentional Logic* (1952a) and in a number of articles, published in the nineteen-fifties, which were known to von Freytag when he published his *Logik*. Von Freytag holds (1961/212), that Veatch, starting from a "realistic" metaphysical point of view, in fact brings forward the same arguments against "logistic pretentions" as he himself does, but that Veatch does not explicitly describe the logic which he wants to defend. The most recent work by Veatch: *Two Logics. The Conflict between Classical and Neo-Analytic Philosophy* (1969) contains no exposition of his logic either. In the first of his articles (Veatch 1950), however, it is made clear that he is an adherent of the logic of John of St. Thomas (1589 – 1644). It then becomes interesting to note that in the opinion of von Freytag, Veatch's logic is essentially the same as his own (*loc. cit.*).

Veatch takes logic proper to be a study of secondary *intentiones*; this is also brought out by the title of his first book. He reproaches contemporary logicians for not occupying themselves with *entia rationis* but only with *entia,* and consequently not with secondary but only with primary *intentiones* (1950/53f., 1952b/*passim*). This is in our opinion quite wrong: modern logicians deal with secondary as well as with primary *intentiones*, an opinion which is shared by Angelelli (1967/199). The study of modal and other intensional logics since the 1920s offers strong support for this statement, and there are also other important arguments (cp. X-12). But this question is of less importance for our present discussion. Since later on we shall have occasion to deal mostly with idealist philosophers it is of importance to note that Veatch's logic is, in his own opinion, not idealistic, but realistic (1953/197f.).[8]

In his most recent work, Veatch maintains that there are two different logics in circulation (Veatch 1969). He calls them respectively "what-logic" and "relating-logic", the former being the "classical" logic.

Veatch regards the traditional logics as in principle essentialistic; in his terminology, a *what* is the same as an essence (277) and a substance (44). He blames the practitioners of modern logic, the relating-logic, for being incapable of answering such questions as: "What is Quine?", "What is a modern logician?"[9] These are questions about the essence or substantial nature of the individual man W.V.O. Quine and of *the* modern logician respectively; or, to put it in another way, of *a* modern logician in general ("as such") ....

Quine is one of the logicians who, for fear of individual "Aristotelian" essences, have rejected quantificational modal logic, a field of research that originated with the work of R. Barcan Marcus (Quine 1963/156, 1966/175f.). Veatch's example is therefore remarkable, like some of his other examples (cp. VII-30); but it may nevertheless have been chosen unintentionally, for he gives the impression of never having heard of modal logic.

Possible answers to the questions he takes as examples are formulated by Veatch in sentences beginning with an indefinite article (68):

> A modern logician is a formalist,
> A modern logician is a man who doesn't read Aristotle and doesn't care,
> A modern logician is a rigorist.

This shows clearly that one cannot hope to treat the philosophico-linguistic problem of the logic of the articles adequately without discussing the relation between traditional and modern logic in general.

It is noteworthy that in this most recent book by Veatch, the last in a series of books and articles in defence of an essentialist traditional logic, there is no attempt at a description of such a logic, nor, in particular, of its theory of inference. Although he obviously rejects relational logic, and hence polyadic predication, Veatch would be prepared to give up the traditional *S is P*-form of essentialist propositions if somebody knew another and better propositional form which would also be adequate for other than Indo-European languages (65). But this does not get us any further. An exposition of the logic to which he adheres is nowhere to be found in his works. Our hope that his last book might contain a supplementation of the extremely incomplete expositions given by von Freytag was in vain. More than two decades ago this hope, that Veatch or someone else would "found a reactionary and *articulate* school, dedicated to clarifying the issues and bringing them to a head", was expressed by Clark (1952/96). So far, Veatch has disappointed our hopes. Equally disappointing in this respect is Wick (1952). Since Pfänder, only von

Freytag and to a lesser extent also Albrecht have shouldered this task.

Maritain. The French philosopher J. Maritain long ago introduced the expressions "la Logique de l'inhérence ou de la prédication" and "Logique de la Relation" (1933/121). Maritain's textbook on logic, the *Petite logique*, was first published in 1920 and went through almost twenty editions. It has been, and still is, very influential on the continent. This book will be one of our most important modern sources for traditional logical theory.

Maritain refers again and again to the works of John of St. Thomas.[10]

It is likely that Veatch owes his distinction between a what-logic and a relating-logic to Maritain. Veatch's what-logic is clearly Maritain's logic of inherence. He refers explicitly to Maritain several times in his first book (Veatch 1950).

Van den Berg. In the years after their publication, the two textbooks by the Dutch author I. J. M. van den Berg (1946, 1952) were frequently read by philosophers and theologians in the Netherlands. Van den Berg, too, refers to Maritain as one of his sources. Below, I shall only sporadically have occasion to mention van den Berg's books, which are very untechnical even compared to the work of Maritain.

5. "PURE LOGIC" AS THE THEORY OF IDENTITY

Philosophers who are antagonistic towards the development of modern logic, with the sole exception of von Freytag, no longer have the courage to expound publicly the logic to which they adhere. It is therefore of importance to point out right away that von Freytag in his book (1961) uses a number of expressions which are seldom found elsewhere in contemporary logical literature, and that he seems to attribute a meaning to certain other expressions which differs from that which is usual nowadays. Some of his expressions we have already met with. I shall list these and also a number of other characteristics of his logic, from which it will immediately become clear that there is more to his endeavour than just an attempt at a rehabilitation of certain syllogistic rules:

(1) He refers to his logic by means of the term "reine Logik" ("pure logic").[11]

(2) He describes it as a doctrine or theory of identity.[12]

(3) The most important logical relation in his system is a relation which he calls "Teilidentität" ("partial identity").

(4) This logical relation of partial identity is founded in an ontological relation which he calls "Teilhabe" ("participation").

(5) He assumes *degrees* of abstraction and concreteness.

(6) He obviously considers 'der Mensch' as an example of a special kind of concept; such concepts may occur as subjects of judgments, which then are expressed by sentences with prenex definite articles, if the language is German.

(7) He characterizes some concepts, like for instance 'der Mensch' and 'die Rose', as *Allgemeinkonkreta,* "konkret" being a synonym in his terminology for "ontologically independent" or "ontologically self-supporting" ("ontologisch selbständig").

(8) He accepts and defends *definitio rei* and definitions *per genus proximum et differentiam specificam.*

(9) He assumes only monadic predication, *S ist P*, and seems to hold that more complicated propositions, such as relational propositions, can be reduced to combinations of propositions with monadic predicates by means of "either... or" and "if... then".[13] Nor does he want any special logic of modalities.

(10) He rejects the identification of concepts with sets.[14]
(1961/11, 15, 33, 39, 26, 38, 26, 49f., 82f., 44, 189).

It is remarkable that those philosophers who oppose the modern developments and who in so doing use the inadequate argument that a philosophical concept is not a class or set, mostly do not know the theory of sets at all, which is a far more realistic theory, in the philosophical sense, than most people realize. Nor do these recalcitrant persons show the least interest in systematic investigation into intensions and their logical functions. Contemporary logical research contains a great number of widely varied approaches to and attempts at defining and describing intensional aspects of logical thought. The philosophical logician R. Carnap, also A. Church, A. Naess, and R. M. Martin have, among very many others, each in his own manner made important contributions to the field of the intensional aspects of logic, and they have done so within the framework of the modern science of logic. In these pursuits they were preceded by G. Frege and C. I. Lewis. In recent years, "philosophical logic", "modal logic", and "intensional logic" have been used as synonyms.

Apart from the difficulty of modern philosophical logical theories, a possible explanation for this aloofness may be found in the fact that those theories are certainly not aimed at the formulation of a purely "ego-centric" – consciousness-centered – intensional logic. In some way or other all these recent theories of the intension (content, connotation, comprehension, sense, or meaning) of terms and of statements are connected with the rest of logic, with predicate logic as well as with the theory of sets and relations. Intensional logic is treated either as a supplementation (and possibly modification) of this already existing theoretical nucleus, as is the case in Carnap's well-known work (Carnap 1960), or it is formulated in terms of this nucleus (R. M. Martin 1963), unless it is dealt with in an entirely empirical manner, as in some works of Naess (1953).

In this context it is worth while mentioning that von Freytag considers his logic to be neutral in the old controversy between a logic of content or comprehension and a logic of extension (*op. cit.*/43), but he nevertheless frequently talks about his logic as the logic of *das Meinen*, which we may be allowed to read as "the logic of intensions" and, as we have seen, he is not willing to treat intensions as classes or sets. Elsewhere in his book we read: "pure logic is a logic of concepts and as such primarily a logic of content" (196). On this inconsistent basis he has made a plea for what he himself takes to be a number of "classical" (10) points of view in logic which taken together presumably form a complete and ready-made system of logic. He sees no need for any improvements to or supplementations of this classical logic. "The system of this simple science" (10) is, in his own words, "the [science called] pure logic". And he maintains that this system includes those principles which "have been cultivated since Aristotle and above all in scholasticism" (*loc. cit.*).

Von Freytag here pictures traditional logic as a homogeneous science with no internal strifes. He thereby excludes from the view of his readers the possibility that traditional logic had its own problems, which were discussed by logicians throughout the centuries, he himself being the spokesman for just one of the factions rather than for all of them.

More than once he stresses that the pure logic which he is defending is a doctrine of identity. In later chapters I shall show that important traditional logicians of the last century, also in Germany, protested against those among their colleagues who maintained that logic was a theory of

identity in the sense of von Freytag. At this point suffice it to say that
logic as a doctrine of identity in von Freytag's sense is by no means necessarily opposed to the notorious logic of G.W.F. Hegel. It is true that
according to this philosophical coryphaeus a logical subject does not
simply stand in a relation of "identity" to itself, but obeys some more
complicated laws. Now, while maintaining that his logic is a theory of
identity, von Freytag does not primarily wish to say that an entity is
identical *with itself*, although this is not denied either.[15] He means that
the relation between logical subject and predicate in a true judgment
is always a kind of identity relation. And this, as we shall see, is assumed
by Hegel as well (VII-7). When someone judges that *S is P* is true, then
a *Teilidentität*, a "partial identity", obtains between '*S*' and '*P*'. "Partial
identity in this sense should be understood as the identity of a part",
he explains (33). This terminology indicates a subject-predicate relation
which is not symmetric, [16] but which is reflexive and transitive (von
Freytag *op. cit.*/30), just like the relation of inclusion between classes. In
so far as he is consistent, therefore, his logic stands in opposition to logical
systems wherein predication is not always taken to be a reflexive and
transitive relation between the contents of two concepts.

In holding that his pure logic is a doctrine of identity, then, von Freytag
defends a kind of identity theory of the copula, that is to say an identity
theory of the meaning of the word "is" in a categorical proposition *S is P*.

It is of considerable interest to observe that Veatch speaks of identity
without qualification, whereas von Freytag talks only of *partial* identity.
The 'is', the copula, in a judgment (expressed by) *x is y* he calls "the intentional relation of identity". He then goes on to ask: "What precisely is
the nature of an intentional relation? The answer we would like to suggest, albeit with some diffidence, is that this relation is always a relation of
identity... And such intentional functions... can only be effected through
this peculiar relation of logical identity... But what is this relation of x to
y in "x is y", if not a relation of identity?... the syllogism with its middle
term would appear to involve a relation of triple identity... Accordingly,
in a proposition the concept of what a thing is (the predicate) is identified
with the thing itself (the subject) – x is y –... any and every predicate
concept in an affirmative proposition is identified with its subject – x *is*
y" (1953/186–191).

Two years before Veatch's most recent book, I. Angelelli, a student of

Bocheński, published a serious study based upon thorough investigations, entitled *Studies on Gottlob Frege and Traditional Philosophy* (1967). In contrast with Veatch and von Freytag, Angelelli is not an adversary of contemporary developments. Nevertheless he, too, is of the opinion that in the traditional theory of predication, as compared to the modern theory in which various meanings of "is" are distinguished, "we have *another theory*, not merely a primitive form of a more refined one" (116). The traditional theory makes no distinction between the "is" in "Quine is a man" and the "is" in "man [a man, *der Mensch*] is [a] vertebrate". About this lack of distinction Angelelli says: "I intend to make it plausible that the absence of that distinction is to be understood in the context of *another* theory of predication where quite peculiar systematic elements are at work (essence)" (108f.). He sums up the traditional doctrine since Aristotle in the following words: "a predicate is identical with the subject when it provides the essence of the subject" (117).

I willingly subscribe to the opinion that there are two logics in existence today, a modern, non-essentialist one and also a fundamentally different traditional and essentialist logic. But, as Angelelli would certainly agree, this standpoint does not exclude in advance that from the point of view of cultural history, the relation of the traditional to the modern non-essentialist structure is comparable to that of astrology to astronomy or of alchemy to modern chemistry.[17] Of astrology as well as of alchemy it can also be said that in these theories quite peculiar systematic elements are at work.

6. THE STATUS OF ARISTOTELIAN SYLLOGISTIC FROM THE POINT OF VIEW OF CONTEMPORARY LOGIC

From the logical literature after the second world war von Freytag further mentions only A. Menne as an author who has something to say of relevance to *Logik*. He describes (11) Menne's *Logik und Existenz* (1954) as a grand rehabilitation of the classical Aristotelian logic by means of the logistic propositional and class "calculi". Menne holds, as von Freytag does, that only the syllogistic form Bamalip (Bramantip, *i.e.* AAI-4) should be removed from the system of valid classical forms in the theory of syllogisms (Menne 1954/128, von Freytag 1961/211). What von Freytag subsequently has to put on record is of no less importance: "Menne, with his great rehabilitation of classical logic, believes he has proved that the theory of inference of classical logic constitutes only a fraction of the wealth of the logical calculus" (211). That opinion is a far cry from von Freytag's own, but is quite consistent with the present state of affairs in logic. The reader will realize that this difference of opinion between Menne and von Freytag is of far greater importance than their agreement concerning Bamalip and the other three disputed syllogistic forms. It is not without importance to observe this, since it shows that von Freytag is practically the only contemporary author who not only advocates a traditional logic but who also tries to describe it, which makes it clear why it will be necessary to refer frequently to his book.

A philosopher, whether or not he considers himself a logician, cannot confine himself to the argument forms in Aristotle's theory of the syllogism, with or without Bamalip, and certainly cannot do so if in practice he uses other quantifiers and operators than those which explicitly occur in Aristotle's syllogistic. In that theory, as is well known, the only logical constants are "all" or "every", "some", "no", "is" and "not". Von Freytag oversteps the boundaries of Aristotle's syllogistic theory of forms by using the important *definite article*. Aristotle in fact occasionally uses it himself, as we shall see later on. It then becomes important to see how von Freytag treats the definite article in the framework of the syllogistic theory of inference.

Present-day logicians do not dispute the validity of Aristotle's syllogistic. Only those inference rules or forms which the Schoolmen called "Darapti", "Bamalip", "Felapton", and "Fesapo" are no longer considered as universally valid (cp. von Freytag 1961/110). The rules Darapti and Bamalip have two universal affirmative categorical premisses, *i.e.* two premisses of the form *every A is B*, and a particular affirmative categorical conclusion, *i.e.* a conclusion of the form *some A is B*. Contemporary logicians hold that inferences of this kind cannot be validly drawn on logical grounds alone,[18] since they reckon with the possibility that terms can be empty (in extension). For Aristotle, whose scientific interests were primarily in zoology, such terms were understandably enough of little or no importance. And as soon as at least the middle term of a syllogism is assumed not to be empty, Darapti will be valid. The schema Bamalip, however, is valid only on condition that the extension of the predicate *P* in the conclusion *some S is P* is not empty. This rule, therefore, can be used only if there is (in the universe of discourse) at least one thing with the property P (cp. Rescher 1964/144). The rules Fesapo and Felapton are both valid under the same condition that validates Darapti, *i.e.* on condition that the middle term M must not have an empty extension (with respect to the universe of discourse). This explains why Menne and von Freytag except only the fourth-figure form Bamalip from the set of syllogistic forms valid in Aristotelian logic (assuming the much debated fourth figure to be added to Aristotle's three). From a formal logical point of view the question is unproblematic, since we are quite clear about the conditions under which these rules can be used as valid argument forms, given the usual representation of the four categorical forms in the syntax of modern logic, and given the inference rules of "orthodox" modern logic. We do not want to contest Strawson's opinion that Aristotle had his reasons for disregarding terms with an empty extension (Strawson 1950). But we shall not consider in detail in this book the informal discussion on this question, nor the new variants of formal logic to which it has contributed, the so-called "free" or presuppositionless logics. The reader is referred to Lambert (ed.) 1969 and 1970. The reason for not going into these developments is that, in the opinion of the present writer, there are important questions about the *reine Logik* ("pure logic") of von Freytag and others which are not successfully dealt with when problems about the validity of the above-mentioned syllogistic rules, or problems about existential presuppositions in general, are taken as the point of departure. I believe that there is a much better approach to the present logico-philosophical conflict between traditional and modern logic. In Chapter III it will become clear in what this approach consists.

As we shall have reason to observe later on (cp. V-18, X-5), the unconditional, universal validity of the three rules Darapti, Fesapo and Felapton is a necessary condition for accepting the logic of von Freytag. It is not, however, a sufficient condition. At least as important for the evaluation of the relative powers of Aristotelian syllogistic and modern logic is the fact that Aristotle restricted himself to monadic predicates and also the fact that neither he nor later traditional logic offered much in the way of a propositional logic or logic of the propositional connectives. Aristotle hardly considered relational statements or statements with polyadic predicates, such as the dyadic predicate "is a brother of". Among the polyadic predicates are the important functional predicates,

which can be illustrated by the predicate "is the father of". Another and
still more important example is $x = f(t)$ where "t" is a time-variable. Such
predicates cannot be ignored when one wants to formulate statements
about many kinds of relationships between entities, for instance when one
wants to discuss properties that depend on time. In any case it is impossible to treat movement and, in general, change at all thoroughly if one
does not have access to many-place-predicates. This of course does not
imply that all kinds of relations are functional.

7. THE LOCKE-BERKELEY PROBLEM AND FREGEAN LOGIC

The ideas which led to the arguments laid down in this book have issued
partly from a study of E. W. Beth's essay: *Über Lockes 'Allgemeines
Dreieck' (On Locke's 'General Triangle')* which was published in 1956.
The main contents of that essay are also to be found in *La crise de la
raison et la logique* (Beth 1957), in his *Foundations of Mathematics* (1965/
190–194), and in the collection of some of his articles called *Aspects of
Modern Logic* (1967), where the Locke-Berkeley problem is treated in the
fourth chapter. Stated in very general terms, this problem is the problem
of the inferential relationship between individual and general statements.
More specifically the Locke-Berkeley problem concerns the following
question: what is the role, in a proof of a general theorem about the
properties of triangles, played by the individual triangle which is drawn
on paper by the person carrying out the proof? This relationship was a
serious problem for generations of scholars and philosophers throughout
the centuries. It was a problem which could be seen as a philosophical or
as a logical problem, in some more exclusive sense of "logical", according
to the taste of the philosopher. Beth maintained that this famous historical
problem is now solved.

   The Locke-Berkeley problem was a serious preoccupation of many
philosophers, among them Immanuel Kant. It can be related to epistemology and methodology and to the philosophy of mathematics, and is
indeed usually treated as an epistemological problem or as a problem in
the philosophy of mathematics. Beth's study shows, however, that the
traditional problem is also, and maybe even wholly, a logical and semantical one. The semantics and logic of the form of expression *an (arbitrary)
M* are at stake. We use expressions of this form when proving a mathe-

matical theorem with a general import, but not only then. For instance, let *M* be the predicate "triangle in a plane", and suppose that we want to prove: for all plane triangles it holds that the sum of the angles equals 180°. This is a universal proposition: *all M's are P*. One usually says that in order to prove a theorem like this, one has to prove, for an arbitrary M, that M has this property P. If and only if one is capable of demonstrating this, is the theorem proved.

Beth solves this problem, which was handed down in the tradition, in terms of quantification theory, *i.e.* the logic of the words "all" and "some" or "there is" which dates from the *Begriffsschrift* of G. Frege (1879) and later works. In our century this theory has become even more persuasive since the appearance of the so-called natural methods of deduction, including Beth's own method of semantic tableaus, which he successfully uses in order to solve the Locke-Berkeley problem. Beth, then, obviously credits the theory of quantification (as it is misleadingly called) with a great philosophical importance. He was not the only one to do that. In his preface to J. Clark's *Conventional and Modern Logic*, the philosophical logician W. V. O. Quine writes (Clark 1952/vii): "But if it is deplorable to exaggerate the cleavage between the old and the new logic, it would be more deplorable to under-estimate the novelty and importance of the new. 1879 did indeed usher in a renaissance, bringing quantification theory and therewith the most powerful and most characteristic instrument of modern logic. Logical and semantical problems with which Aquinas and others had grappled admit of simpler and clearer treatment in the light of quantification theory."

It is easy to verify, however, that many past and present philosophers do not consider the theorem about the sum of the angles in an arbitrary triangle as a universal statement, but rather as a statement of the form *the/a/an M is P*, or *le (la) M est P* in French, or *der (die, das) M ist P* in German, for instance in the following formulation: "Das Dreieck hat eine Winkelsumme von 180°", or in Dutch: "De som van de hoeken in de driehoek is 180°". Nevertheless, Beth gives no place at all to the articles in his study.

Now one of two things must be the case. Either Beth's exposition is indeed of relevance to the problem of the logic of the articles as they are often used, both in philosophy and elsewhere (especially the definite articles in continental philosophy) – although he did not himself make this point, or else his exposition is of no importance at all for the logic of the articles and hence neither for the definite articles in German, French, and Dutch philosophy nor for the "zero articles" in Eastern European philosophy. In that case one might hold that he never treated the real Locke-Berkeley problem, and that his appeal to quantification theory did not bring the solution of that problem any closer. In the second case there is a logical problem left over, which can be formulated by means of the term "a triangle", or "the triangle", or "das Dreieck": this is the problem of describing the relationship of propositions and judgments about *the* triangle ("*das* Dreieck").

It seems likely that philosophy might benefit from a quite general study of the logic of the articles in older forms of philosophy and, as far as continental western European languages were concerned, of the definite articles in particular.

8. The dialogical formulation of fregean elementary logic

Since Quine wrote the words quoted above, quantification theory has been through yet another development which is of the greatest importance from a philosophical point of view. I am referring to the dialogical formulation given by P. Lorenzen to elementary logic, to which the logic of the quantifiers belongs. Lorenzen formulates every logical problem of the type: does the statement C follow as a conclusion from the premisses $P_1, P_2, ..., P_n$?, as a dispute between two parties:

Opponent	Proponent
P_1	C
P_2	
$\vdots$	
P_n	

(dialogical tableau)

These dialogical tableaus bear a strong similarity to the semantic tableaus and even more to the deductive tableaus introduced by Beth and used by him to solve the Locke-Berkeley problem.

One might say even that these three methods are as many interpretations of one single tableau-method, and that the interpretation is expressed mainly by the headings of the two columns of the tableau. For the semantic and deductive tableaus these headings are:

True	False		Premisses	Conclusion
P_1	C		P_1	C
P_2			P_2	
$\vdots$			$\vdots$	
P_n			P_n	
(semantic tableau)			(deductive tableau)	

In recent years the eyes of more and more philosophers, especially on the continent of Europe, have been turning towards these tableaus of Beth and Lorenzen. Thus the tableau method in the dialogical version of Lorenzen is discussed in a work by H. Lenk, *Kritik der logischen Konstanten* (1968), of which I did not become aware until the present work was almost completed. Lenk especially focusses his attention on the propositional connectives ("and", "or", etc.) and on the quantifiers and their treatment in the philosophical literature from the beginning of German idealism until our own time. He does not discuss the logic of the articles at all. Apel only mentions Lorenzen's work in a note (1967/32 n. 54), but Eley, in his attempts to discuss the philosophical basis of what he calls formal logic, is more explicit and is very favourably inclined towards this dialogical point of departure.

Modern logic, which does not know that usage of the articles which can be observed in discourse about *the* triangle, has not only brilliantly stood the test of a dialogical analysis, but a dialogical formulation of this logic leads to further clarification. The philosophical world does not seem to have realized that through Lorenzen's tableaus the meanings of the word "arbitrary", in German: "beliebig", "willkürlich", which is used in expressions of the form *an arbitrary M*, have become even clearer than in earlier formulations of modern logic, and no longer conceals any secrets. Beth did not seem to realize that either; at any rate he never pointed it out explicitly. It is particularly revealing to survey the logic of this form of expression with reference to dialogical situations (cp. III-2 below) and, to the extent that yet other articles and other uses of articles might be necessary, it ought also to be possible to describe what their logic is in dialogical situations. These are sufficient reasons for focussing especially on a dialogical setting in our investigation into the logic of the articles in philosophical tradition.

9. THE CRITICAL STUDY OF APPLIED CONCEPTUAL STRUCTURES: A LOGICAL DISCIPLINE

In this section I shall try to outline briefly an important field of research: the logical foundations of philosophy. As compared to the amount of work done in the foundations of mathematics, this is a sadly neglected field. It may appropriately be regarded as a crucial part of *the study of applied conceptual structures*. It does not coincide with the study of the history of logic as commonly understood and pursued, since usually only those changes in logical theory which are regarded as valuable in one way or another are considered in any detail. In the first place, contributions to logical speculation and logical systematization are in general allotted a greater or a smaller place in a history of logic according as posterity judges them to be, or not to be, valid or otherwise useful contributions, and not according to their philosophical and other cultural influence. Old and new works on logic that are regarded as bad are shunned by practically all professional logicians, by historians as well as by theoreticians. Even the connections between various untenable, though influential, pieces of logical theory and other problems in logic are almost never discussed, to say nothing of their connection with problems of other kinds. As long as this attitude is academically dominant among logicians, logic as a discipline cannot fulfil its tasks either of cultural clarification or of cultural criticism. Second, in as far as contributions to this field of study exist, they tend to be piecemeal, often from praiseworthy modesty, but also because it is the prevailing fashion.

The discipline which I have in mind is of course closely related both to the history of ideas as conceived by Lovejoy (1936) and to the "archaeological" intentions of Foucault (1970). Other authors who have had related intentions could easily be mentioned (R. G. Collingwood, S. Toulmin). Several of them, however, present their ideas about this broad field of research into human intellectual patterns and "molecules", as Lovejoy called them, as if their own ideas about it were completely original and bore no relation to the ideas of other students of intellectual structures. This is a tendency which I strongly regret.

On the other hand, Lovejoy was not a logician, nor is Foucault. It is therefore necessary that their ideas and their work should be supplemented by logicians. Incidentally, it does not seem to have occurred to either

of them that the investigation of codified logics and also of more speculative works on logic offers an excellent entry into past and present conceptual structures, which are often there described in *optima forma*.

These structures often come into being at the very moment of their
codification, and only become culturally influential afterwards, but not
always. In any case, while Foucault speaks of the *episteme* underlying the
(various) modes of thought in a certain age, I prefer to speak of the *logos*
within which the intellectual and, often, great parts of the practical life
of an age, is confined. This *logos* may be disclosed through accurate descriptions of *logical* detail. If they are willing to take part in this investigation of already existing conceptual structures, which is not a simple one,
logicians may discover that they are at a point of vantage.

It tells us something about the lack of intellectual impact of the discipline called "history of logic" that it has not been able to prevent a number of contemporary linguists (Lakoff 1970; McCawley) from postulating
the existence of a What, called "Natural Logic", which is presumably
embedded in another What, called "Natural Language", from which, if
all goes well, Natural Logic will in due time be extracted by the efforts of
linguists. Had they read pp. 9–14 of Beth's *Foundations of Mathematics*,
this notion of Natural Logic, and perhaps even that of Natural Language,
might never have been re-instated. In any case, the whole idea of a variety
of conceptual deep structures runs counter to the spirit of modern linguistics.

For the purposes for which slogans are coined, we may very well say
that logic is nothing but *critical linguistics*. In Veatch's terminology: a
modern logician is simply a critical grammarian. It follows that the logical
study of conceptual structures and, more especially, of the logical foundations of philosophy, will be a critical one. To criticize a theoretical logic
or conceptual structure does not here mean to measure its shortcomings
against the qualities of Absolute, or of Natural Logic – it obviously
cannot mean that – but to try to assess its relative merits as compared
to those of modern logic. I have developed this notion of the relative or
comparative evaluation of logics in some detail elsewhere (1971).

The critical study of applied conceptual structures as a task for logicians was, one might say, inaugurated by Bertrand Russell. His contributions to it are found scattered in a number of his works (*e.g.*, in his 1900,
1903, and 1946 publications).

Leonard Nelson's masterly *Fortschritte und Rückschritte in der Philoso-phie (Progress and Regress in Philosophy)* may be regarded as the next great contribution. It was published posthumously by Julius Kraft in 1962. This work was one of the early sources of information and inspira-tion for my own investigations, which led to the present book.

Notwithstanding its obvious merits, Nelson's book has a certain defect that comes clearly to the fore when it is compared with another work which has the same defect, if it may be called that, but in another garb. I am re-ferring to Aebi's original and equally masterly study of Kant, which caused a heated debate on the European continent when it appeared in 1947. In each case the reader is likely to acquire a false perspective, al-though he may not learn anything which is substantially wrong and a great deal which is extremely true. Both books range too narrowly, focussing too much upon one or two philosophers in order to give an entirely realistic picture of its subject-matter. Nelson's work carries the sub-title: *From Hume and Kant to Hegel and Fries.* In his opinion, the person who is responsible for the pitiable state of logic in German philosophy is, above all, Fichte, followed by Hegel. According to Nelson the situation in theoretical philosophy was quite acceptable up to and including Kant. This may be admitted, but only in a relative sense. Aebi, on the other hand, calls her book: *Kant's Begründung der deutschen Philosophie (Kant as the Founder of German Philosophy)*, and directs her whole critical energy at Immanuel Kant who, again not unjustly, is blamed for perpetrating a *fallacy of four terms* in the heart of his philoso-phy.[19]

Is Kant, then, rather than Fichte or Hegel, *the* philosopher who must carry the blame for the total confusion resulting from the transcendental, or philosophical, or pure, logics which are still taken for granted and often openly admired by very many contemporary philosophers? A recent analysis of Kantian "transcendental" logic whose first three chapters no philosopher-logician can afford to miss (Swing 1969) may perhaps be read as a moderate corroboration of Aebi's claims in this respect (cp. Swing, *op. cit.*/52, though Swing seems wholly ignorant of her work).

It is not difficult, however, to argue that the focus of criticism and blame might just as well be pushed back several hundred years; a study of Ong's brilliant exposition of Ramism will soon convince anyone that the impact of Peter Ramus was another main source of logical decay in European thought (cp. Ong 1958).

These authors, Nelson, Aebi, and Ong, may be right in many and even in the majority of their analyses. Even so, it is easy to show that many of those features of the thought of Fichte and Hegel, or of Kant, or of Peter Ramus, against which Nelson, Aebi, or Ong have fulminated, are not only rooted in the works of Plato and of the neo-Platonists, but often in the works of Aristotle and later Aristotelians as well.[20] Upon closer inspection the important difference between Fichte and Kant, or between Ramus and Aristotle, may even become less interesting than what separates all of them from contemporary logicians. Their more specific characteristics in matters of logic turn out to be as many different answers to some very difficult problems in the construction of logical theory (that is to say, of logical syntax and theories of inference). Some of these problems are now solved, *i.e.* they have been given solutions which seem clearly superior to all the earlier ones, though others are still regarded as open (*vide* the problem of induction). It seems to me that the importance of the influence of the former problems and of the various solutions proposed is gravely underestimated.

This shows, I think, that the policy of concentrating on the logical characteristics of one or two philosophers, or of a short period of logical history, or even of a whole school, however valuable and necessary it may be as a logical preamble, is a policy that will not suffice in the long run. It must be supplemented by more problem-oriented procedures, which are likely to become diachronic.

Beth's study of the Locke-Berkeley problem is an excellent example here. Beth points out that there are systematic connexions between

(a) the problem of generalization and instantiation in the theory of the quantifiers;

(b) the method of proof by *ekthesis* which Aristotle used in the *Prior Analytics*;

(c) the opinion, held by Aristotle's commentator Alexander of Aphrodisias, that proofs of general statements in which reference is made to properties which are first attributed to an individual entity (for instance, to an individual triangle) are not *syllogistic* proofs, and hence – for Alexander – not *logical* proofs, but proofs δι' αἰσθήσεως;

(d) the scholastic doctrine of expository syllogisms: here Beth acknowledges his indebtedness to I. Bocheński; and

(e) the problem of Locke, Berkeley, and Kant about the role of the figure drawn on paper in a proof of a theorem in geometry.

To this we may add, hypothetically to start with:

(f) the logic of the articles in traditional philosophy, especially of certain uses of (possibly zero) articles which will be outlined in II-4.

A closely related cluster of problems is treated, both historically and systematically, in the previously mentioned work by Angelelli (1967), which may be regarded as one of the best recent examples of a whole book devoted to the study of the logical foundations of philosophy, and hence to the study of applied conceptual structures. His approach is, however, ontologically-semantically oriented, while in the present study I shall choose an inferential approach, in accordance with the view that the very centre of a logical system is its theory of inference. I think Angelelli underestimates the importance of the transition from monadic to polyadic (relational) predication. Furthermore, though he is interested in the problem of the uses of articles, Angelelli does not discuss ekthesis at all.

Another example is Lenk's scrutiny of a great number of philosophies of propositional form (Lenk 1968). It contains a vast amount of information, but he leaves out the category of indefinite propositions (Aristotle's ἀδιόριστοι), which are sentences with a (zero) prenex article.

The notion of ekthesis is discussed at length in Hintikka's most recent book (Hintikka 1973), in which Beth's analysis of the Locke-Berkeley problem is one of the main issues.[21] However, Hintikka does not connect the Locke-Berkeley problem and the ekthesis-procedure with the problem of the logic of the articles. I hope that the present attempt to do precisely that may be found to contain something of value.

The main text of this book will deal with certain features of some conceptual structures which are found in systematic philosophies, hence with the logical foundations of philosophy, while references to applied conceptual structures in the wider sense will be relegated to the footnotes.

NOTES

[1] British ordinary language philosophy must be placed in the first of the two groups. It is of importance to point out here that J. L. Austin translated Frege's *Grundlagen*

into English (Frege 1959), and that P. T. Geach and M. Black are responsible for the appearance of English translations of a number of Frege's philosophical writings (Frege 1960). These are good examples of the positive interest in, and knowledge of, modern logic in these circles.

[2] Augustus De Morgan wrote in his famous article *On the syllogism: IV, and on the Logic of Relations:* "... I affirm that all the difference between Aristotle or Occam and the lowest of the noble savages who ran wild in the woods is only part, and I believe a very small part, of the development of human power. If the logician could leaven his own mind with a full sense of what his foregoers did for thought and for language, a spontaneous admission would grow that if these same foregoers had worked themselves into the same familiarity with relation in general which they obtained with what I call *onymatic* relations [internal semantical relations, E.M.B.], and still more if they had cultivated those yet wider fields which lie beyond, the common language would have now possessed facilities on the want of which he founds his assertion of the sufficiency of the old logic. Though satisfied that the educated world is in advance of the current system of logic, I feel equally sure that a more extensive system would work a still greater progress".

[3] "For it can be observed that often there is a division within one and the same philosophical school, due to the employment of different methods, *i.e.* on the one side mathematical-logical analysis, on the other phenomenological procedure. ... It is most remarkable how mathematical logic is capable of bringing about an appreciative union of philosophers of the most different and even antagonistic schools: Platonists, Aristotelians, nominalists, even Kantians and some pragmaticists; while on the other hand the abyss between the adherents of this method and the adherents of phenomenology often seems so deep that no rapprochement seems possible any more at all" (quoted from Tranøy 1954).

[4] About the works of von Freytag, F. Schmidt says: "Vertritt die traditionelle Logik, steht der Identitätslogik Leibnizens sehr nahe" (in Leibniz 1960/543). Cp. the remark by Scholz (1967/46) about B. Erdmann, whose heir von Freytag seems to be in many respects. Cp. also IX n. 10.

[5] See for instance the survey of the titles of the most important works by W. V. O. Quine, on the back cover of Quine 1966: *A System of Logistic, Mathematical Logic, Elementary Logic, Methods of Logic, From a Logical Point of View....*

[6] The Aristotelian syllogistic corresponds to (two-valued) first order class logic, provided that certain assumptions of existence are added to the latter. That is to say, Aristotelian logic is limited to monadic (non-relational) predication.

[7] The logical work of De Morgan, which dates from the period from 1846 to 1868, was published in one volume only recently by P. Heath (De Morgan 1966). Cp. n. 3.

[8] The meaning of "realism" is vague; see Peirce's use of this word (Peirce 1967, IV/34).

[9] Cf. Russell 1947/66. See also Apel 1967/29f. about Skjervheim. For the role of "whats" in psychology, see Woodworth 1952/11ff.

[10] The logical writings of John of St. Thomas were re-issued in 1955 (here referred to as John of St. Thomas 1962, 1965). In a critical essay on Veatch 1950, Doyle disputes Maritain's interpretation of John of St. Thomas (Doyle 1953).

[11] The subtitle of Frege's *Begriffsschrift* also contains the expression "reines Denken". In the philosophical tradition, however, the word "pure", or "rein", has a special meaning and in this meaning these words are primarily predicated of concepts.

[12] Compare this position with the completely different view expressed by C. S. Peirce in *The Logic of Relatives* (1897): "Moreover, it must be acknowledged that the illative

relation (that expressed by "therefore") is the most important of logical relations...
It can be demonstrated that formal logic needs no other elementary logical relation
than this... It can further be shown that no other copula will of itself suffice for all
purposes. Consequently, the copula of equality ought to be regarded as merely derivative" (1967, III/198f.).

[13] Compare Kant's *Urteilstafel* in his *Kritik der reinen Vernunft* (1968/140). The Dutch
author K. Popma has not discovered the difference between the Kantian and the
modern logic of relations (1963/158), yet he is strongly antagonistic towards modern
logic because it is not sufficiently static (160), an argument one cannot very well deploy
against the logic of Kant. We will return to this matter in IX-16.

[14] In the opinion of the present author sets are concepts, and the theory of sets a
theory of concepts, but the meaning of "concept" is so extremely vague that there is
room for several kinds of concepts and theories about them, be it with different fields
of application or as mutual complementations.

[15] Von Freytag justly points out that it is a serious, if a very common mistake, to
believe that logical identity is the same as *Genidentität*, or invariability during a certain
period of time. On this point his opinion coincides with that of modern logicians.
"The law [of identity] says "A is A", not "A is always A", for it does not mention
any properties of the object, not even its temporal properties. If it is temporally determined, then the temporal data belong to it[s defining properties] and partake of
its identity. The statement does not say that the A at a given moment is identical with
the A at another moment" (1961/17). Bolzano was the first writer to stress the importance of taking references to time as going with the subject term, and not with the
predicate term as was usual in philosophy before his time; co. VII-11, VII-n. 44.

[16] Cp. Jacoby 1962/18.

[17] The reader may be interested in consulting Foucault (1970/17–30) on the Four
Similitudes and the theory of signatures in Renaissance thought.

[18] This is true providing that Aristotle's universal propositions are interpreted in this
way: $(x) [Ax \to Bx]$. In this interpretation universal categorical propositions *every A
is B*, where the term *A* has an empty extension, are always *true*. Particular categorical
propositions are then given the following formulation: $(Ex) [Ax \& Bx]$. The problem
of the validity of the disputed syllogistic forms is precisely the same problem as that
of the soundness of the principle of *conversio per accidens* (*cpa*), that is, the principle
that allows us to draw the conclusion *some B is A* from a premiss *every A is B*. If the
universal and the particular categorical forms are interpreted in this way, then *cpa* is
not a valid inference. For the universal proposition is then true and the particular
proposition false if the extension of the term *A* is empty.

The *cpa*, and with it the four disputed syllogistic forms and also the inference form
ad subalternatam, were considered to be valid by the scholastic proponents of the extensional theory of identity (VII-2). In order to be able to say that the logical subject
and predicate have identical extensions, these logicians considered it a necessary condition that both subject term and predicate term refer to something (Moody 1967b/
312). From this point of view if "the subject" *every A* has an empty extension, this
is a sufficient reason for saying that the proposition is false. Whether the extension
of the predicate is empty or not then makes no difference. Take as an example the
proposition: "All Centaurs are hoofed beings". Those schoolmen regarded a proposition like this as *false*, whereas according to present-day logicians it is a true statement when interpreted with respect to a non-mythological domain. The position of
the schoolmen implies that one can always conclude from the truth of *every [all] A is*

[*are*] *B* to the non-emptiness of the extension of *A*, hence to the truth of *some A is B* (Moody 1953/50, 1967a/531).

According to Łukasiewicz, however, the modern symbolic rendering of the categorical forms is not in agreement with Aristotle's intentions; if we assume another, hypothetical interpretation, the contested syllogistic forms, *cpa* and inferences *ad subalternatam* become valid (Łukasiewicz 1957). We shall nevertheless not discuss this hypothetical interpretation of Łukasiewicz here, because it does not throw any new light on von Freytag's "pure" logic; consequently, Jacoby rejects that interpretation (1962/72f.). For a recent contribution to the problem of the relation between syllogistic and modern logic, see Vieru 1971.

Von Freytag formulates the four categorical forms in the plural manner: "alle A sind B" etc. (1961/*passim*). It may be assumed that this is for the sake of euphony in the German language; compare IV-8.

[19] One review of her book contained the following lamentation: "What a pity that the men who might have measured their strength against hers seem to be extinct, or at least to keep silent" (R. F. Beerling in *Algemeen Nederlands Tijdschrift voor Wijsbegeerte en Psychologie* 41 (1949)).

[20] Cp. van der Meulen (1951/50) in his polemic against Aebi's book.

[21] When Hintikka's book was published, the English translation of my own book was already almost completed, which will explain why I have not discussed his arguments and conclusions here.

CHAPTER II

NAMING WHAT IS

1. A FIRST ACQUAINTANCE WITH CONCRETE UNIVERSAL TERMS

Von Freytag draws an ontological distinction between ontologically concrete and ontologically abstract things.[1] According to his usage of language, entities are ontologically concrete when they enjoy an independent manner of being; what has no independent being is ontologically abstract. Everything which is ontologically independent, or ontologically self-supporting, hence the totality of ontologically concrete entities, is said by von Freytag to be "real teilbar und abtrennbar" (1961/24; cp. Whately 1859/90f.). He does not explain, however, what "real" means, since this topic belongs to the ontology of spiritual Being ("die Ontologie des geistigen Seins", 24). Consequently we are given no criterion stating the conditions that have to be fulfilled in order that something be really divisible and separable. His conception of ontological independence therefore hangs in the air, but this need not detain us from a further investigation into the logic which he defends.

After having drawn this ontological distinction, von Freytag proceeds to another distinction, this time not based on an ontological criterion but entirely of a logical nature: the distinction between individual and general ("allgemeine") concepts. According to von Freytag, each thing, whether concrete or abstract, can occur as an individual as far as logic is concerned; this is a statement to which no one will wish to take objection. The construction ("Bauart") of a violin by Gagliano, for instance, so he remarks, can be an individual concept with respect to the use of language. As soon as he comes to logic proper, however, von Freytag considers the construction of a Gagliano as being a group of characteristics which taken together form the essence ("Wesen") of the topic under consideration (24f.).

These logical and ontological distinctions are mutually independent and so are capable of pairwise combinations. This is illustrated by von Freytag by means of the substantive "rose". Drawn up systematically his exposition is as follows (compare Angellelli 1967/12f.):

	The concept	is
(1)	'this rose'	individual, concrete
(2)	'the red of this rose'	individual, abstract
(3)	'the (a) rose' (as a general concept)	*universal, concrete*
(4)	'the nature of the rose' ("Rosennatur")	universal, abstract

A term which is being used to express a concrete universal concept we shall call "a concrete universal term".

2. THE LOGOPHORE

In this study we shall concentrate on the third type of concept distinguished by von Freytag: the universal concrete concepts or concrete universals, as they are usually called. It is noteworthy that neither he nor kindred authors are consistent in their use of articles. German, French and Dutch authors mostly put a definite article before the general term and speak in terms of "der allgemeine Begriff 'die Rose'". At other times they do not, so that "Rose" alone has to indicate what they call a concrete universal concept. In some languages in which substantives do not as a rule begin with a capital letter, one often uses a capital letter in order to indicate that the term refers to a concrete universal, in the sense of von Freytag. In English, for instance, "der Mensch" is often translated by "man" if it is taken to refer to a concrete universal. But as we have seen above, in English the indefinite article is very often also used where a German or French or Dutch or Scandinavian author would use a definite article.

In connection with his observations on so-called concrete univerals, von Freytag introduces an expression which we shall later adopt in our own exposition and which we shall use in order to formulate reasonably precise hypothetical statements about the logic which von Freytag tries to defend.

"One might say", he writes, "that such concrete universals as "rose" contain a *logophore*, an ontologically self-supporting carrier of the properties in question, which apart from this characteristic of ontological independence remains *indefinite* ["unbestimmt"], an empty place into which a suitable individual can enter, when one proceeds from the "rose" to "this rose". The logophore is the concreteness of the individual deprived of its individuality, but not of its ontological independence. The "red of this rose" [example no. 2] had no logophore, no empty place, and as only the red and not the rose carrying it is meant, it was abstract. Here [in 'the red of this rose'] a concrete individual ['this rose'] is deprived of its ontological independence, but not of its individuality" (26, translation and italics mine E. M. B.). Clearly a concept containing a logophore is a *carrier of Logos.*

One might also ask whether conversely the concept 'the rose' is deprived of its individuality, but not of its ontological independence. Von Freytag does not mention this.[2]

According to von Freytag this classification of concepts into independent and dependent is made on ontological grounds, which in his opinion are external to the science of logic.[3] There is no immediate reason for us to object to that. But it is nevertheless of great interest to know the logic which determines the inter-subjective use of concrete universal terms and also to know the conceptual background which is required for the use of such expressions. This question is all the more important as von Freytag's language, in which "die Rose" occurs as an expression of a concrete universal concept is unusual from the standpoint of modern logic and suggests an orthodox Platonism or Aristotelianism. In any case it must be based on some kind of concept-realism.

Now this is certainly no answer to the question of what the logic of such terms is, since many modern logicians also adopt a kind of concept-realism, albeit a kind that differs from orthodox Platonism and (orthodox?) Aristotelianism in important respects. It is admittedly difficult to list everything there is in the world in a manner which is sufficiently comprehensive for logical purposes, without ascribing some irreducible kind of existence, "abstract" but ontologically independent, to any classes or sets. And it will be clear that what we choose to *call* general concepts is of no importance, whether: "concrete universals", "independent general abstractions", or "sets"; these various kinds of concepts can be characterized and distinguished only by the logical rules people follow in dealing with the expressions used for naming them. In the present phase of our investigation we are, therefore, still a long way from being able to decide whether the logic of von Freytag's four kinds of concept is acceptable or not.

3. HUSSERL ON DEFINITE ARTICLES

E. Husserl is one of the best-known writers on *reine Logik*. Unfortunately Husserl never personally wrote a systematic account of what he understood as logic, but concentrated on the problem of providing an undefined logic with a philosophical foundation. With regard to his rejection of polyadic predication and a logic of relations[4] it is of importance to observe straight away that we do find in his works expressions and concepts that are clearly related to von Freytag's "logophoric" concepts and to Veatch's "whats". In his *Logische Untersuchungen*, published in 1901, Husserl distinguishes as three different *Formen*

$$ein\ A, \quad alle\ A, \quad das\ A\ \ddot{u}berhaupt,$$

and gives as examples:

$$ein\ Dreieck, \quad alle\ Dreiecke, \quad das\ Dreieck.$$

"Das Eins und das Das, das Wenn und das So, das Alle und das Kein, das Etwas und das Nichts, die Quantitätsformen und die Anzahlbestimmungen usw. – all das sind bedeutende Satzelemente" (1968 II, I/147). He explicitly draws a distinction between the meaning of "das" and of "alle": "Die Formen *das A* und *alle A* (ebenso: *irgendein A überhaupt* – gleichgültig welches) sind nicht bedeutungs-identisch; ihre Verschiedenheit ist keine "bloss grammatische" und am Ende gar nur durch den Wortlaut bestimmte. Es sind *logisch* Unterschiedene Formen, wesentlichen Bedeutungsunterschieden Ausdruck gebend" (149).

From a systematic point of view it is of little importance that the difference between '*alle A*' and '*das A*' is characterized by Husserl as a *logical* and by von Freytag as an *ontological* difference. Whether a system or a principle is called logical or ontological makes no difference to the further treatment. What is logically relevant in the expositions of both authors is above all that '*das A überhaupt*', in short: '*das A*', exists and that an expression *das A überhaupt* cannot be replaced by an expression *irgendein A*. It is also unimportant whether '*das A überhaupt*' is said to be a concept or an object ("Gegenstand"). Of primary importance is that, according to Husserl, '*das A überhaupt*' exists and is of relevance to logic. His use of a variable "*A*" indicates that Husserl took this existential statement to hold in general, whatever predicate *A* one would like to consider. This rule, if it is one, is seen to be in agreement with current usage in the works of a great number of philosophers.

I take Husserl's expression *irgendein A überhaupt – gleichgültig welches* as a synonym for the expression *an arbitrary A* ("*ein beliebiges A*"). As Husserl himself says, this last form of expression is in a sense only a variant of *alle A*. None the less, it is an expression which is of the greatest value when we want to formulate logical principles and rules.[5]

Finally, we notice that not only does a logically relevant concept '*das A überhaupt*' seem to exist, in Husserl's opinion, for every predicate *A* of every language, but it can also be referred to in the language of our choice, for instance by means of an expression *an A* or *the A*, or *A*, in English, and by *das* (*der, die*) *A* or *das A überhaupt* in German.

The definite article is also frequently used, however, in order to refer to a unique individual, as in the expression "the Prime minister of Britain", or the (empty) expression "the present King of France". Sometimes one finds the expression *the (arbitrary) A*, where *an arbitrary A* is obviously meant.[6] For these reasons it would be highly preferable to make use of one or more different signs or modes of expression wherever the definite – or, in English, the indefinite – article is not employed in either of the two latter senses but, for instance, in the sense of Husserl. This would allow us to prevent confusion where it is possible to do so.

Conversely, many "general" or "universal" entities are frequently referred to without the use of any article, a habit which contributes still further to the confusion.

4. DEFINITION OF "LOGOPHORIC USES" OF THE ARTICLES AND OF OTHER OPERATORS BY WHICH SUBSTANTIVES ARE FORMED

As a consequence of our observations in the first three paragraphs of this chapter, we shall introduce the following terminology for the purposes of the present investigation:

A proposition

(a)   *the/an M is P*

will be called "a logophoric proposition" and will be said to express a logophoric judgment *iff* the article is not used anaphorically and the proponent of (a) does not, by this proposition (a), mean

(b)   *the only entity in the domain possessing the property M, also has the property P,*

nor does he use (a) as a stylistic variant of the proposition

(c)   *all M's are P,* or *every M is P,*

nor of

(c')   *an arbitrary M is P,*

nor of

(c")   *the class (*or: *set) of M's is included in the class (set) of P's,*

nor of

(d)   *some M (M's) is (are) P,*

and if, furthermore, the proposition

(e)   *the/an M is M,*

in which *the/an M* has the same sense as in (a) and "is" is used as a copula of predication, is regarded by the proponent as (tauto)logically true.

The clause that (e) be regarded as a logical truth is the semantic correlate of that principle which was called by Beth "Plato's Principle".[7] Beth's formulation of this principle is as follows: "If any entity u has the property A, *the* corresponding absolute entity a [corresponding to the property A] has the property A" (Beth 1946; italics mine). The use of language which is based upon traditional logic is in conformity with this clause. Thus Kant, for instance, states that "Der Mensch ist Mensch" is tautological (1801/174). W. Wundt takes the same proposition, being an example of *das formal identische Urtheil,* to be an application of *A ist A,* a formula he does not distinguish from $A = A$ (1893/193). And A. Pfänder holds that *die Pflanze überhaupt* falls under the concept '*Pflanze*' (1963/145).

An expression *the/an M* will be called "a logophoric expression" when it is used in some logophoric proposition. Even if it does not occur within a proposition but is mentioned separately, we shall speak of a logophoric expression if it is mentioned with reference to possible logophoric uses of it in some proposition. An article which is a part of a logophoric expression will be said to have a logophoric use in that expression.

In order to make it clear that a logophoric meaning is intended and not one of the meanings of (b)–(d), we might replace *the/an M* by another token sequence, for instance $\imath M$,[8] to be read as: *id M*. By this choice of "id" as a verbal form for "$\imath$", we take into account the expressions "identity", "idea" and "ideal", as well as the connotations of the homonymous Latin pronoun, especially as it is used in psychoanalysis. A logophoric proposition *the/an M is P* can then, for the purpose of distinguishing it from (b)–(d), be rendered as *$\imath M$ is P*, or as $P(\imath M)$. A judgment which can be expressed only by means of a logophoric proposition will be called "a logophoric judgment".

The differences between the German articles "der", "die" and "das", or between the French articles "le" and "la", are purely grammatical, that is to say that these differences are irrelevant to a logical investigation. The same seems to hold for the English articles "the", "an" or "a" when used logophorically. We shall therefore render all logophoric uses of propositions *the M is P, an M is P, M is P, der M ist P, die M ist P, das M ist P*, and so forth, by *$\imath M$ is P*.

Everything we have said in this section so far is equally applicable, without any change whatsoever, to those languages in which no use is made of a "definite", or of an "indefinite", article.[9] The expression *M* of a language T will be said to occur logophorically in

(a′) $M \ is_T \ P,$

where "is_T" is the translation of "is" in T, if and only if (a′) may not be interpreted as (b)–(d) and the proposition $M \ is_T \ M$ is a tautology.

If we also want to characterize the language of the English philosopher Cook Wilson[10] and others as logophorical usage by means of these definitions, then it becomes necessary to add: an expression *M* with the grammatical structure *A-ness, A-heit* or a similar structure occurs logophorically in (a′) if and only if (a′) may not be interpreted as (b) or as (c) or as (d), with *A* instead of *M*, while *M is A*, that is to say the sentence *A-ness (A-heit) is A*, is regarded as meaningful and (tauto)logically true.

This last addition, that *A-ness is A* must be meaningful and true is a necessary addition if we want to prevent Quine's "squareness" from falling under our definition of a logophoric term.[11] For the proposition

(α)   squareness is a concept

is not synonymous, materially or in any other respect, with the proposition

(β)   the unique entity with the property square is a concept

(compare (b) above), nor with

(γ)   any square thing is a concept,

nor with

(δ)   some square things are concepts.

But "squareness is square" is not meaningful and true, if "squareness" means the class of all square things, nor even if it means the "heap" of all square things. The occurrence of "squareness" in (α) is therefore not logophoric.[12]

William of Ockham's opinion that "albedo est alba" is incorrect, is quite remarkable (*Summa Logicae pars prima*, Chapter 63). As we saw earlier in this paragraph, Kant, Wundt, and Pfänder expressed themselves in a manner so as to allow us to think that they would disagree with William. Similarly, G. Martin offers the German proposition "das Rechteck ist ein Rechteck" as an example of *das absolut identische Urteil* in the logic of Leibniz. Martin takes this to be an application of the formula *A est A*. It is hard to see how Leibniz, Kant, Wundt, and Pfänder could have denied that whiteness is white, given the logical truth of their own examples. And von Freytag does indeed say of 'die Rose' that this is an independent entity, "nicht nur Rosennatur, sondern etwas, was Rosennatur hat" (1961/26). In the same manner, 'der Mensch' must be something that has human nature.

The definitions in this section are entirely negative. We have left undecided, therefore, whether there are several different logophoric uses or only one. The above negative definitions delimit a logical *milieu*, within which there may very well be important differences of many kinds, including logical ones, between different authors. But from what has already been said it will be clear that anyone belonging to this logical *milieu* uses a different logic from what is nowadays usually called modern logic.

It is of importance to observe that "der Mensch ist Mensch" is true on formal logical grounds when the expression "der Mensch" does not indicate a concrete universal but is being used as a stylistic variation of "ein willkürlicher Mensch" or of "irgendein Mensch, gleichgültig welcher". In other words: *an arbitrary M is M* is tautologically true, for every noun or adjective *M*. Modern logicians do not say, however, that the term *an arbitrary M* refers to an ontologically independent entity.[5]

5. The H-thesis

Aristotle certainly did not accept (e) of the last paragraph as a logical truth for all predicates *M*, as some Platonists seem to have done. Aristotle's attitude to this principle of Platonic predication is neatly summed up in Owen's essay *The Platonism of Aristotle* (in: Strawson (ed.) 1968). "There is", Owen says of Aristotle's logic of predication, "one sort of predication that does not seem to imply the Self-predication Assumption: White is not white in the sense in which Socrates is white. But there is another sort, represented by the predication of *man*, which for convenience I shall call 'strong predication'; and this sort does *seem* to imply this Assumption" (Owen, *op. cit.* 159; last italics mine). The crucial place in the Organon is *Posterior Analytics* I 22 82a. Here Aristotle himself speaks of strong, or strict, predication. Of this kind of predication he says that the predicate signifies a substance or essence, and that "predicates which signify substance [essence] signify that the subject is identical with the predicate or with a species of the predicate. Predicates not signifying substance which are predicated of a subject not identical with themselves or with a species of themselves are accidental or coincidental; *e.g.* white is coincident to man, seeing that man is not identical with white or a species of white, but rather with animal, since man *is* identical with a species of animal" (I 22 83a 24–30).

This strongly suggests that Aristotle's strict, or essential predications are either incomplete definitions or else expressions of strict identification, and hence that strict predication obeys the principle of self-predication (reflexivity). Our hypothesis is somewhat weakened by Aristotle's subsequent assertion that in strict predication "A and B cannot be predicated reciprocally of one another". This objection can be met by the assumption that Aristotle disregarded the possibility of giving the same value to his variables "A" and "B".

In *Metaphysics Z*, Aristotle holds – contrary to his earlier view, and more in line with Plato – that the primary subjects of discourse are species, not individuals (cp. Owen, *op. cit.*/160). A great many later philosophers of whom some called themselves "metaphysicians", some simply "philosophers", have been (or are) primarily or even exclusively interested in the *logically* (rationally) *primary* subjects of discourse. Aristotle allows them to think that these are always species.

"Philosophical logic" was traditionally understood as *the logic of discourse about* the primary subjects of all discourse. That is to say, traditional philosophical logic is the logic of discourse about species and their essential predicates. Consequently little attention is paid to Aristotle's "secondary" accidental predication and to the logic of discourse about individuals, and self-predication is seen as, philosophically, the normal thing. For a student of traditional philosophical logic the influence of Aristotle's theory of predication becomes indistinguishable from that of Plato's theory.

N. Hartmann discusses the Platonism of Aristotle from an ontological point of view and reaches parallel conclusions. For the philosophers who came after Aristotle it was not necessary, he says, "to ascribe to essences a being "before the things" and "above" them; one could also conceive of them in Aristotelian fashion as substantial forms present "in the things". One thereby avoided the difficulties of a duplication of the world, but without abandoning the basic idea" (*i.e.*, the basic idea of Plato's philosophy, that there *are* "essences"; N. Hartmann 1942/204).

Hartmann clearly holds, and rightly so, that it was to a great extent assumed in the philosophical tradition that a basic entity lay concealed behind every general term. "It is certainly true," says Hartmann, "that everything which is basic is general ["allgemein"], but it does not follow that everything which is general is also basic" (*op. cit.* / 207). This criticism is valid but becomes clearer and more effective when it is formulated as a criticism of assumptions about fruitful uses of language.

Concerning von Freytag, it is not clear whether he considers *all* terms *der* (*die, das*) *M* as names for ontologically independent entities, or only some of these terms. His classifications have to do with concepts and not with terms of a language. Concerning the meaning of "concept", he says: "Alles Meinbare kann also für die Logik Begriff sein, und meinbar ist bekanntlich alles, Wirkliches wie Unwirkliches, Vorstellbares wie Unvorstellbares, Widerspruchsfreies und sogar gelegentlich Widerspruchsvolles [!]. Man kann nichts angeben, was nicht meinbar wäre, denn gäbe man es an, so hätte man es bereits gemeint" (*op. cit.* / 23). This does not yet answer our question whether he considers *all* general terms as referring to ontologically self-supporting general concepts. He makes no attempt at formulating a *rein-logisch* comprehension-axiom[13] for concrete universals, so that it might become clear which concepts may be treated as ontologically independent or "concrete", and which concepts should be regarded as abstract concepts only. Nor does one find in his book existence postulates corresponding to the axioms of existence in modern axiomatizations of set theory.

The following rules (i)–(iv) represent an attempt at summing up some very important features of a philosophical use of language which is found in the works of a great number of authors. I have especially in mind philosophers in the idealist and the Thomist traditions, and among others Hegel, Husserl, Heidegger and Habermas, in whose honour I shall speak of *the H-thesis*. It thereby should become easier for logically trained readers to make a comparison with modern logic, and, if possible, to achieve some semantic clarification. I take idealist, "dialectical" and neo-Thomist philosophers to base their thought and their use of language on the following syntactic rules:

(i)   If M is a predicate of a language T, then ιM is a logophoric term in T, *i.e.* a term which can be used logophorically.

(ii)   Every logophoric term in T is an acceptable (meaningful) term in T.

(iii)   If P is a predicate of T, and if ιM is a logophoric term in T, then the word sequence ιM *is* P is a well-formed (meaningful) sentence in the language T. This sentence may also be written in this manner: $P(\iota M)$.

From (i)–(iii) together with clause (e) in the definition of "logophoric uses" of terms it follows that:

(iv) ιM *is* (a, an) M, or: $M(\iota M)$, or: ιM *is M-like*, or: ιM *has* (or: *carries*) *the M-nature*, is a logophoric tautology of T for every predicate M of T.[14]

I shall call the conjunction of (i)–(iv) "the H-thesis".[15]

The condition that there must exist at least one thing with the property M in order that $\imath M$ *is* M may be regarded as a tautology, which is a part of Plato's Principle, as formulated, rightly or wrongly, by Beth, has deliberately been omitted from the H-thesis.

Whoever adheres to this thesis will regard *the/an M is P* as meaningful and well-formed from a logical point of view, whichever predicates M and P are chosen. We still do not know what meaning one is supposed to associate with such sentences. As was said at the end of Section 4, we have so far given only a negative definition of such judgments. For we have only said that a term *the/an M* is being used logophorically when for some purposes of reasoning and communication it may *not* be replaced by *the only M*, nor by *all M*, nor by *some M*, nor by *an arbitrary M*, if this latter expression is used in the manner of contemporary logic. In subsequent chapters we shall try to develop a more positive theory of logophoric usage. The expressions "concrete" and "abstract" will not play an important role in the working out of this theory although the expression "allgemein Konkretes" or "concrete universal" will serve us as a first indication that the author may have had a logophoric use of language in mind.[16]

6. EXAMPLES

Many traditional logicians take their examples from mathematics surprisingly often. A favourite example is the sentence: "In the triangle (or: a triangle) the sum of the angles is 180°."

Examples are also frequently chosen from the taxonomic systems of biology: "the rose is a plant", "the dog is a vertebrate", "das Neuholländische Schnabelthier besitzt Milchdrüsen" (Wundt 1893), "der Hund säuft" (Lotze 1912/80). Traditional logicians and philosophers never tire of the sentence "Man is mortal", "der Mensch ist sterblich", as an example from a human universe of discourse.

We shall choose another example, containing a psychological-anthropological predicate in order that our example may be relevant at least to philosophical anthropology (as it is called), to the social sciences, to theology, and to ethics.

Since "Dutchman" and "diligent" are predicates of the English language, the word sequence "The Dutchman is diligent" – or "der Holländer ist fleissig" – expresses a logophoric judgment for anyone who accepts the

H-thesis.[18] We could, therefore, choose this sequence of words as an example.

It will be quite clear, however, that the scientific and the philosophical importance of a sound logic can be illustrated just as well, and its ethical importance [19] much better, by means of an example that expresses a property which is socially and ethically less attractive. For this reason we shall from now on use as our standard example the following sentence, which may be used as a logophoric judgment:

(1)   the Bantu is Primitive.

We shall use a capital letter in the predicate, since the standard form of a judgment in traditional logic is: '*S is P*'. We assume

(a)   that the judging person is not alone in the universe,

and

(b)   that the judging person formulates the judgment 'the Bantu is Primitive' in some language, either in writing or in speech.

The judgment in question can more appropriately be characterized as follows:

(2)   'ι Bantu is primitive';

the choice of language and of language form is thus unimportant, but we shall use the English formulation (1) as an expression of (2).

Von Freytag points out that his *reine Logik* treats of the judgment and not of the assertion. One may of course accept this, but not without re-marking that as soon as he uses his logic not just for communication with God or for other strictly private metaphysical purposes, but expresses these judgments in a language which is accessible to other people, then it must unavoidably have consequences for the logic of propositions and in general for the logic of the chosen language as a means of expression. Whately characterized a proposition as a 'judgment expressed in words" (1859/41).[20]

NOTES

[1] The use of the expressions "concrete" and "abstract" has undergone great changes throughout the years. These terms are therefore quite uninformative without a

preceding nominal definition and we shall not attribute any systematic meaning to them taken alone. Cp. Whately 1859/81, 89; Wundt 1893/111; Husserl 1968, II, 1/4; Pfänder 1963/150; Adorno 1956/126.

[2] Compare Angelelli 1967/117, 229 n. 33.

[3] A. Pfänder, to whom von Freytag frequently refers, says exactly the same (1963/120, 139).

[4] Cp. VII n. 52.

[5] The meaning of "arbitrary" is explained in III-2.

[6] Cp. Tarski 1964/4 on the misunderstanding about "variable numbers"; Beth 1967/72. Anyone who speaks about "variable numbers" might as well speak about "indefinite numbers"; the importance of this remark will become clear later on.

[7] Beth discussed Plato's Principle as early as 1946, but the best exposition of the setting in which this principle belongs is found in Beth 1965/9–14, where the principle itself is stated on p. 12: "Suppose we have an entity u which has the property A; then there is an entity a which has the following property: for any entity x distinct from a, we have (i) the fact that x has the property A presupposes the fact that a has the property A, and (ii) the fact that a has the property A does not presuppose the fact that x has the property A. (I do not give a statement by means of logical symbols, as this would involve the introduction of modal operators.)" Beth shows how this principle can be derived from Aristotle's Principle of the Absolute (cp. VII-14, sub (iv)). Plato's Principle should not be identified with the formula of first order predicate logic which in Beth 1962/141 and Beth 1967/20f., with little justification, is called "Plato's law".

[8] This notation, which we borrow from G. Peano, is often used in contemporary logic to indicate the *unit class* or *singleton* of an entity m, that is to say the class or set which has m as its only element. As names for unit classes we shall use token sequences of the form: $\{m\}$, whereby token sequences of the form: $\imath m$, or $\imath M$, become available as names for logophoric entities.

[9] Compare VI-30. See also Kraak and Klooster 1968/109f.

[10] Cook Wilson uses the expression "Aness" in connection with entities which he calls "concrete universals". About the sentence "This flower is a hyacinth" he says: "In this sentence what the nominative case stands for is a particular in the strict sense of the word, and a particular substance. It is a particular of the universal represented by 'Aness'. Aness is not merely a universal, *it has the specific quality or character corresponding to the symbol A*" (quoted from Foster 1931/4; my italics). Comparing Cook Wilson with Whately we see that the usage of language with respect to classification of terms fluctuated within English, too. For Whately mentions "holiness" as an example of an *abstract common term*, these being names for attributes, and he adds to this that no general terms, neither *concrete common terms* like "man" nor abstract common terms like "humanity" or "holiness", can refer to a real thing existing in nature. It is unclear what Whately understands by the term "nature", but even if this word is taken in a narrow metaphysical *hic-et-nunc*-sense, it is obvious that for Whately, 'holiness', in any case, is not what von Freytag calls "real teilbar und abtrennbar", whatever he may mean by that. To Whately, then, 'holiness' is not a concrete universal, while Cook Wilson takes "hyacinthness" to name a concrete universal. We see, then, how all kinds of substantival forms, among them – restricting ourselves to English – the use of capital first letters, of the definite article, of the indefinite article, or of one of the suffixes "-ity" and "-ness", have been classified as "concrete" common terms.

[11] W. V. O. Quine formulates the problem of conceptual realism in terms of "-ness" forms. He calls the word "square" a "general term" (1963/76); here Whately would have

48   THE PROBLEM

spoken of "a concrete common term". They formulate the logical-ontological problem differently, however. For Quine, this is the problem how far we really need, in addition to terms like "square", another kind of term like "squareness". He calls the latter "abstract singular terms", while Whately called them "abstract common terms" and Cook Wilson took them as names for *concrete universals*. Quine arrives at the concept 'squareness' by the application of "a fundamental operator 'class of', or '-ness'"; thereby, starting from the propositional function "x is square" he arrives at "the attribute *squareness*, or, what comes to much the same thing, *the class of squares*" (*loc. cit.*). Quine, then, unlike Whately, treats the *attribute*, to which like Whately he refers by means of the word "squareness", as that ontological entity which is in question in the discussion about conceptual realism. Or perhaps it might be better to say (expressing the situation in realistic terms) that Quine refers to the ontologically problematic entity by the name "attribute". The distinction between conceptual realism and nominalism cuts across the distinction in traditional and modern logic and is therefore useless for a study of the semantics and the logic of the definite (in English often the indefinite) articles as used in logophoric thought and language. Cp. Angelelli 1967/16 (Section 1.3), 29 n. 41.

[12] Cp. Stegmüller 1965a/59, 1965b/61; Kneale and Kneale 1964/19; Fearnside and Holther 1963/60; Behn 1925/60, 120. See also III-2, about the Rescher-Copi discussion. – Quine observes (1963/69, 73) that while the "heap" (not: the class) of red things is a red thing, the heap of squares is not square.

[13] Cp. Beth 1964/94, 1965/465f., 1967/93f.

[14] *Every* general term is then "predicable" (cp. Russell 1964/101f.; Beth 1948/72). In contemporary logic this is far from being the case. Whether a theory of logical types is introduced or not, there is in any case a consensus that *not every* general term can be predicated of itself (or: of its reference). The connection assumed by the Norwegian philosopher Skjervheim between the irreflexivity of the predicative "is" in type theories on the one hand and a supposed neglect of general philosophical "problems of reflection" on the other hand (Skjervheim 1964/166f.) may look like no more than a particularly bad pun, notwithstanding his reference to Fitch (1952/217–225); but I believe that it goes much deeper. What Skjervheim says strengthens my belief in the over-all systematic importance in traditional logic of clause (iv) of the H-thesis, that is: of Plato's Principle, upon which the greater part, and maybe all, of those philosophies of life and philosophical anthropologies rest which still compete with the social sciences. Cp. Angelelli 1967/194.

[15] If we omit the word "logophoric" from the H-thesis, we may say that Frege, in *Über Begriff und Gegenstand* (in: Frege 1967), accepts clauses (i) and (ii), but not (iii) and, still more important, not clause (iv). Cp. also Stenius 1969/16.

[16] That an author calls a term "general", "common", or "universal", and "concrete" is not a sufficient condition for taking him to mean it in a logophoric sense, for Whately's concrete common terms are hardly logophoric terms according to our definition in Section 4; cp. Whately 1859/89, where "man" is said to denote "merely *any* man, viewed *inadequately*, *i.e.* so as to omit, and abstract from, all that is peculiar to each individual; by which means the term becomes applicable alike to any one of several individuals, or (in the plural) to several together". Whately represents, one must conclude, a transitional period in theoretical logic. Calling a term by those names is not a necessary condition either: see Edmund Husserl, who speaks about *abstract* general terms *das A*. The crucial question is whether or not Husserl held *das A ist A* to be well-formed and tautological.

17 Even Merleau-Ponty, certainly no philosopher of mathematics, takes a great
interest in judgments about triangles (Merleau-Ponty 1945/440f.). Cp. also Husserl
1948/449–451; Bakker 1964/73.

18 A modern logician may call 'Dutchman' an object or *Gegenstand*, but of this object
"diligent" cannot be predicated.

19 "It often goes unnoticed, however, that the definite article does not only possess
logically unattractive properties, but that it also has a quality that belongs to the field
of ethics. It is doubly unattractive that this quality is a negative one, and that as the
victims of its immoral behaviour the definite article has chosen the philosophers. This
negative property consists, in short, in the fact that wherever the definite article is
employed, the impression is given that only one single object is spoken of, in an unambiguous manner… The history of philosophy offers strange examples of the
mischief that can be caused in this manner" (Stegmüller 1960/171). I owe this quotation
to professor J. B. Ubbink.

20 A recent attempt to formulate a language-independent definition of "proposition"
within the frame-work of a possible-worlds model theory is found in Cresswell 1973/42:
"A proposition is a set of heavens", "heaven" being a term for a special hypothetical
construct.

CHAPTER III

THE SEMANTICS OF THE LOGICAL CONSTANTS

1. INTRODUCTION AND ELIMINATION OF LOGICAL CONSTANTS

When logophoric terms *the/an A* (for instance, Husserl's *das A über-haupt*) are taken up into someone's use of language and are burdened with great philosophical significance, these terms must then be assumed of importance for reasoning. In that case, it must be possible to connect assertions of the form *the/an M is P* logically with statements of other grammatical forms, such as *S is (an) M, all Ms are P,* or *S is (a) P.* It would be very strange indeed, if *the M is P,* or *an M is P,* could be in no way rationally related, in a communicable manner, to other judgments and propositions. Practice also shows unmistakably that persons who make use of a logophoric language do assume such rational connections as well as the possibility of rendering these connections by means of the properties of natural languages. Our problem is to describe them.

It would be preferable to expound these connexions in the form of *introduction* and *elimination* rules for the logophoric articles "the", "a" or "an", etc. The exposition should at the same time make clear how these articles are to be dealt with dialogically. We do not want to limit ourselves to those natural languages in which a definite or an indefinite article is available and frequently used.[1] Let us generalize these requirements as follows: we set ourselves the task of formulating such rules of introduction and elimination for the symbol "*ι*" in logophoric terms *ιM*. We cannot know in advance whether one introduction and one elimination rule will be enough to characterize the logophoric usage of articles. It is very possible that we shall have to distinguish between several logophoric articles. But in that case, each of these articles should have a clear semantics of its own. Some systematic clarification is needed in any case, if the position that logophoric uses of language are understandable is to be maintained.

Some preliminary remarks about such rules of introduction and elimination are in order; we shall take the universal quantifier, the expression "(for) all...", or "(for) every...", as an example. To the existential quantifier, the expression "there is ...", corresponds the expression "some" in traditional logic. The introduction and elimination rules for these quantifiers are clear enough, thanks to Frege, G. Gentzen, S. Jaśkowski, and numerous authors of modern textbooks. The elimination rule for the universal quantifier can be formulated in the following manner: whenever a proposition *for every individual* (thing in the universe of discourse) *it is the case that F* is asserted, *F* may be predicated of x whichever element of the universe of discourse x may be; this is to say that the proposition *Fx* may be asserted. Idealist logicians will want to replace "individual" by "individual concept". If this change is made in the above formulation of the rule of elimination, then it will be acceptable from an idealist point of view, too. Idealists will not be inclined to describe their logic in terms of a universe of discourse, however. The expression "universe of discourse" was introduced by Boole. We shall denote it by the letter "D".

We can illustrate the rule of elimination for the universal quantifier by an example: if a (true or false) proposition

(1)   *all S's are P*,   or: *every S is P*,

for instance,

(2)   all Bantus are Primitive,

is given, that is to say:

(2′)   for every individual (possibly: individual concept) x (in D): if x is a Bantu, then x is Primitive,

then we may subsequently *eliminate* the universal quantifier "for every..." from the discourse by asserting:

(3)   if Saul is a Bantu, then Saul is Primitive.

In doing so the reasoning person who assumes the truth of (1) need not in any way put any restrictions upon his choice of an individual (or individual concept), provided that it belongs to D. It can be shown empirically that this logical rule holds for words like "all", "every", "each", etc., when these are used in propositions (sentences in the indicative mood).

This elimination rule was called by the schoolmen the *dictum de omni*. Nowadays the formulation of an introduction rule for the universal quantifier offers no problem either, although it is more difficult. The solution is based upon Frege's formulation of the logic of the quantifiers in 1897. Logicians today agree as to the conditions which have to be fulfilled in order that we may validly *introduce* the universal quantifier into a discourse which, so far, does not contain this quantifier prefixed to the propositional function *Fx* in question (for instance the function: *if x is an S, then x is P*). And the same holds for the existential quantifier. The "int-elim-rules", as they are called, for these quantifiers can be found in every not-too-elementary text-book of theoretical logic.

2. THE MEANINGS OF "AN ARBITRARY S"

It is often said that, given (1), one may introduce "an arbitrary individual term" (we chose "Saul"), and it is also said that one may select "an arbitrary individual", being "an arbitrary element of D". If the extension S of the general term *S* is taken as our universe of discourse, we arrive at the statement that, given (1), one may select "an arbitrary S", *in casu*: an arbitrary Bantu. The universal quantifier is then, of course, *eliminated*.

The following locution is also current: if one wants to *introduce* a universal proposition, *for every x, Fx*, one should be able to show for one "arbitrary" individual in the domain D that it possesses the property called *P*. As soon as someone thinks that he can assert a proposition of the form (1) above, for example, the proposition (2), or the proposition

(4)　　　　in every triangle, the sum of the angles is 180°,

he will have to show for one "arbitrary" Bantu, or for one "arbitrary" triangle, that this entity possesses the property in question. This Bantu, or this triangle, must be "completely arbitrary", otherwise the proof is not valid.

This terminology raises the question: what is meant by an expression *an arbitrary S* (German: *ein beliebiges S*; Dutch: *een willekeurige S*; Norwegian: *en vilkårlig S*; French: *un S quelconque*).

This terminology is improved by speaking, not of *an arbitrary S*, but of *an arbitrarily chosen S*. The choice of an individual term, and thereby of an individual that has this term as its proper name, must be made in an arbitrary manner. It is the choice which is then said to be or not to be arbitrary, not the chosen term or individual. But this improved terminology does not yet solve the problem. It is clear that there are two senses of "arbitrary", which are not, however, distinguished in textbooks on logic. For making an arbitrary choice of an individual with the property S, given the truth of a universal premiss *every S is P*, or *all S's are P*, in order to conclude that P belongs to this arbitrarily chosen thing that is S, is definitely not the same as arbitrarily choosing an individual with the property S so that from the truth of *that S is P* we can conclude to the truth of *all S's are P*, nor as introducing an arbitrarily chosen thing that is S on the strength of a premiss *some S is P*, or *some S's are P*.

Rescher holds that if the old notion of a variable as a name of a variable entity, for instance of a variable number, or of a variable triangle, or of a variable human individual, is objectionable (as explained in Tarski 1964/4), then the expression "any/an arbitrarily selected individual" is objectionable (1968/134–137). I think he is wrong on this point. Rescher criticizes Copi, who in his textbook *Introduction to Logic* says the following: "We shall use the small (hitherto unused) letter '*y*' to denote any *arbitrarily selected* individual" (quoted from Rescher, *op. cit.* / 134). As it stands, this sentence is indeed very unclear, but it does not follow that "*any* talk of... 'arbitrarily selected individuals' is thoroughly inept", as Rescher says that it does.

The meanings of an expression *an arbitrary S*, or *an arbitrarily chosen S*, become intuitively clear only when we formulate the rules of logic in terms of rules for a critical rather than a deferential attitude towards communicated language.[2]

(a) If somebody has asserted *every S is P*, or *all S's are P*, then he has thereby taken upon himself the burden of proving *X is P*, where *X* is a name for an individual which is chosen from among the elements of the set of S's in accordance with the preferences of a critic who disputes the truth of *every S is P*. This debater has a *free choice* among the S's.

(In conformity with the usage in traditional logic we shall from now on use capital letters for individual terms, too.)

He will usually, however, not make his choice blindfold, but will generally have reasons for making a very special choice. Another opponent will as a rule make another choice.

(b) When someone has asserted a particular proposition *some S is [are] P* and this is doubted by another person, then the first person need do no more than to establish the truth of *X is P* for something that is S, called *X*, this time chosen at will (following one's preference, as a free choice) by himself, and not by his critic as would have been the case if instead of the word "some" he had pronounced the word "all", or the word "every".

These two simple rules form the base of the logic of the quantifiers. We can put them up schematically in the following manner (adapted from Lorenzen 1969/25):

Speaker	Critical listener	Reply (defense)
Rule a: *every S is P*	(*You admit that*) *X is an S.* *Is X, then, P?*	*X is P*
Rule b: *some S is P*	?	*X is an S and X is P.*

In the course of a debate between a proponent of a thesis which we may call *C* and an opponent of the former person with respect to that thesis, each party may make statements of the form *every [all] S is [are] P* as well as of the form *some S is [are] P*, with the other party as a critical listener. If the contested thesis[3] *C* is itself a statement of the form *every S is P* (or *all S's are P*) or one of the form *some S is [are] P*, the above rules can be stated in the form of a dialogical tableau (cp. Section 3 of Chapter I):

Opponent	Proponent	Opponent	Proponent
	C ⋮ *every S is P*		C ⋮ *some S is P*
X is an S; *is X, then, P?* (X chosen from the S's (in D) according to the preference of the opponent)	(no objection) *X is P*	?	*X is an S* *and X is P* (X chosen from the S's (in D) according to the preference of the proponent)

In borderline cases, like monological discourse (reasoning, logical thinking) and the production of proofs, the proponent must take upon himself the roles of every conceivable opponent at the same time.

The difference between the introduction and elimination rules now rests upon the order in which the inscriptions in these two last schemas are read. If we start at the upper left and read the schema in this way: ⅄ ⅄, we obtain the introduction rule for the logical word in question; if we start at the upper right and read the schema in the following manner: ⅀, we get the elimination rule.

1°. Anyone who wants to prove *every S is P* must reckon with every possible choice that can be made by his opponents. The most difficult opponent is the one who for some reason or other chooses an individual S about which the proponent is given no other information than that it has the property S, together, of course, with all those properties which are common to all Ss. An opponent may choose whichever individual S he wishes (rule (a)); a self-critically ("logically") reasoning proponent will therefore have to choose an individual S about which he has minimal information (defined as above), and will have to try to show for such an individual that it has the property indicated by P. This is the introduction rule for "every", or "all".

2°. Anyone who wants to prove *some S is [are] P* may do this by demonstrating the truth of *X is P* for an X, chosen in accordance with his own preference (rule (b)). This is the introduction rule for "some".

3°. Suppose the "monologically" reasoning person wants to demonstrate (not necessarily in a mathematical sense!) the truth of some statement C ("C" for "concludendum"). A "given" truth or premiss has for him the logical value of a proposition which will be accepted as true by any opponent. A premiss *every [all] S is [are] P* can be exploited by the proponent of C, by choosing an individual S in accordance with his own preference (rule (a)). For he may of course assume the same critical attitude towards the "every"-propositions of his opponents as they may do towards his "every"-propositions. If communication is to be possible, the logical particles must have the same meaning for all listeners. The choice of something, X, which is S, is therefore made in accordance with the preferences of the proponent of C, who wants to use this premiss in his argument for the truth of C. A reasoning person does not have to act critically in *this* choice of an individual X-thing. This is the elimination rule for "every" (or "all").

4°. If the premiss is *some S is [are] P*, we have a situation which is analogous to that in which the thesis to be demonstrated, C, is itself a universal proposition *every* [*all*] *S is* [*are*] *P*. A logical, *i.e.* a maximally self-critical person has to reckon with the most troublesome and embarrassing critic as a possible source of the premiss and with the original choice of an S by the latter (cp. rule (b)). He will therefore have to choose an S about which he possesses minimal information. This is the elimination rule for "some".

If we understand by "a critical choice" of an entity X, or of a term *X*, a choice which is made while taking into account every possible opponent of the thesis *C*, the truth of which is to be demonstrated on the grounds of, among other things, the premiss at issue, then we can formulate the rules of elimination *sub* 3° and 4° in the following manner, here adapted to the symbolism of traditional logic:

Elimination rule for "every":
every S is P
X is an S

X is P
(*X* can be chosen freely)[4]

Elimination rule for "some":

some S is P

X is an S and X is P
(*X* must be chosen "critically", and the conclusion may not be the last statement in the argument)

The introduction rules which we formulated *sub* 1° and 2° can be given the following schematic forms:

Introduction rule for "every":
X is an S
X is P

---

every S is P
(*X* must be chosen "critically")

Introduction rule for "some":
X is an S
X is P

---

some S is P
(*X* can be chosen freely)

The occurrences of the word "critically" in two of these four rules indicate that valid reasoning *from a particular proposition* as well as reasoning *to a universal proposition* involves considerable difficulty. Reasoning from a universal proposition as well as reasoning towards a particular proposition is, however, extremely simple.

Nowadays the rules in which the word "critically" or an equivalent must be included also belong to what is called the rules of "formal" logic.

Without the additional remarks in brackets these schemas are not, however, universally valid, *i.e.* schemas yielding true conclusions whenever the premisses are true, whatever predicates are chosen for "*S*", "*P*" and "*X*". People who are unfamiliar with the clause in brackets may want to say that the (simplified, bracket-free) argument forms for the elimination of "some" and the introduction of "all" are not "formally" valid. This was, I believe, the common outlook in the logic current before Frege's formulation of the logic of the quantifiers. Arguments from particular judgments as well as demonstrations of universal judgments were considered as resting upon a logical procedure, *ekthesis*, which was held to fall outside the scope of *formal* logic, that is to say outside that logic which can be formulated by means of variables. Salmon (1973) draws an interesting comparison with the contemporary situation in the theory of induction.

3. THE PROBLEM OF THE ELIMINABILITY OF LOGOPHORIC ARTICLES
AND OF THEIR MEANING IN DISCUSSIONS

To begin with, we shall concentrate on the question of the rule or rules of elimination of articles which are used logophorically. Assuming that someone has arrived at a judgment *the/an M is P*, without entering into the question of *how* he arrived at it we can formulate the above question as follows: can one subsequently, on rational grounds, relate this judgment to individual M-things or individual concepts and, if the answer to this is affirmative, which means of expression are available, suitable, obligatory or permitted in order to communicate this relation?

Our question can also be delimited in the following manner. Has the proposition any consequences which can be formulated in another proposition, in which the logophoric "the" or "a", or a logophoric zero article, does not occur? Or are the "the"- and the "a"-judgments always the end result of thought, a goal in themselves?

In order to find an answer to these questions we would like to establish, by investigating the literature on this and related topics, whether

	Example:
the/an M is P	the Bantu is Primitive
S is (an) M	Saul is a Bantu
S is (a) P	Saul is Primitive

is a logically valid mode of inference, in the sense that the terms M, P, and S can be freely chosen and the conclusion will be true provided the premisses are true. This is a question which forces itself on every student of logophoric uses of language. If the answer to this question is "yes", then we shall have found the elimination rule we were looking for. If the answer is "no", we can of course go on searching, but the answer we have already reached will nevertheless be of great importance. A straightforward answer to this question is of the greatest value, quite especially for ethics and for the social sciences and for so-called philosophical anthropology.

The problem can also be given the following formulation. Suppose that somebody has made a logophoric assertion *the/an M is P*, and suppose that some critical listener subsequently says: *S is an M*; *is S, then, P*? The question at issue now is whether or not the proponent of the logophoric assertion, by expressing this judgment, has semantically bound himself to defend *S is P*; and if that is not the case, what kind of logical obligation did he then take upon himself?

That is to say, we want to have the second and the third rules of the following schema filled in:

Opponent	Proponent
	(1) *the/an M is P*
(2) ?[e.g.: *S is M*; *is S, then, P*?]	
	(3) [??]

An introduction rule for the logophoric "the" or "a (an)" can then be obtained from such a schema by reading it in the order (2)–(3)–(1).

4. VON FREYTAG'S TREATMENT OF *das M ist P*

We should have liked to have been able to settle, on the strength of von Freytag's book, whether the inference-schema which was described at the beginning of the last paragraph is universally valid in his logic or not. It is quite astonishing to discover that he fails us completely and that he tries to circumvent and bypass this question in every possible way. That *Lehre von den Schlüssen* (theory of inference) of which his book treats turns out to be restricted to the Aristotelian syllogistic, expressed in a symbolism of his own. This is to be found in the sixth chapter of von Freytag's book. But the terms *der* (*die, das*) *M* do not occur in his treatment of the theory of syllogisms at all.

As a proof of this, consider the fifth chapter of his book, called "Die Lehre von den Urteilen" (the theory of judgments; 1961/58–91). A treatment of logophoric forms of judgment is missing. Anyone who looks, in this chapter, for a theoretical account of the concept 'the logophore' which he introduced earlier (26, 39f., *i.e.* in the chapter "Die Lehre von den Begriffen", the theory of concepts), will have his hopes dashed. When the logophore is mentioned at all, this is done only indirectly and without any connexion with the systematic treatment given in the fifth and the sixth chapters. One also searches in vain in both chapters for a connexion between the logophore and the form *der* (*die, das*) *S ist P*, nor is any distinction between individual and general judgments of this grammatical form to be found.

In his chapter on judgments, von Freytag says: "In pure logic one has always emphasized that every judgment has the form *S is P*" (61). Nevertheless, in discussing various kinds of judgments or assertions von Freytag turns out to be prepared to distinguish between universal judgments 'all ...', particular judgments: 'some...', and *individuelle* judgments: 'Sokrates ist Mensch'. But he holds that, logically speaking, this last judgment has the same *Struktur* as the quantifier-free judgment 'Europäer sind Menschen', 'Europeans are men', which makes it impossible to understand what he means by "Struktur" in this connection. For the purposes of our investigation it is very important to point out that *judgments with a concrete universal 'der (die, das) M' as subject do not explicitly occur in his enumeration of judgment forms.* The distinctions he does make he calls distinctions of *quantity*; thereby we are led to the guess that logophoric judgments will be treated in a classification according to *quality*. That guess turns out to be wrong, for his distinctions according to quality consist in the unusable Aristotelian-Kantian distinction between "positive" and "negative" judgments (61). It throws no light upon the differences between "all", or "every", and the logophoric operator or operators. We must be allowed to conclude that von Freytag does not dare to tackle this topic.

However, what he has to say about the quantity of judgments is not entirely without significance for the problem of the logic of logophoric judgment forms. We find in his book the following: "Quantity concerns only the subject, and as it usually confronts us, it does not concern the subject as a general concept, but as an individual or a class of individuals. Here one speaks in terms of "all", "some", and "the". Linguistically we are in the heart of class logic, concerned with the simplest relations between classes. "All" is to say "the whole set", "some" means "a sub-set of", "*the*" ["*der* (*die, das*)"] *means "the element"*" (65, my italics). This passage strongly suggests that his logic of concrete universal concepts '*the/an M*' is the same as his logic of '*this M*'. He explicitly declares that if the subject refers to an individual, then this subject-concept is to be treated with respect to its relationships of identity and diversity *as a genus-concept* (65). It is difficult to reconcile this with a statement he makes elsewhere in his book, that the distinction between individual and general concepts is "um so wichtiger für die Logik" (24). That distinction turns out to be, in the first place, of importance in his logic for the theory of negation; for von Freytag assumes two kinds of negation of judgments with general concepts as subject, *viz. privative* and *limitative* negation (27).[5] But he maintains that this distinction between two kinds of negation does not apply to judgments with individual subjects (27).[6] As the reader will agree, there is considerable conflict between his various statements.

Now the inference form *the/an M is P, S is an M, ergo S is P*, about which we want to know whether or not it is valid in von Freytag's logic, does not contain the word "not" at all. As far as our question about the validity of this inference form is concerned there is therefore no reason for von Freytag to distinguish between an individual interpretation and a general interpretation of the term *the/an M*.

In the chapter about judgments he says: "The logic of concrete entities is the same as that of abstract entities" (24). A separate theory of the logic of concrete universals would therefore not be needed in his logic.

But this conclusion is really most surprising. From the truth of "this Bantu [for example Saul] is primitive" one can of course not conclude that the Bantu John is primitive. If we assume a general concept 'the Bantu' to be subject to the same logic as the concept 'this Bantu', then the judgment 'the Bantu is Primitive' is of practically no relevance for the speaker's own conception of individual Bantus. The judgment 'a triangle has angles whose sum amounts to 180°' would then equally bear no relevance to our ideas about individual triangles. This cannot possibly be von Freytag's intention.

He almost goes as far as to say that the use of concrete universals does not or need not find expression *in language*: "Quantity concerns only the subject, and as it usually confronts us, it does not concern the subject *as a general concept*, but *as* an individual or a class of individuals" (my italics). It is a most obscure passage, the more so if we bear in mind logophoric uses of the articles. Von Freytag continues: "However, these relations of class logic are *based upon relations of identity and diversity belonging to pure logic*; though in practical thought and therefore also in the linguistic expression less attention is paid to the latter" (65, my italics).

If we treat judgments with concrete universals as subjects as if they were individual judgments we arrive at very strange results, as will be shown in Chapter VI.

5. ARE LOGOPHORIC JUDGMENTS IN SOME CASES REAL DEFINITIONS?

The traditional logician W. Wundt, whom we mentioned earlier, and who is better known as a psychologist, classifies as *Subsumtionsurteile* not only the individual judgments 'this is a house' and 'the sun is a fixed star', but the judgment 'the wolf is a beast of prey' as well, the latter obviously being meant by him as a general judgment (Wundt 1893/197). Erdmann speaks of "generelle Urteile" and offers as one among several examples: 'der Mensch hat reflektierendes Wahlbewusstsein' (Erdmann 1907/484). To the generic ("generelle") judgments belongs according to Erdmann a special group, to wit the real definitions (486).[7] These are judgments "which circumscribe *the* content of an *object* systematically" (my italics). We notice, first, that he ascribes content to an object or *Gegenstand* and, second, that he simply assumes *the* content of a *Gegenstand* to be a definite, or definable, entity. Pfänder also distinguishes between different kinds of *Arturteile* with the form '*der (die, das) M ist P*'.

Returning to von Freytag, we search in vain for such expressions as "generelles Urteil", "Subsumtionsurteil" and "Arturteil" and for a treatment of the various kinds of judgments which Erdmann, Wundt, Pfänder and others called by those names. He does, however, speak about real definitions and he does defend this notion (52, 60), but without

offering a single example of a real definition. Von Freytag takes them to
be judgments which differ from "mere assumptions" in that in the language
form of the latter the word "sei" is used, not "ist". It is not made clear
whether this means that he considers *every* 'ist'-judgment '*der* (*die, das*)
M ist P' in which the subject is a concrete universal, to be a real definition.

The formulation of real definitions is no task for pure logic, von Freytag
thinks (49, 52). Philosophy, however, must in his opinion conceive of
certain definitions as real definitions, and "this results in different attitudes towards terminology and in the whole manner of thinking, differences
which have often led to ineffective polemics" (48f.). But in as much as an
adherent to his pure logic might consider the expression "real definition"
to stand for an important class of judgments, he is under an obligation to
clarify the logic of such judgments, their "monologic" as well as their
"dialogic".

6. ARE LOGOPHORIC JUDGMENTS IN SOME CASES A KIND OF MODAL JUDGMENTS?

Von Freytag treats of modalities in his fifth chapter, albeit in a manner
which is surprisingly naive considering the year of publication, 1955.
There is, he thinks, a close relationship between the logical doctrine of the
modality of judgments and what he calls "die logische Systematik der
Begriffen", the logical systematics of concepts (81).[8] This systematics
turns out to be the doctrine of definitions in terms of *genus proximum* and
differentia specifica and the ordering of concepts in "concept-pyramids".
Von Freytag describes (80) the relationship in the following terms: "Art
muss Gattung sein" and "Gattung *kann* Art sein", read: (*species*) *must be*
(*genus*) and (*genus*) *may be* (*species*). Among his examples one finds the
following judgments: 'In a plane Euclidean triangle the sum of the angles
must be 180°, and 'A ["das"] triangle *may* be obtuse-angled' (79).

It is important to notice that in his exposition of modal "muss"- and
"kann"-judgments von Freytag makes mention of *definiteness* and *indefiniteness*: "We see that here the definiteness or indefiniteness of what
is expressed in the judgment, and which is due to certain presuppositions
that are not mentioned explicitly in the judgment itself (in our example
something like the axioms of Euclidean geometry), are at stake" (79).

One might think that this indefiniteness would have to be of another kind

than the quantitative indefiniteness which is meant when indefinite propositions like "Europäer sind Menschen", "Europeans are human beings",
are taken as examples. Aristotle called such quantitatively indefinite judgments "ἀδιόριστοι". As a classical point of contact with one of these two
kinds of indefiniteness, *Analytica Priora* I, ch. 13, 32b 10 may be considered.
Here Aristotle is saying of the expression "to be possible", "ἐγδέ-
χεσται", that in one of its senses it can mean "the indefinite ["ἀόριστοι"], which can be both thus and not thus" (Jenkinson).[9]

The forms *the/an M is P* in English and *der (die, das) M ist P* in German,
however, do not contain any modal words. Von Freytag does not even
broach the question of the logical relations between "ist", "sei", "muss"
and "kann sein", from which one might hope to develop a theory of
inference.

The logic of logophoric judgments with concrete universals '*the/an M*'
as subject has still not appeared.

7. SOME INTRODUCTORY REMARKS ON THE DEFINITE ARTICLE IN CONTEMPORARY LOGIC

We shall devote this paragraph to a survey of standard contemporary[10]
conceptions of propositions like

(1)   The Bantu is Primitive.

Present-day logicians, whether conceptual realists or not, are usually
prepared to regard a proposition like (1) as meaningful and as completely intelligible from a logical point of view, provided it is made clear
that it may be interpreted in one of the following ways. (Though nothing
has as yet been said about the *truth* of this proposition.)

(a) In the first place one may explain such a proposition as a careless
rendering of

(2)   Every Bantu is Primitive.[11]

When the set of all human beings is taken as universe of discourse, (2)
may be a testable proposition. A necessary condition for this is that criteria be given which determine the circumstances under which an individual is to be considered as primitive. When criteria are chosen that do not
entirely conflict with the lexical definition of "primitive" in natural language, then (2) will be false and refutable.

(b) Secondly, it is conceivable that a philosopher or another participant in a discussion asserts (1) and that, when asked what he means by that sequence of words, he will answer:

(3)   The vast majority of Bantus are Primitive.

This mode of speech is, however, very unprecise. In a scientific context, the proponent of (3) would usually be requested to make his assertion more precise. He may then, for instance, make use of the quantifier "most" which, in contrast with the expression "the vast majority", does have a logic.[12] He might therefore answer that according to a certain known criterion for primitivity and for a certain given value of the time variable "t",[13] the following holds true:

(3′)   Most, but not all, Bantus-at-t are [were] Primitive.

He may, however, also use a statistical quantifier and assert:

(4)   At least 80% of the Bantus-at-t are [were] Primitive.

At this point it should be stressed that no conclusion whatsoever about any individual Bantu can be drawn with certainty from proposition (4), nor from (3). It is neither implied in these propositions that a certain Bantu, S, is primitive, nor that he is not, nor does (4) imply that the said Bantu, S, has a *degree* of primitivity, which in some way is expressed by the percentage 80. The percent-operator does not modify the copula in the assertion: "The Bantu is Primitive", nor does it modify the predicate. Nevertheless both these propositions, (3) as well as (4), have grammatical forms which allow us to criticize them systematically and hence perhaps to refute them.

(c) Again, the expression *the M* may be used in order to indicate one special individual:

(5)   the (present) Minister of Finance.

With respect to (5) we may assume that the English or the Dutch or some other people form the universe of discourse. The expression (5) then has a well-determined sense whenever there is precisely one minister of finance in this domain, and assertions beginning with "the Minister of Finance" will therefore as a rule be meaningful. If an expression like (5) is held to be a *term* with the following "form":

(5') $\imath x M x$,

to be read: *the one and only x* (in D) *such that x is M*, one might as well
omit both "*x*"-es, however, and write simply $\imath M$.

This is one of those senses of the/an M, which were excluded by the
definition of "*the/an M* is a logophoric term". The symbolic rendering of
the proposition

(6)   The Minister of Finance is intelligent

is the schema

(6') $\imath x M x$ *is P*,   or: $P(\imath x M x)$,

to be read: *the one and only x such that x is M has the property P* (with
implicit reference to a universe of discourse, D). The sign "$\imath$" in (5') and
in (6') which is an abbreviation of "the one and only" was first used with
this meaning by Russell (Whitehead and Russell 1962/173). I have deliberately chosen a different though graphically similar sign in order to
express the logophoric "the" or "a (an)". The intention will be obvious:
the use of these two signs makes it clear that we are dealing with two
(or more) uses of the definite article which are logically radically different
from each other. For neither Husserl nor von Freytag require in their
logic that the universe of discourse contain at most one M in order that
der (*die, das*) *M ist P* be meaningful and true. Although there are many
roses, the assertion "the rose is a flower" is, for Husserl, first, both
meaningful and true, and, second, a proposition with a completely different meaning from that of the proposition "every rose is a flower", and
also different from the meaning of "an arbitrary rose, whichever you like,
is a flower".

When it is argued that a term *the M* is meaningful if and only if there is
precisely one entity in the universe of discourse with the property called
M, then of course we get into trouble in case there is no M in the universe
of discourse at all. An assertion like

(7)   The Dutch Emperor is tyrannical

for example, is hard to evaluate in a subject-predicate logic, because the
so-called subject term has no reference. In order, among other things, to
get around this difficulty – there were also other motives present – Russell

proposed his Theory of Descriptions (Russell 1905, Whitehead and Russell 1962/173 f.). This theory implies that a proposition *the M is P* with a *definite description*, an expression *the M*, as its grammatical subject should be interpreted as

(7′)   *there is one and only one entity in the universe of discourse which has the property M, and that entity also has the property P.*

To accept Russell's Theory of Descriptions is to regard the proposition *the M is P*, when used in the sense *the one and only M is P*, as an abbreviation of the longer proposition (7′). When interpreted in this manner, (1) is synonymous with the existential proposition

(7″)   *there is* one and only one individual in the universe of discourse which has the property of being a Bantu, and that individual is Primitive.

The assertion (1) is then meaningful and can be judged true or false with respect to any universe of discourse, including universes of discourse which do not include any Bantus; but as soon as the universe of discourse contains more than one Bantu, or none at all, this proposition is false, quite irrespective of the nature of these Bantus. In Russell's theory, schema (5′) is not regarded as the form of a term with a meaning of its own, not even when the universe of discourse contains precisely one M. Only complete sentences like (6) or (7) are said to have a *logical form*, by which is meant that they can be coupled with a certain elucidating standard grammatical form, the symbolic rendering of which looks as follows:

(7‴) $(Ex) [Mx \& Px \& (y) (My \rightarrow I(y, x))]$

to be read: *there is an x which is M and which also has the property P, and which is such that if anything, y, possesses the property M, then y is precisely the same entity as x.* The condition for the meaningfulness of the term *the M* is taken up into the logical form (7‴) of the propositional form (6′) which has a more usual grammar.

One often says that the latter form (6′) and thereby also the propositions (6) and (7) are logically "analysed" with the result that they possess the logical form (7‴). The expression "logical analysis" in this context is not in every respect a happy one. We have to do, in the first place, with the fact that the standard grammatical form (7‴) is much *more precise than* the grammatical form *the M is P*.[14] It is even more precise than *ıM is P*. This grammatical form, (7‴), is therefore of great value for communication, especially for critical dialogic behaviour. It is more precise than *the M is P* because when the form (7‴) is used the reader or listener is given a well-nigh complete survey of the implications of the proposition so formulated and of the conditions which must be satisfied if one wants to introduce it into a discourse; the same cannot be said when the form *the M is P* is used.[15]

These conditions and implications are, however, not expressed by means of special introduction and elimination rules for the definite article or for the symbol "*ı*". We do not need any such rules. For in Russell's analysis, the meaning of *ıM is P* is formulated by means of the existential quantifier, the universal quantifier, the identiy sign and two propositional connectives. The logical constant "*ı*" is therefore not a primitive logical constant.

It will be clear that both inference schemas

$$\frac{\begin{array}{l} \text{ıM is P} \\ \text{S is (an) M} \end{array}}{\therefore \ \text{S is P}} \qquad \frac{\begin{array}{l} \text{ıM is P} \\ \text{S = ıM} \end{array}}{\therefore \ \text{S is P}}$$

are logically correct (universally valid).[16] The implications of propositions *ıM is P* present no problem when one assumes Russell's Theory of Descriptions (or a related theory, cp. V-19). Both inference forms are just as valid as the inference form *every S is P, S is (an) M, hence S is P*, and on precisely the same condition: a domain (universe of discourse) must be given or presupposed in every application, otherwise neither "*ı*" nor "every" (or: "all") can be interpreted.

Unfortunately, similar formulations of logophoric judgments *ıM is P*, from which the logical implications of such judgments become, at least to a certain extent, apparent, are not at our disposal.

8. IS THE "PURE LOGIC" A COMPLETED SCIENCE?

In modern "quantificational" logic, accurately formulated int-elim-rules are given for the universal and the existential quantifiers. In consequence propositions in which such words and maybe also propositional connectives occur are *dialogdefinit*, although since Brouwer it is a much debated issue whether they are *wertdefinit*, *i.e.*, whether they always have a truth-value, "true" or "false", and, since Gödel and Church, whether they are *beweisdefinit* – decidable –, *i.e.*, whether there is or is not a proof-procedure, so far as mathematical propositions are concerned (Lorenzen 1962/15, 21, Kamlah and Lorenzen 1970/157, 163; Lenk 1968/572, 574f., 592f.).

If we adopt Russell's Theory of Descriptions, for example, for com-municational purposes, the definite article, when used in presumably uniquely referring propositions (propositions with the meaning "the one and only..."), is as clear logically as the existential and the universal quantifiers, at least as long as it is not embedded into a modal context. It must be admitted, however, that there is more than one theory of descriptions in modern logic (*e.g.* Hintikka's), but they are all construed with the purpose of clarifying the logical relations between propositions of this kind and other propositional forms. Russell's theory is very practical for most purposes and is also very often taken for granted, for instance in the philosophy of science.

The logophoric operator (or operators) "the/an" or "das... über-haupt", by means of which, according to a great many philosophers, a concrete universal or, as some say, an ontologically self-sufficient (inde-pendent) concept is formed from a noun or adjective M, does not satisfy the same requirement of logico-semantic clarity, at least not at the present moment. This is why we have raised the problem of formulating the se-mantic conditions for the introduction and the elimination of the logo-phoric operator.

Our investigation into the possibilities of explaining the semantics and the logic of logophoric judgments will not take the form of an epistemo-logical investigation into the way in which one may reach correct judg-ments about '*the M*'. The semantic rule (or rules) of introduction does not coincide with such a prescription for obtaining knowledge either. Anyone who maintains the contrary must nevertheless admit that an ac-count of how to "know" an entity '*the M*' does not coincide with an in-vestigation into the *eliminability* of the logophoric operator.

Von Freytag holds his logic to be neutral with respect to the controversy between a logic of content or intension and a logic of extension (1961/43).[17] But many authors regard what von Freytag refers to by means of the expressions "die Meinung", "das Gemeinte" or "das Meinbare" as the sense, meaning, intension, intention, *Gegenstand*, or content of the judgment or of the concept. Indeed, von Freytag does so himself (23).

In his review of von Freytag's book Lorenzen, therefore, laconically re- marks that whatever his intentions may have been, von Freytag has done no more than to formulate an intensional analogue of certain very ele- mentary laws of first order class logic (Lorenzen 1956). This description is in accordance with what von Freytag himself says about the explicitly formulated laws of inference relations in his logic: he speaks of "das [vollendete?] System dieser einfachen Wissenschaft", *the* (completed) system of this simple science (11).

The really salient point about his logic is not its limited stock of laws of rules – simplicity ought to be welcomed wherever it can be achieved – but rather the fact that the nature of his logic leads him to devote a very limited space and often none at all to a number of essential problems. Thus in his book the semantics and logic of the definite article is, in the words of Scholz quoted earlier, *völlig vernachlässigt*.[18]

NOTES

[1] Cp. Kraak 1966/122–125.

[2] Cp. Aristotle, *Topica* VIII 14 164b 1–5: "For it is the skilled propounder and objector who is, speaking generally, a dialectician".

[3] Cp. Naess 1966.

[4] Beth formulates the elimination-rule for "all" by means of the following addition: the chosen individual term "may be *any* individual parameter" (1962/49, my italics). Thus he expresses the same logical connection between "any" and "all" as was formulated in the same year (1962) by the linguistic philosopher Z. Vendler, namely that "any" is used in order to express freedom of choice. The object of his discussion is "this very peculiar aspect of the use of *any*, which ... succeeds in blending indetermi- nation [NB!] with generality, freedom of choice" (1967/80). "Any raven you may select will be black" (93). An imperative "take any" is no command: "*Take any* is not an order which is obeyed or disobeyed, but an offer, which is accepted or declined" (85f.). Vendler shows that "while *each* and *every* always connote existence, *all*, by itself, does not" (91). One *might* say, therefore, that the universal categorical proposi- tions of traditional logic were of the form *every S is P* or *each S is P*, and not of the form *all S are* [or: *is*] *P*. Traditional logicians indeed preferred the singular to the plural form of the copula (cp. IV-8). But it is more than doubtful whether present-day adherents to some mode of traditional logic will accept the "every"- or "each"-form as correctly rendering the judgments expressed by universal categorical propositions. – Vendler also distinguishes between speaker and listener in his analyses of the logical constants; the proponent is by him called "the sponsor". (Section 3.8 of Vendler's article (85f.) can be improved upon; see Lorenzen 1969/28.) Thomas Hobbes wrote in his *Leviathan*,

Part I, Chapter 4: "... there being nothing in the world Universall but Names; for the things named, are *every* one of them Individuall and Singular. One Universall name is imposed on many things, for their similitude in some quality, or other accident: And whereas a Proper Name bringeth to mind one thing onely; Universals recall *any* one of those many" (my italics). I owe this quotation to my former student Mr. I. Kisch Jr.

[5] Cp. IV-18 XII-6; also von Freytag 1961/80 f.; Kalinowski 1972/26.

[6] The expressions "privative" and "limitative Verneinung [negation]" should be brought into connection with the difference between *internal* and *external* negations of quantified propositions, modal or otherwise, and with the history of this difference; see XII-6. Cp. the *Prior Analytics* 26b 14, 27b 20, 27b 27, 28b 28, 35b 11.

[7] See Robinson 1965, Chapter VI, especially pp. 189f., for a survey of the meanings of "real definition".

[8] Cp. Trendelenburg 1862, II/246–248 for a lucid exposition of the metaphysical interpretation of traditional logic.

[9] Cp. IV-18.

[10] As non-standard, or unorthodox, modern logic I count the recent investigations into free logics, and the identity theories of quantified modal logics.

[11] Frege held, in "Über Begriff und Gegenstand", that the interpretation of the proposition "das Pferd ist ein vierbeiniges Tier" as a universal proposition "alle Pferde sind vierbeinige Tiere", or perhaps rather as "alle wohlausgebildete Pferde sind vierbeinige Tiere", is probably the best one (1967/170). There would then be no special need for general propositions "das Pferd..." (cp. VIII-23). Angelelli doubts "whether this interpretation is absolutely valid" (1967/103) with respect to the tradition. As we shall see, the history of the traditional form *the/an S is P* shows that his doubts are not unfounded. Cp. VIII-20. Kamlah and Lorenzen (1971/168) assume that a general proposition of the form *der (die, das) M ist P* is always a necessary universal proposition *it is necessarily the case that all M's are P*. Montague (1970/397) reads "a unicorn is an entity such that..." as "some unicorn...", and adds: "I have made no attempt to capture the ambiguity, felt strongly in this sentence, according to which the indefinite article **a** may sometimes have the force of universal, as well as the more usual existential, quantification." Montague wrongly takes such sentences to be simply ambiguous, expressing either of two perfectly clear propositions, *viz.* either a universal proposition or else an existential proposition.

[12] The logic of "most" is described by Rescher (1964/258f., 1968/170f.). One might say that Rescher has here realized the intentions of Hospinianus (cp. IV-16).

[13] Bolzano pointed out that in an *S-P*-logic, time-determinations ought to be taken up into the subject term and not into the predicate term, in order not to create any problems where they can be avoided. Cp. note 4 of Chapter I. In this connection Scholz remarks: "Es ist nicht schwer sich vorzustellen, wie viel weiter wir sein würden, wenn statt der unüberlegten Dialektik, welche die philosophischen Auseinandersetzungen mit der wirklichen Welt wie eine schwere Krankheit beherrscht, die besonnene Art *Bolzanos* gesiegt hätte. Man kommt auch in der wirklichen Welt als Philosoph viel weiter mit dem "bisschen Verstand", wenn man sich etwas mehr Mühe gibt, als den Dialektikern zugestanden werden kann" (1941/179f.). But see also Patzig 1965/45.

[14] See Naess 1966/39 for a definition of "the formulation *U* is *more precise than* the formulation *T*".

[15] Hintikka's theory is discussed briefly in V-20.

[16] Cp. Geach 1968/51. The equality sign here stands for strict, or strong, complete identity; for this notion we shall later use the sign "I".

[17] Compare De Morgan's attempts (1862) at formulating a general logic of terms, an "onymatic" system, which can be interpreted intensionally as well as extensionally. See also Hintikka 1969/87, where the opinion is defended that "the theory of reference is... the theory of meaning for certain simple types of language".

[18] Angelelli refers (1967/135 n. 105) to Hillebrand 1891/37 as an example of the traditional lack of logical interest in the use of the articles. In the course of one page Hillebrand uses the four expressions "einen Thaler", "Thaler", "der Thaler", "der Thaler selbst" as mutually interchangeable names for one objective *Gegenstand*. Cp. VIII-1.

PART 2

HISTORICAL SURVEY

CHAPTER IV

FROM THE HISTORY OF THE LOGIC OF
INDEFINITE PROPOSITIONS

1. INTRODUCTORY REMARKS

We have already mentioned, in Section 4 of the last chapter, that von Freytag ascribes the same *Struktur* to quantitatively undetermined judgments in the plural like 'Europäer sind Menschen' as to individual judgments 'Sokrates ist Mensch'. The former judgment lacks an indication of whether all Europeans are meant, or only some. According to him, individual judgments have, logically speaking, no quantity (1961/65). From this he infers: "Individual judgments like "Socrates is (a) human being" *hence* have logically the same structure as "Europeans are human beings" or "Sinners are human beings"" (65f., my italics).

Von Freytag does not give any name to the kind of judgments to which 'Europeans are human beings' and 'Bantus are Primitive' belong, possibly because he does not regard them as representing a special kind of judgment at all. Considering his classification of concepts it seems likely that his special interest is in judgments in the singular like 'Der Europäer ist...', 'The Bantu is....'. But he does not introduce any special name for these judgments either.

Aristotle called quantitatively undetermined judgments "ἀδιόριστοι". The German traditional logician B. Erdmann held the opinion that the class of general judgments which are formulated in the singular, like 'Der Mensch hat reflektierendes Wahlbewusstsein', coincides with the class of ἀδιόριστοι (1907/484). When we remember that von Freytag makes no distinction, in his doctrine of judgments and in his theory of inference, between individual and general judgments '*der (die, das)*...', even though, according to his doctrine of concepts, '*der (die, das) M*' may be general concepts, we come to the following conclusion: for the purposes of the theory of inference the judgments 'Saul is Primitive', 'Bantus are Primitive' and 'the Bantu is Primitive' all belong to one and the same logical category.

Von Freytag in fact admits this. Traditional logic – or as he says:

"classical logic" – has often been reproached, he says, for not distinguishing between these kinds of judgments "especially in the theory of inference". But he maintains that given his definition of logic this is rightly not done.

From the standpoint of contemporary logic, however, it is most surprising to learn that it is held to be possible, with respect to the theory of inference, to put all sorts of quantitatively undetermined propositions and individual propositions together into one class. This surprise is not lessened by the fact that it is impossible to understand from von Freytag's book what that theory of inference is like. It will clearly be necessary to consult other sources in order to check whether it is really true that in the course of the long history of theoretical and practical logic individual judgments and ἀδιόριστοι were not distinguished in the theory of inference. It is now also clear that the problem of the logic of the articles in traditional philosophy cannot be solved unless we thrash out the logic of these two kinds of judgments, *i.e.* the individual and the indefinite judgments. Without attempting an exhaustive treatment, we shall therefore start with a digression into the history of the logic of indefinite general judgments, that is to say, of quantitatively undetermined judgments with some kind of general import. It seems likely that the key to the riddle of the concrete universal terms *the/an M* in the logics of von Freytag and others has to be sought in the traditional treatment of indefinite propositions. Angelelli, too, thinks that "in the "propositiones indefinitae" we have the universal subject in all its purity" (1967/95).

2. ARISTOTLE'S ἀδιόριστοι

It seems wise to ask first what kinds of judgments or propositions were recognized by Aristotle. In the *Prior Analytics* there are three kinds: universal propositions, particular propositions, and indefinite propositions, the ἀδιόριστοι. By a universal proposition he means a proposition in which something (B) is said to belong "to all or none of something else"; by a particular proposition he understands a proposition in which something (B) is said to belong to "some or not to some or not to all" of something else (A). The indefinite propositions finally – the ἀδιόριστοι – are those in which it is said that the characteristic in question "does or does not belong, without any mark to show whether it is universal or particular" (An.Pr. I 1, 24a 15–20).[1] Aristotle gives the following examples of indefinite propositions (24a 21–22):

		Translation by Edghill (ed. Ross)	Translation by Rolfes[2]
(1)	τῶν ἐναντίων εἶναι τὴν αὐτὴν ἐπιστήμην	contraries are subjects of the same science	das Konträre fällt unter dieselbe Wissenschaft
(2)	τὴν ἡδονὴν μὴ εἶναι ἀγαθόν	pleasure is not good	die Lust ist kein Gut

In the *Prior Analytics*, Aristotle expressly ascribes the same logic to affirmative propositions as to particular propositions. His words are: "So there will be a perfect syllogism [Darii and Ferio, E.M.B.]. This holds good also if the premiss BC should be indefinite, provided it is affirmative: for we shall have the same syllogism whether the premiss is indefinite or particular" (*An. Pr.* I 4.26a 28–30), and: "It is evident also that the substitution of an indefinite for a particular affirmative will effect the same syllogism in all the figures" (*An. Pr.* I 7. 29a 27–29). This must mean that he takes the logic of a quantifier - free proposition *M is P* to be the same as the logic of a particular proposition *some M is P*. The question whether a negative indefinite proposition *M is not P* is the same as the logic of *some S is not P* is not explicitly raised.

In the seventh chapter of *De interpretatione* Aristotle offers a classification according to that which is named in the subject. He then distinguishes two main kinds of proposition, *i.e.* (i) propositions concerning a universal subject, understanding by a universal subject something that may itself be predicated of many subjects, and (ii) propositions concerning an individual subject.

Aristotle mentions the four following propositions as being examples of the first kind of proposition: "Every man is white", "No man is white", "Man is white", and "Man is not white". The two latter are said to have a universal subject but not to be of a universal character. Examples of particular propositions are not given here. As examples of the latter sort of proposition we find "Socrates is white" and "Socrates is not white".

The expression "ἀδιόριστος" is not found in *De interpretatione*. This provoked the following remark from H. Maier: "The non-general propositions about general [entities] would accordingly fall apart into two subdivisions, for which, however, no particular designations are found in *De interpretatione*" (1896/159). Using some terminology which Aristotle introduced later, we may characterize these sub-classes as the class of particular and the class of indefinite propositions.

Łukasiewicz holds that Alexander of Aphrodisias (*floruit* ca. 200) was the first to assert the *equivalence* of indefinite and particular propositions (Łukasiewicz 1957/5). Theophrastus, Aristotle's pupil and successor, already held particular propositions to be indefinite (Bocheński 1956/115). Aristotle, however, says explicitly only that for syllogisms with indefinite premisses the same forms are valid as when the premisses are particular, although in the fifth chapter of the *Prior Analytics* he mentions in passing "the indefinite nature of the particular statement". Maier nevertheless sees no reason "to mingle the particular with the indefinite judgment" (163). We will return to the connection between indefinite and particular propositions in Chapter X, where this connection will be discussed in some detail.

Łukasiewicz is of the opinion that indefinite propositions are of no importance in the logical system of Aristotle. But in the seventh chapter of *De interpretatione* it is said that the propositions "man is white" and "man is not white" are each other's *contradictories*, although they may *both be true*. This means that we are probably up against one of the sources of subsequent confusions, which ought certainly not to be overlooked. From this chapter of *De interpretatione* it becomes clear that there is a direct connection between *the problems of negation* on the one hand and *the problem of the articles* on the other. The problem of the negation of indefinite propositions *the/an M is P* of course disappears if we construe indefinite propositions as particular propositions, the way Alexander probably held that we should, provided that the logic of particular propositions is sufficiently well worked out.

We have, I think, every reason to ponder over the ways in which indefinite propositions are treated in philosophical tradition. One cannot but sympathize with Ackrill, who remarks that "it is a pity that Aristotle introduces indefinite statements at all" (1963/129). Ackrill goes on to say that since Aristotle treats indefinite propositions as equivalent to I- and O-propositions, "*he might as well have dispensed with them altogether and* confined his attention to A, E, I, and O forms" (*loc. cit.*). But Aristotle did not do so, nor did later tradition. If we want to become acquainted with the traditional logic of the articles as used by philosophers, *i.e.* in Dutch and in German especially the logic of the definite articles, then we shall have to find an answer to the question *why* Aristotle and later thinkers gave a place to indefinite propositions in their logic.

3. *Propositiones indefinitae* IN THE LOGIC OF THE SCHOOLMEN

Von Freytag refers to the logic of the schoolmen, among his other sources (1961/10). But none of the detailed theories about the *proprietates terminorum* developed in the thirteenth century are found in that nineteenth-century logic which takes a theory of concepts, like the one propounded by von Freytag, as its point of departure. On the other hand the Latin language has no articles. For these reasons one might feel inclined to ignore medieval logic altogether in this study, were it not that the schoolmen wrote extensively on *propositiones indefinitae*. Moreover the way in which some of the most prominent scholastic logicians used the expression "propositio definita" is an important indication that problems which were originally considered to be distinct have become conflated in the traditional logic of the nineteenth and the twentieth centuries.

In the entirely unoriginal albeit influential *Summulae logicales* by Peter of Spain (*ca.* 1205–1277; cp. de Rijk 1970), probably written around 1232, we find a division into four kinds of categorical proposition: "Propositionum categoricarum alia est universalis, alia particularis, alia indefinita, alia singularis" (Section 1.08). The author then gives an example of an indefinite proposition: "Indefinita est illa in quia subicitur terminus communis sine signo, ut "homo currit"; vel aliter: propositio indefinita est quae significat inesse vel non inesse" (Section 1.09).

This sentence, "homo currit", we find in the theory of *suppositio* of William of Sherwood (1200/1210–1266/1271) as well in the theory of *suppositio* of Peter of Spain himself. This gives us sufficient reason to enquire into that theory. This famous scholastic theory of *suppositio* is a theory of the senses of terms in different kinds of propositions. Since this theory starts out from and studies combinations of terms with other terms, Moody considers the doctrine of the *suppositio* of terms as a syntactic theory (1953/22). Geach, however, often translates "*suppositio*" by "mode of reference" (1968/56). We might speak of a syntactic theory of meaning. De Rijk holds that "the Mediaeval logicians based the *suppositio* on the *capacity* of a term, when used in a proposition, to be interpreted for one or more individuals" (1967/571). This capacity is the *significatio* of the term, not its *suppositio*. But in the opinion of de Rijk, there is, in the theories of authors like William of Sherwood and Peter of Spain, in fact no sharp difference between the *significatio* (being the first *proprietas termini*) and the *suppositio* of a given term.

4. *Suppositio simplex* AND *suppositio personalis:*
WILLIAM OF SHERWOOD

William of Sherwood distinguishes nine *suppositiones communes* which may be ascribed to a general ("communis") term. According to his theory of *suppositio* the word which is written in capitals has another *suppositio communis* in each of the following propositions (quoted from Kneale and Kneale 1964/252):

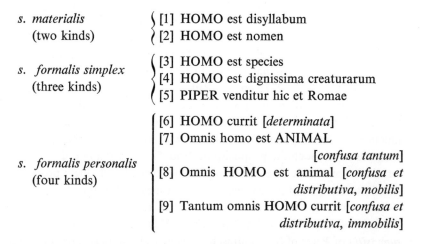

s. materialis (two kinds)	[1] HOMO est disyllabum
	[2] HOMO est nomen
s. formalis simplex (three kinds)	[3] HOMO est species
	[4] HOMO est dignissima creaturarum
	[5] PIPER venditur hic et Romae
s. formalis personalis (four kinds)	[6] HOMO currit [*determinata*]
	[7] Omnis homo est ANIMAL [*confusa tantum*]
	[8] Omnis HOMO est animal [*confusa et distributiva, mobilis*]
	[9] Tantum omnis HOMO currit [*confusa et distributiva, immobilis*]

Comparing this list with the work of von Freytag, we must conclude that in the latter no theoretical distinction is made between propositions [3] to [6] inclusive. It would seem that following the theory of *suppositio* of William of Sherwood, the subject term "Bantu" in our example "the Bantu is Primitive" may be said to have *suppositio* [4] as well as *suppositio*

[6]. In full, the fourth *suppositio* is characterized as follows: *suppositio formalis simplex pro significato comparato ad res, solum cum reduplicatione speciei*. This is a different kind of *suppositio* from that in the proposition

The Bantu is dispersed over large parts of Africa,

in which the term "Bantu" clearly has *suppositio* [5], and also from that in the proposition

The Bantu is a species [or: a genus],

in which it has *suppositio* [3]. But [3], [4], and [5] are all variants of *suppositio simplex*.

By the word "mobilis" in the name of *suppositio* [8] the following is meant: the predicate in that proposition may also be predicated of an individual under the subject "homo". Provided that the qualification "in quantum" is inserted, the same is said by W. Kneale to hold of [4]: whatever may be predicated of the subject *homo* in [4] may also be predicated of an individual entity *as far as* it is human: "Iste homo, *in-quantum homo*, est dignissima creaturarum" (255f., my italics). This is the meaning of the words "solum cum reduplicatione speciei" in the name of this *suppositio*. We shall return to the problem of the logic of reduplicating propositions, *i.e.* to the logic of the expression "in quantum", in Section 30 of the present chapter.

In William's theory the subject of [5] was said to be *vaga*, that is to say erratic or wandering, and it was *immobilis*. This kind of sentence is also discussed by Kraak. (1966/126).

The *suppositio* of the subject term in Peter of Spain's example of an indefinite proposition, "homo currit", is, however, a variant of *suppositio personalis*, namely, *suppositio personalis determinata*.

5. *Suppositio simplex* AND *suppositio personalis*:
PETER OF SPAIN

Peter explains *suppositio simplex* as follows: "Suppositio simplex est acceptio termini communis pro re universali significata per ipsum, ut cum dicitur "homo est species" vel "animal est genus", iste terminus "homo" supponit pro homine in communi et non pro aliquo inferiori, et similiter de quolibet termino communi, ut "risibile est proprium", "rationale est differentia"'" (1947/58, Section 6.05). A term having *suppositio simplex*, then, is being used to designate a universal, and not to designate the individuals falling under this universal. Bocheński translates "homo est species" into German as "*der* Mensch ist eine Art" (1956/193, italics mine). He stresses that the problem of the semantical correlate of a term in *suppositio simplex* was a difficult philosophical problem, about which the schoolmen disagreed. To Peter of Spain he ascribes the opinion – which was not shared by William of Ockham – that a term having *suppositio simplex* is a name for the *Wesen* or the nature of those things whereof this term may be truly predicated (1956/196; see also de Rijk 1967/588f., Moody 1965/42).

Peter's classification of kinds of *suppositio* may be rendered schematically as follows:

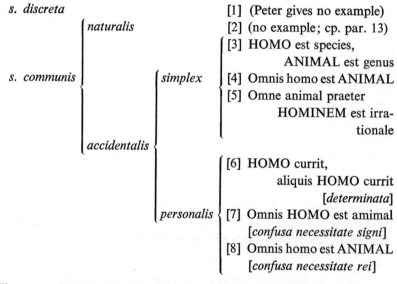

s. discreta — [1] (Peter gives no example)

s. communis
- *naturalis* — [2] (no example; cp. par. 13)
- *accidentalis*
  - *simplex*
    - [3] HOMO est species,
      ANIMAL est genus
    - [4] Omnis homo est ANIMAL
    - [5] Omne animal praeter
      HOMINEM est irra-
      tionale
  - *personalis*
    - [6] HOMO currit,
      aliquis HOMO currit
      [*determinata*]
    - [7] Omnis HOMO est amimal
      [*confusa necessitate signi*]
    - [8] Omnis homo est ANIMAL
      [*confusa necessitate rei*]

There are clear agreements with the doctrine of William of Sherwood. Thus both of them offer "homo est species" as an example of *suppositio simplex* and "homo currit" as an example of *suppositio personalis determinata*.[3] But there are important differences as well. First, the distinction between *suppositio materialis* and *suppositio formalis* which we encountered in William's theory has disappeared. Secondly, Peter shows no interest in the difference between William's [4] and [5]. It is even unclear in which category Peter would place the *suppositio* of the subject in William's examples of [4] and [5]. *A fortiori* the important distinction that William draws between these propositions is lost in Peter's theory. Finally we observe that Peter here ascribes two different *suppositiones* to the term "animal" in "Omnis homo est animal". We seem justified in drawing the conclusion that, in practice, *suppositio personalis confusa necessitate rei* coincides with a form of *suppositio simplex*.[4] This remark is also made by Kneale and Kneale (1964/263). However, Peter himself finally rejects the notion of a *suppositio personalis confusa necessitate rei* (cp. de Rijk 1970/86ff.).

Of the *Summulae logicales* by Peter of Spain, who became Pope John XXI, there have been no less than 166 editions, although it must be said to be less profound than the work of William of Sherwood, especially when we consider the absence of William's distinction between [4] and [5]. It was *"the* introductory textbook for the next three hundred years" (Kretzmann 1967b/126). We may therefore certainly infer that what is said in that book about *propositiones indefinitae* and about *suppositio simplex* has influenced European thought concerning logic and philosophy in general.

It has certainly been no less significant that besides *suppositio personalis* and *suppositio simplex* these thirteenth century logicians, William of Sherwood and Peter of Spain, also introduced *suppositio naturalis*. This topic will be taken up in Section 14.

6. THE THEORY OF *suppositio* AS A THEORY OF TYPES

Bocheński (1959) regards the doctrine of *suppositio simplex* as an elementary theory of types.[4] This is in part correct. The difference between the first and the second types in Russell's type theory may indeed be illustrated by means of the terms "homo" and "species"; *homo ε SPECIES*, and it is incorrect to infer from this that *Socrates ε species*, although *Socrates ε homo*. The theory of *suppositio* and especially the doctrine of *suppositio simplex* is therefore an important supplement to a logic containing the H-thesis which was formulated in II-5.

Now the quantifier-free example which Peter gives us concerning immobility of a subject term is not "homo est SPECIES", but "HOMO est dignissimus creaturarum", and also "ROSA est pulcherrima florum". *A fortiori*: "HOMO est digna creatura", "ROSA est pulcher flos". These propositions are grammatically comparable to "The Bantu is Primitive", and it would seem that they are logically of the same kind as well. These three propositions all differ from the propositions "Homo est species", "animal est genus" in this way, that "Socrates est digna creatura", "this rose is a lovely flower", "this Bantu is primitive" are meaningful propositions which may even be true, whereas "Socrates est species" and "Blackie is a genus" are meaningless (or false). Von Freytag reports that it was not uncommon in traditional logic to regard individuals as *ultimae* (or: *infimae*) *species*, but that an individual is never considered as a genus (1961/36, 39).[5]

It is therefore impossible to accept Bocheński's appeal to a stratification of concepts as an exhaustive explanation of the conceptual structure or structures behind the theories of *suppositio*. The difference between predicates like "genus" and "species", on the one hand, and predicates like "digna creatura", "pulcher flos", "primitive", on the other, disappears completely in that perspective.

Bocheński notes a similarity between the principle of *typical ambiguity* (type-ambiguity) which Russell used in his theory of types and the doctrine of "analogous" expressions in scholastic logic (1959/128). Taking up this suggestion we may perhaps say that there is a typical ambiguity between the use of "pulcher" as a predicate for second-order entities like 'rosa' in the sense of a *res* or *forma* and as a predicate for an individual rose. The traditional technical term is not "typical ambiguity", but "analogy".

This does not, however, solve the problem of the semantics and the logic of general terms with different kinds of *suppositio*.

7. MOBILITY AND THE LOGIC OF GENERAL TERMS

There is no denying that the theory of *suppositio* and the doctrine of *suppositio simplex* are relevant to the problem we have set out to investigate. This is certainly so if all quantifier-free propositions are *ipso facto* counted as indefinite propositions, for instance as propositions in which subject terms are said to have *suppositio simplex*; but they are still relevant to our problem even if only those propositions are called "indefinitae" whose subject terms have *suppositio personalis determinata*. Our example "the Bantu is Primitive" is to be compared with the proposition "homo est dignissima creaturarum", the subject term of which has *suppositio simplex* according to William of Sherwood's theory of *suppositio*, rather than with the proposition "homo currit". But it should perhaps not be forgotten that certain forms of *suppositio simplex* may coincide with certain forms of *suppositio personalis*, at least in Peter's theory.

I cannot agree with Moody (1953/25) and de Rijk (1967/589) when they say that the various accounts of *suppositio simplex* are of little relevance for formal logic. To be sure, the "formal" (with respect to communicated language) aspects of the theory of *suppositiones* were not worked out, with the result that from a dialogical point of view the theory left much to be desired; but this should be regarded as an example of uncultivated logical ground, not as a proof of irrelevance.

It must especially be regretted that the distinction between various kinds of *suppositio* was neither pursued as a problem in the theory of inference nor even very clearly related to it, either to the theory of syllogisms or to scholastic studies of *consequentiae*. It is not sufficient for logical purposes to distinguish five or nine *suppositiones* of the word "homo" and to characterize them by means of a number of suggestive expressions which are not themselves related to the proposition in question. We should like to know which conclusions, mediate or immediate, we may draw for each of the various kinds of *suppositio* of the terms, and we should also like to know from what premisses a proposition with terms having a given kind of *suppositio* may *in principle* be inferred. Ideally, the validity or invalidity of an argument should be determined as a function of the *suppositiones* of the terms in the premisses and in the conclusion.

Peter takes the term "homo" in the proposition "Omne animal praeter hominem est irrationale" to have a form of *suppositio simplex*. He writes

Unde non sequitur: "omne animal praeter hominem est irrationale, ergo omne animal praeter hunc hominem", sed est ibi fallacia figurae dictionis, procedendo a simplici ad personalem. Similiter hic: "homo est species, ergo aliquis homo"; et hic: "omnis homo est animal, ergo hoc animal"; in omnibus his fit processus a simplici suppositione ad personalem (6.06).

Therefore: "every animal except man is void of reason, hence every animal except this man", is not a valid inference, but is a fallacy *figurae dictionis*, since we are shifting from simple [*suppositio*] to personal. The same holds here: "Man is a species, hence some man is"; and here: "every man is [an] animal, hence [he is] this animal": in all these cases there is a shift from simple to personal *suppositio* (6.06).

In his description of personal *suppositio* we read, regarding the proposition "Omnis homo est animal":

Unde iste terminus "homo" dicitur supponere confuse mobiliter et distributive, sed confuse et distributive tenetur quia supponit pro omni homine; *mobiliter* autem quia licet descensum fieri sub eo pro quolibet suo inferiori, ut "omnis homo est animal, ergo Socrates vel Plato". Sed iste terminus "animal" dicitur *confundi immobiliter* quia non licet fieri descensum sub eo, ut "omnis homo est animal, ergo omnis homo est hoc animal", sed ibi est processus a simplici suppositione ad personalem sicut hic: "homo est dignissima creaturarum, ergo aliquis homo"; et "rosa est pulcherrima florum, ergo aliqua rosa". Sed in hoc differunt quia in his est suppositio simplex a parte subiecti, in alia vero a parte praedicati (6.11).

We say, therefore, that the term "man" has a confused, mobile and distributive *suppositio*, confused and distributive because it stands for every man, but *mobile* because it is allowed to descend below this [term] to an arbitrary thing which falls under it, like in "every man is [an] animal, hence Socrates or Plato". But this term "animal" is said to be joined [to the other parts of the proposition] in a way so as to make it *immobile*, because it is not permissible to descend below it, for instance, "every man is [an] animal, hence every man is this animal"; here there is a shift from simple to personal *suppositio*, just like here: "man is the most worthy among the creatures, hence [so is] some man"; and "the rose is the most beautiful flower, hence [so is] some [individual] rose". But they [these propositions] differ, because in the latter [two] it is the subject which has simple *suppositio*, but in the former the predicate [my italics, E.M.B.].

The general term at issue, S [or: P], is clearly called "mobilis" when the other term in the proposition, $P[S]$, may be ascribed to every entity which has the property called $S[P]$. Now what exactly is the condition which has to be satisfied in order that a general term be *mobilis*? This is nowhere clearly explained. As long as one remains within the confines of scholastic terminology, this condition will have to be formulated by means of the expressions "simplex", "personalis", "confusa", "confusa tantum", and "distributiva". The *suppositio* of a term $S[P]$ is called "confusa" if this term refers to more than one individual. If it refers to every individual which belongs to the class of the S-things [P-things], then the *suppositio* is distributive and the term distributed; if not, then the schoolmen spoke of the term as "confusa tantum". As is well known, later writers concentrated on *distribution* as a systematic explanatory concept and the term "distribution" became the only technical term in theoretical use (cp. Rescher 1964/116f.; see also our IX-10).

8. THE LOGIC OF THE GRAMMATICAL SINGULAR

For what is to come, it is of great importance to notice that the schoolmen in their logical analyses operated with propositions containing singular predicates: "est ...", "currit ...". This is the case not only in propositions [3] to [6] inclusive in William of Sherwood's theory of *suppositio*, of which it may be said that they are all indefinite propositions in Aristotle's sense, since they contain no quantifiers, but also in the "every" – or "all" – propositions [7]–[9]. The schoolmen, writing in Latin, discussed "Omnis homo est animal" and not "Omnes homines animalia sunt". The same holds for Aristotle himself: "πας ἄνθρωπος λευκός". In the twentieth century Maritain emphatically asserts that the logically correct formulation is "Tout homme est mortel", and not "Tous les hommes sont mortels" (1933/54, n. 37, 147). Nidditch, on the other hand, considers the retention of the grammatical singular in all kinds of propositions as one of the reasons why theoretical logic did not reach a higher level within its long history (Nidditch 1962/35). The grammatical form "Tout homme est mortel" as well as "Every man is white", already invites confusion with all kinds of individual propositions because of their grammatically singular character. This danger becomes even greater when the operator "πας", "omnia", "every" or "tout" is omitted and propositions are considered beginning with "ἄνθρωπος", "homo", or "man". We have seen above (Section 1) that this is no illusory danger. Some kind of additional theory, of *suppositio* or something else, is clearly necessary.

We know that Albert of Saxony regarded some kinds of *suppositio* as the result of a *modification* of the subject term, a modification effected by the prenex quantifier, if any. The argument by which, in his *Perutilis logica*, he justifies this position contains three premises. He asserts, first, that two mutually contradictory propositions have the same logical subject, and, second, that *every S is P* and *some S is not P* are mutual contradictories.

Third, he says: if the word "every" (or "all") were *part* of the logical subject of a proposition *every S is P*, then this proposition could not possibly have the same logical subject as the proposition *some S is not P*, its contradictory. For one cannot very well say that "every" forms a *part* of the logical subject of the latter proposition. Consequently, "every" cannot be a *part* of the logical subject of *every S is P* either.

His solution is to say that the word "every" ("all"), or "some", is a modifier or functor: "Rather, it is a modification of the subject and signifies the manner of *suppositio* in the subject itself" (quoted from Boehner 1952/23). This shows that he takes some *suppositiones* of terms to be determined by the prenex quantifiers, which he regards as functors with the subject term S as argument and with 'the logical subject', being the term (the word) S with a certain *suppositio*, as its value, the outcome of the operation. The logical form of a universal categorical proposition then is: *all, applied to S, is P,* or: *all (S) is P.* It is natural to complete the analysis by saying that P is ascribed to the meaning of *all (S):*

(all (S)) is P, read: *(all of S) is P.*

If this be the philosophically correct analysis of the universal propositional form, then the particular form will be analysed as follows:

(some (S)) is P.

Assuming Albert to be representative of the schoolmen in general as far as his analysis of the use of quantifiers is concerned, this explains why they formulated universal and particular propositions in the singular. In any case, Ockham also considered quantification to be an operation on terms (cp. Matthews 1964/95, 98), contrary to Frege, who takes propositional functions as the arguments of the "quantifiers" (cp. Potts 1973). As Matthews says, Ockham's view that corresponding A- and I-propositions differ *only* in "quantity" is plausible only if we take him to be thinking of the subject and predicate terms as quantified (Matthews, *loc. cit.*). We shall return to this important issue in Chapter VII (Section 3) and also in Chapter X. From now on I shall often write *all S is P,* as a translation of *omnis S est P.*

9. INDICATING *suppositio* BY MEANS OF ARTICLES

"If Aristotle had not started a queer fashion in the use of general terms, there would have been no theory of *suppositio simplex*. And similarly if Latin had posessed an indefinite article there might have been no theory of *suppositio determinata*", W. Kneale suggests (1964/274). We may add that if Latin had possessed even more articles, then the theory of *suppositio simplex* might have taken on the form of a doctrine of the correct use of these other articles. De Rijk says that the various kinds of *suppositio* in scholastic logic "are mainly determined by the use of quantifying prefixes (*signa quantitatis*) or of demonstrative pronouns", but also that "quantification was thought not to apply to terms taken in material or simple supposition" (1967/569). It then becomes desirable to apply some kinds of prefix (or suffix), which we may call articles. The linguist Seuren has argued for the use of operators, such as quantifiers, in the theory of grammatical deep structures (1969/20, 116). From a dia-logical-logical standpoint it is of no less importance, however, that such operators be used in surface structures, and thus in vernacular language as well as in more technical philosophical or scientific usage.

Let us indicate a *suppositio simplex* by the letter "s" and a *suppositio personalis* by a letter "p":

(1) s_1 homo est species,
(2) s_2 homo est dignissima creatura,
(3) s_3 piper venditur hic et Romae,
(4) p_1 homo currit,
(5) p_2 homo est animal,
(6)   omnis homo est s_4 animal,

etc. These operators "s_1", "s_2",..., may be regarded as an extended system of articles. "s_1" is related to the operator of class abstraction. One might say that Aristotle tried to describe the logic of "p_2" and of "s_4" (cp. XII-2, sub (ii)), but not the logic of "s_2", "s_3", and "p_1". In fact "p_1" corresponds to several different operators, some of which are equivalent to certain complex uses of the quantifiers "for all..." and "there is..."; this will be discussed in Section 11. "s_2" may be replaced by existential quantifiers.

Provided that $s_2 M\ est\ M$ is regarded as a tautology, "s_2" may be replaced by "$\imath$". We already know that *S is P* does not follow from $s_2\ M\ is\ P$ and *S is (an) M*. Peter of Spain does not tell us what conditions must be satisfied in order that we may *introduce* this article into a discourse, nor does he explain how this article may be *eliminated* from a discourse in which it has been used. As long as these rules are not clear, propositions of the form $s_2\ M\ is\ P$ may be considered to be meaningless, unless they are analysed, as *exponibiles*, in terms of other operators for which we do possess such rules. We will return to this topic in Section 14.

10. *Indefinita – determinata – indeterminata* AND THE CONTEMPORARY
UNCERTAINTY WITH RESPECT TO DEFINITE AND
INDEFINITE ARTICLES

It is surprising to learn that the subject in a *propositio indefinita*, which according to Kneale and Kneale may be translated into contemporary English
by means of an *indefinite* article, was characterized as *suppositio determinata*. In fact William of Sherwood himself remarks that we might as well
have spoken of a *suppositio determinata* (cp. Kneale and Kneale 1964/258).

Kretzmann defines *suppositio personalis determinata* as the *suppositio*
of a general term used to indicate an individual *which is not identified*. He
then translates "homo currit" by "a man is running" (Kretzmann 1967a/
372), and so does W. Kneale (1964/257). Assuming that Kretzmann is
right in his interpretation of *suppositio determinata* we conclude that
the indefiniteness on which William of Sherwood and Peter of Spain
focussed their attention was at least sometimes and perhaps even primarily
the indefiniteness of the *identity* of that unique subject which is spoken
about in "homo currit", for instance, rather than the indefiniteness of the
so-called *quantity* of (the subject of) the proposition, caused by the absence of an operator "every" (or "all"), "some", or "no". Of course the
latter indefiniteness will in general entail the former, but as the converse
does not generally hold (see next section), we have to do with two different
kinds of indefiniteness. If indefinite and even definite articles are at the
logician's disposal, it is not likely that one of these kinds of indefiniteness
will be overlooked and left out of theories. In our opinion the schoolmen,
however, shifted their attention from the indefiniteness of "quantity" as a
logical (or, as most of them said, dialectical) problem to the indefiniteness of the identity of the individual which is meant, as when someone says:
"a man is running below there on the street".

Depending on the context, the sequence of words "homo currit" may
be interpreted in at least four different ways:

(a)   An unidentified man is running.
(b)   An arbitrary man is running.
(c)   Men in general, *i.e.*, most human beings, run or walk (as most
        birds fly).
(d)   Man ("der Mensch"; *suppositio simplex* or *naturalis*) runs
        (and a bird flies).

In his German translation of Section 6.09 of Peter of Spain's *Summulae logicales*, Bocheński unwittingly illustrates how uncertain is the relation between modern linguistic usage and *suppositio personalis determinata*. He translates one occurrence of "homo currit" by "ein Mensch läuft" and another by "der Mensch läuft" (1956/197). In the second case he chooses a definite article, while Kneale and Kneale here use an English indefinite article. This may perhaps be explained as a systematic difference between the grammar of German and of English, but it remains a surprising fact that Bocheński, writing in German, first uses an indefinite and later a definite article. In all probability he has not understood "der Mensch läuft" as an essentialist proposition; more likely the expression "determinata" also tempted him to use the definite article. For a reader or listener this is most confusing as the number of functions that are ascribed to the one small word "der" becomes much too large. One can then no longer expect the articles to have any logic at all, and consequently they may as well be omitted altogether. In English the situation is no better, for as we have seen the *indefinite* article is used not only in translations of propositions with terms having *suppositio determinata*, but many authors use them, like Veatch, in essentialist propositions as well ("a modern logician is a rigorist"). If the same article is used in both these cases one may easily obtain the impression that [4] ("homo est dignissima creaturarum") and [6] ("homo currit") in William's list of *suppositiones* are indefinite propositions having the *same* logic. This result is particularly to be expected when the theory of *suppositio* sinks into oblivion, as it has done in more recent traditional logic. This points in the direction of a conflation of the problem of the ἀδιόριστοι as a whole not only with the doctrine of *suppositio personalis determinata* but at the same time with the doctrine of *suppositio simplex*. And indeed we shall see later on that in the more recent tradition the two kinds of indefiniteness, the lack of a quantifier and the indefiniteness of individual identity, which we have encountered in the works of Aristotle and of Peter of Spain respectively, were not systematically distinguished.

11. *Suppositio determinata* AND STATEMENTS OF EXISTENCE

In section 6.09 of his *Summulae*, Peter of Spain discusses the propositions "homo currit" and "aliquis homo currit". In both of these propositions he ascribes *suppositio determinata* to the term "homo". His own words are: "licet in utraque istarum iste terminus "homo" supponat pro omni homine tam currente quam non currente, tamen pro uno solo homine currente verae sunt". That is to say, not only the particular but also the indefinite proposition "homo currit" is true as soon as at least one man is running. In other words: the particular and the indefinite proposition have, as Aristotle also taught, the same truth-conditions. Since Russell and Tarski, logicians may be expected to recognize the importance of the difference between "is true of" and "refers to", which is so rightly emphasized by Bell (1971), but I think we may safely assume that this difference was not understood by the schoolmen, nor in any other period of traditional logic. Church shares this assumption when he defines *suppositio personalis determinata* as the use of a general term for an element of the corresponding class. Nowadays, he says, one gives in such a case either a description, in English beginning with "the", or else a proposition beginning with an existential quantifier corresponding to the indefinite article "a" (in: Runes 1962/307). Church is of the opinion, then, that "homo currit", in which "homo" has *suppositio determinata*, always means one of two things: (a) the (one and only) man is running, or (b) a man is running, in the sense of

(1)   (Ex) [Man x & Runs x].

This prompts Angelelli to ask the following question: "Did Boethius think in that way?" (1967/130). Concerning the distinction between essential and accidental properties, Boethius wrote: "Nam si quis dicat, homo sedet, quod est accidens separabile, cum quicumque singulum hominem, id est individuum, sedere viderit, tunc id *et de specie predicat*, ut dicat: quoniam Cicero sedet, Cicero autem homo est, homo sedet" (quoted from Angelelli; my italics). This means that Boethius also predicates an accident, such as *sedet*, of the species *homo*, although the subject term in the proposition "homo sedet" was held by the schoolmen to have *suppositio personalis*.

Angelelli signalizes the same theory of predication in the works of Thomas Aquinas (1224–1274). There are four ways, Aquinas teaches, in which something may be predicated of a universal. One of these he illustrates by means of the example "homo ambulat". The relevant passage runs as follows: "Quandoque autem attribuitur ei aliquid ratione singularis in quo invenitur, puta cum attribuitur ei aliquid quod pertinet ad actionem individui; ut cum dicitur *homo ambulat*" (quoted from Angelelli 1967/131).

Angelelli's question is a rhetorical one; it must indeed be answered in the negative. Boethius and Thomas analysed "homo ambulat" into a logical subject expressed by a subject term "homo" and a logical predicate expressed by a predicate term "ambulat".

The theoretical distinction between accidental and essential properties occasionally leads an author to introduce certain qualifications or modifications which are supposed to indicate that accidental properties (*currit, sedet, ambulat, albus*) belong primarily to individuals, although they ascribe these properties in a derived sense to *homo* (to 'homo'?) as well.[6] In Angelelli's view these modifications, "quoniam", "quia", etc., are not much in evidence in traditional textbooks. Their effect is "*to confirm that it makes sense to say* "man (not: this man) is sitting", whereby the fact that the predicate *sitting* applies primarily to this man and only secondarily to *man*, is irrelevant" (1967/116).

In (1), on the other hand, the logical subject – if one wants to preserve this technical term – is the domain (universe) of discourse, of which it is said that it contains things which satisfy the propositional function "Man x & Runs x".

Church is right, however, in saying that a proposition having the truth-conditions which Peter of Spain ascribed to "homo currit" and also to "aliquis homo currit", will be expressed by (1) in contemporary logic, unless a definite description "the..." is meant.

Boethius and Aquinas do not seem to draw any logical distinction between "aliquis homo currit" and "homo currit", the subject term of which, "homo", is said to have *suppositio personalis determinata*. They ascribed the same *suppositio* to the subject term of particular propositions and so does Maritain, though not for every kind of particular proposition.

It may be advisable to point out that I make no claim to idiomatic completeness with respect to the logic of operators like "a (an)", "any", "every", and "each" with respect to the differences between them – a topic which has recently attracted some attention (cp., *e.g.*, Vendler 1962). Propositions in which such words occur in other than prenex position, immediately preceding or forming a part of the grammatical subject, will not be discussed here at all.

12. LOGOPHORIC VALUE JUDGMENTS AND *suppositio simplex*

In Peter's *Summulae* we find nothing to tell us which conclusions a listener is entitled to draw from an asserted proposition "homo est dignissima creaturarum". But the proposition "Socrates est digna creatura", for instance, is certainly not cognitively unrelated to "homo est dignissima creaturarum", and this relation is probably different from that assumed to hold between "Socrates est species" and "homo est species". (One cannot simply say that "Socrates est species" is a meaningless string of words, since in some traditional logics individuals were considered to be the *infimae species*. This is an argument against Bocheński's assimilation of the theory of *suppositio* to a theory of types.)

Angelelli is not able, either, to give a logico-semantic analysis of the refractory proposition "homo est dignissima creaturarum", found, for instance, in the works of Aquinas. The predicate of this proposition, which is a standard example in scholastic logic, cannot in his opinion be a *mark* (German: "Merkmal") of, or in, the concept 'homo'; for if it were it would be applicable to Socrates. Nor can it be a *property* of the concept in the sense of Frege's *Eigenschaft*, since this would fall outside the author's (Aquinas's) intentions, so he thinks (Angelelli 1967/131f.). This is not altogether clearly expressed, but we may assume that Angelelli means the following. By this proposition the schoolmen wanted to express something of relevance to our view of individual human beings. But the proposition at issue does not have that kind of relevance if its predicate qualifies the concept 'homo' in the same way as the predicate "abstract", which is indeed a name of a property of this concept.

The proposition "homo est dignissima creaturarum" according to Angelelli resists analysis even in terms of what he takes to be the *traditional* theory of predication. This proposition is clearly a value judgment. Thanks to the predicate we have chosen, our standard example "the Bantu is Primitive" may also be used to express a value judgment; in fact, it will commonly be so used.

As we have seen, W. Kneale regards "homo" in "homo est dignissima creaturarum" as *mobilis* provided the expression "... in quantum homo" is inserted in the conclusion. This expression, "in quantum", gives the impression that the property *homo* may belong to an individual subject in different degrees, in accordance with the Platonic doctrine of a graded

methexis-relation. The English expression "in so far as", which is sometimes used in similar cases, also has this connotation.

13. ARTICLES ANALYSED IN TERMS OF QUANTIFIERS: FROM *suppositio*
simplex TO *suppositio personalis* BY MEANS OF POLYADIC PREDICATION

In the theory of *suppositio* formulated by William of Ockham (ca.
1285–1349), the normal *suppositio* is taken to be *suppositio personalis*.
He rejected the explanation of *suppositio simplex* given by William of
Sherwood and Peter of Spain. In his opinion not only *suppositio materialis*
but also *suppositio simplex* have to do with propositions in which something is being said about a sign, and not about a *forma* or *res*. Moody is
inclined to regard the difference between Peter's definition of *suppositio
simplex* and that given by William of Ockham as a transition from
Platonism to Aristotelianism (Moody 1965/42).

As long as this difference between their positions is of no consequence
for the theory of inference their disagreement may seem to be merely
verbal. But I believe there is a real and still operative difference of opinion
which remains important for the contemporary use of language, even if
the theory of *suppositio* be locked up in the scientific lumber-room; and
I believe so for the following reason. In contrast with many other scholastic
philosophers, William of Ockham explicitly rejected the proposition
"Albedo est alba" "Whiteness is white", as incorrect: "... nam haec est
simpliciter falsa: 'Albedo est alba', qualitercumque termini supponant"
(*Summa logicae pars prima, cap. 63um* [*De suppositione, quid est*]).[7] The
answer which is given by a particular author to the question whether this
proposition is a tautology or not, a question answered in the affirmative
in the H-thesis (which seems to be accepted by the majority of the contemporary followers of a traditional logic), is, I think, of far greater logical
and cultural importance than an informal exposition of the nature of what
sentences like "homo est species" are about. Ockham rejects, I believe,
(the universality of?) that clause of the definition of "logophoric usage"
which says that *the/an M is M* be a (tauto)logical truth, and which is
contained in the H-thesis, *i.e., he rejects Plato's principle* and so cannot
invoke it in order to solve difficult cases.

Ockham cannot, then, simply analyse the proposition "homo est dignissima creaturum" as a proposition in which the property of being *the*

most dignified being in Creation is being ascribed to a subject 'Man', or 'der Mensch' – for instance to essential human nature. Nor does he want to say that this predicate is being ascribed to the *sign* "homo". In his own words: "... opinio dicentium, quod in ista: 'Homo est dignissima creaturarum', subiectum habet suppositionem simplicem, est simpliciter falsa; immo 'homo' habet tantum suppositionem personalem in ista... illa forma communis, cum sit pars illius hominis, non est nobilior isto homine.[8] Et ita, si subiectum in ista: 'homo est dignissima creaturarum', supponeret pro aliquo alio ab homine singulari, ipsa esset simpliciter falsa" (*Summa logicae I, cap.* 66).

In a literal sense ("de virtute sermonis") the proposition at issue is false, but, Ockham says, understood according to the intention of the speaker it is true ("secundum tamen intentionem ponentium est vera"). For he does not mean that "homo sit nobilior omni creatura universaliter", but only "nobilior omni creatura quae non est homo" (*loc. cit.*).

But if that is so, the grammatical form of this proposition becomes logically redundant as a means of expressing judgments of this kind, for the proposition is in Ockham's interpretation equivalent to

(1) $(x) (y) [(\text{Homo } x \ \& \sim \text{Homo } y) \rightarrow x \text{ Nobilior } y]$.

We have thereby eliminated a logophoric monadic subject-predicate-proposition, making use of polyadic predication instead:

x Nobilior y.

Of course this asymmetrical predicate, "nobilior", is also a value predicate and it may be difficult in a concrete case to decide what is worthier than what. *But there is no logical problem in (1) about mobility and descensus.* If Socrates is a human being, the conclusion follows that Socrates is worthier than every creature which is not human:

(2) $(y) [\sim \text{Homo } y \rightarrow \text{Socrates Nobilior } y]$.

And in this conclusion the expression "in quantum" does not occur.

In Ockham's opinion it is very often the case that philosophical statements are false *de virtute sermonis*, but true according to the real intention of the speaker ("et verae in sensu, in quo fiunt, hoc est, illi intendebant veras propositiones per eas", *loc. cit.*). Or, to put the point in the vernacular of present-day logicians: the logical form of a philosophical proposition does not always coincide with its grammatical form. But Ockham gives no instructions for the formulation of intentions which cannot be expressed in simple categorical propositions when these are understood literally, nor for the interpretation of those traditional forms into which they were in fact cast. He does not seem to realize fully that his own interpretation of (1) cannot be given a structure conforming to Aristotelian syntax such that its consequences will be validated by the Aristotelian theory of inference.

14. "PHILOSOPHICAL" *suppositio: suppositio naturalis*

In the oldest treatises in which the *suppositio* of terms is discussed, the meaning of this expression is explained as a property which a term has when used in some proposition (de Rijk 1967/571). This propositional or contextual approach formed the original background of the whole theory of *suppositio*, which was, in fact, an attempt to formulate the truth-conditions of categorical propositions of various kinds (597). In the opinion of the present author "*suppositio*" may be translated by "mode of reference", or simply by "reference". Clearly the mode of reference of a term will in general vary with the proposition considered and especially with the term's position within that proposition. In Russell's terminology, we can say that medieval logicians regarded every noun or adjective in a proposition as a *logically complete* term, *i.e.*, as a referring term, and that they regarded its reference as being a function of the way in which it is embedded in the proposition under consideration. These functions were called *suppositiones*.

The thirteenth century. Independently of each other, Peter of Spain, William of Sherwood, Lambert of Auxerre, and at least one more (anonymous) thirteenth century logician nevertheless introduced a property or function of terms, called (by Peter) *suppositio naturalis* or (by William) *suppositio habitualis*, which is clearly non-contextual, and as a consequence hard to distinguish from the term's *significatio*.

Concerning this *suppositio naturalis* Peter explains: "Suppositio naturalis est acceptio termini communis pro omnibus de quibus aptus natus est praedicari, ut "homo" *per se sumptus de natura sua* habet suppositionem pro omnibus qui sunt et qui fuerunt et qui erunt" (*Summulae logicales*,

Section 6.04; my italics). The term "homo", and every other term as well, has natural *suppositio* due to "its own nature".

As a part of the theory of *suppositio* this is not easy to understand. Kneale thinks that a scientific proposition like "Omnis HOMO est mortalis" offers a good example of what Peter meant (Kneale and Kneale 1964/263). However, in that case it is hard to understand what the difference is between *suppositio naturalis* and *suppositio simplex* in Peter's logic.

Now Peter's notion of *suppositio naturalis* is clearly opposed to *suppositio accidentalis* (cp. the schema in Section 5). De Rijk interprets Peter as follows: when a term occurs in a proposition it has a *suppositio* with respect to that proposition; this was called *suppositio accidentalis* or *respectiva* and comprised both *suppositio simplex* and *suppositio personalis*. Since "accidens" meant not only accident, but predicate as well (de Rijk 1962, II/585, 669), *suppositio accidentalis* was, for Peter, a mode of reference which the term had *with respect to* the predicate term, *respectu praedicati*. (This becomes comprehensible when the background supplied in Section 8 above is taken into account: quantifiers were often thought to belong to, and in any case to operate upon, terms in order to yield modified terms: subject and predicate, which together made up the proposition.) Since *suppositio naturalis* by definition falls outside this notion it is a *suppositio absoluta, sine respectu ad praedicatum, i.e.* it is a function of the "loose" word or, as Peter says, of the *terminus per se sumptus*.

In yet another treatise, the relevant part of which also dates from the thirteenth century, a *terminus per se sumptus* is furnished with *suppositio absoluta*. For the three indistinguishable notions *suppositio naturalis*, *suppositio habitualis (suppositio secundum habitum)*, and *suppositio absoluta* de Rijk chooses the collective name "virtual *suppositio*".

Why introduce a "natural" *suppositio* at all? Boehner, and, following him, de Rijk, hold that Peter's natural *suppositio* was meant as a term's natural capacity – issuing from its *significatio* or connotation – for having contextual *suppositio* of various kinds (de Rijk 1971/75f.). Maybe we can say that the *suppositio naturalis* of a term sums up the ranges (left fields) of all its accidental *suppositiones* (regarded as functions), with respect to every proposition in which the term can *possibly* occur.

No wonder, then, that de Rijk observes a sliding scale between the notions of *significatio*, *suppositio naturalis* and *suppositio accidentalis* in

Peter's logic, and a similar "fading effect" in William of Sherwood's theory of *suppositio*. In both works a sharp distinction is first introduced between the meanings of the technical terms "*significatio*" and "*suppositio*", but this distinction is subsequently blurred by their introduction of *suppositio secundum habitum* or *naturalis*. This took place at a time when scholars were influenced by Aristotle's *De interpretatione* and *De anima* and by later Greek and Arabic works on philosophy. The result was that in the thirteenth century a kind of "philosophical" logic was born (de Rijk 1967/574, 597).

The fourteenth century wrought a consolidation of the contextual, more especially the propositional, outlook in matters of reference. Nothing, in consequence, is heard of *suppositio naturalis, absoluta,* or *habitualis* in the sense of the logicians of the thirteenth century. William of Ockham opposed the notion of *suppositio simplex*, not the notion of *suppositio naturalis*, which he does not so much as mention. Most of the other four-teenth century logicians do not mention it either.

There are, however, at least two exceptions: the nominalist Jean Buridan (*ca.* 1295–1356) and the realist Vincent Ferrer (*ca.* 1350–1419). Both authors assume a "natural" *suppositio*, though of a contextual kind, in agreement with the spirit of the age. The theory of the former can be summed up as follows: the subject term of a scientific ("demonstrative") proposition has a natural *suppositio in* that proposition in as much as it refers to all past, present and future individual (particular) things to which it may truly be ascribed. In the author's own words: "[Suppositio] Naturalis autem quando [terminus] supponit indifferenter pro omnibus suis suppositis, sive sint praesentia, sive praeterita vel futura. Et hoc modo suppositione utuntur scientiae demonstrativae. Aliter per demonstratio-nem ostendentem quod triangulus habet tres, etc., non haberemus scientiam de triangulis futuris, quod est inconveniens, ut dictum fuit" (quoted from Scott 1965/669). So Buridan ascribes *suppositio naturalis* to the term "a/the triangle" in a proposition like "The sum of the angles in a/the triangle is 180°", so as to cover also *future triangles*. As de Rijk says, his *suppositio naturalis* is clearly *omnitemporal* (1973). His pupils Marsilius of Inghen and Albert of Saxony do not mention *suppositio naturalis*, which in Jean Buridan's sense does not seem to have been very influential in philosophy.

Scott points out that the definition of *suppositio naturalis* given by the Spaniard Vincent Ferrer, in his treatise *De suppositionibus dialecticis* from 1372, differs not only from that given by Peter of Spain (which was non-contextual), but also from the (contextual) definition given by Jean Buridan. The realist Ferrer, having no qualms about universals and essences, makes use of the expressions "essentia" and "essentialiter" in his definition: "suppositio naturalis dicatur, quando terminus communis *accipitur respectu praedicati* sibi essentialiter convenientis, quemadmodum est in ista propositione 'Homo est animal'. Tunc enim *res* per terminum huiusmodi importata sumitur absolute *per suam essentiam seu naturam*, ut ibidem clarius habetur" (quoted from Scott, *op. cit.* 670, my italics). The reference to the predicate distinguishes Ferrer's notion of *suppositio naturalis* from Peter's; secondly, as Scott rightly holds, his definition is clearly essentialist (which is not to say that it is clear), and that distinguishes it from *suppositio naturalis* in Jean Buridan's sense. De Rijk characterizes Ferrer's notion as *atemporal* (1973).

Note that Ferrer speaks of *suppositio naturalis* in "HOMO est animal", whereas William of Sherwood and Peter of Spain ascribed *suppositio personalis confusa* to the subject term of "Omnis HOMO est animal"; Peter adds: *necessitate signi*.

Some modern authors, *e.g.*, Boehner, have thought that *suppositio naturalis* entirely disappeared from the scene after Peter of Spain. That is certainly not true. In the fifteenth century Peter of Spain's type of non-contextual *suppositio naturalis* is taken up by the school of Albert the Great. In the seventeenth century the Spanish logician and theologian John of St. Thomas and other Thomist philosophers take over the essentialist, atemporal, but contextual *suppositio naturalis* introduced by Ferrer, while some of John's opponents understood *suppositio naturalis* in the sense of Peter of Spain's *suppositio* of "loose" terms (see further, Sections 28, 29).

15. ARGUMENTS WITH INDEFINITE PREMISSES IN THE LOGIC OF THE RENAISSANCE: RAMUS AND KECKERMANN

Our discussions in Sections 11 and 13 may be summed up in the following manner:

A number of applications of the grammatical forms *S est P, the S is P,* and *an S is P* may be avoided and replaced by more precise, though more complex, propositional forms. This is to say that a number of *propositiones indefinitae*, whether they contain a prenex article or not, may be replaced by propositions the dialogical logic of which is perfectly clear. A necessary condition for this is that we have an exactly-formulated logic of the quantifiers, perhaps supplemented by a rigorous modal logic, at our disposal; this logic of the quantifiers must be formulated for propositional functions of arbitrary length and composition and, furthermore, polyadic predication must be permissible and with it internal quantification; we must, for instance, be able to express the identity relation by means of a dyadic predicate and to embody this predicate in a quantified propositional function. Then, and only then, may the theory of *suppositio* be dropped.

These possibilities have been developed only in the course of the last hundred years.

Renaissance logicians chose, one might say, the opposite direction. Since they often wrote in their own national languages, *articles* are used frequently in logical works and in philosophy in general; but at the same time the theory of *suppositio* disappears almost completely, while the principle of monadic predication is retained. Multiple and internal quantification is therefore inconceivable for these logicians, too.

We shall hold to our resolution not to delve into a variety of more or less precise statements of intentions and results, but to base our own conclusions on the argument-forms which are actually to be found in the works of logicians from different centuries. We can thus form a picture of their influence on language and argumentation.

The *Dialectique* by Pierre de la Ramée, or Peter Ramus, dates from 1555. In this work (ed. Dassonville 1964) we find a syllogism in the first figure with a prenex plural definite article (Ramus 1964b/128):

Les consuls créez par vertu doibvent grandement pourveoir au faict de la République.
Cicéron est Consul créé par vertu.
Cicéron doibt donques pourveoir au faict de la République.

The grammatical form of this argument is:

The M's [are] P,
S is [an] M,
ergo, S [is] P.

This verbal form is valid only if the major is read: *all M's are P*, or *every M is P*, or if it is at least taken to imply this proposition.

The posthumous edition of the *Dialectique* in 1576 contained the following "syllogism":

Le bon poëte est joyeux, oisif et en seureté.
Ovide n'est pas joyeux, oisif et en seureté.
Ovide donques n'est pas poëte.

The form of this argument is:

The good P is M [perhaps: *the ideal P is M*,
                              or: *the real P is M*].
S is not M.
Ergo, S is not [a good?] P.

This form, too – being of the second figure – is valid only if the major premiss is known to imply: *every M is P*. For it is unlikely that this major premiss is meant as a proposition with a uniquely describing subject term; it is certainly some kind of indefinite proposition. Another example of the same form would be

The real Bantu is primitive.
Saul is not primitive.
Ergo, Saul is not a [real?] Bantu.

In the *Dialectique*, Ramus assumes the following classification of simple propositions (1964b/117):

$$\textit{énonciation simple} \begin{cases} \textit{commune} \begin{cases} \textit{géneralle} \\ \textit{spécialle} \end{cases} \\ \textit{propre} \text{ [individual]} \end{cases}$$

He offers examples of *énonciation spécialle* beginning with "quelque", and examples of *énonciation généralle* beginning with "toute" or "nulle".

In the *Dialecticae libri duo*, however, the particular proposition or *axioma particulare* is counted together with the *axioma proprium* as a kind of *axioma speciale*:

$$axioma\ simplex \begin{cases} generale \\ \\ speciale \end{cases} \begin{cases} particulare \\ \\ proprium \end{cases}$$

(cp. Kneale and Kneale 1964/303). In the English translation of this work, from 1574, the exposition runs as follows: "The proposition is speciall, when it speakethe specially *and of a parte*.... The speciall is eyther *indefinite*, or proper: *Indefinite when it speakethe of no certain thing*: as, *Some man is learned*" (1966/75f., my italics). *The particular propositions, then, are the propositions which are here called "indefinite".*

The theory of syllogisms in this translation contains almost thirty examples and, by contrast with the *Dialectique*, the word "all" occurs but once. *Instead of universal propositions are found indefinite ones, which in this work are called "generall".* The name "generale", or "generall", which is being used here is probably the origin of that expression "generelle Urteile", "general propositions" which is often found in more recent traditional logical literature, such as for instance in the works of Kant and of Erdmann. The translator mostly uses the definite article in the "generall" propositions, but occasionally the indefinite article and sometimes no article at all. Among the examples of valid syllogisms we find the following (1966/84f.):

> The troubled man reasonethe not well
> The wyseman reasonethe well:
> The wyseman therfore is not troubled.

> Mortall thinges are compounde,
> The spirithe or soule is not compounde:
> The soule therfore is not mortall.

Ramus called these syllogisms "generalis" ("generall"), because *both* premisses are called "generalis". If one premiss, the minor, is *a particular or an individual* proposition, he calls the syllogism "specialis". He gives examples of this, too (83, 85, 86):

> A wyseman is to be praised:
> But some wyseman is a poor man:
> Therfore some poor man is to be praysed.

The enuyous is not valyante,
Maximius is valyante:
Maximius therfore is not enuyous.

A daunser is Ryotous:
Murena is not Ryotous:
Murena therfore is no daunser.

These syllogisms are called "speciall", because only one premiss is "generall", the other being "speciall" (particular or individual). A syllogism is called "proper" only if *both* premisses are "proper" (individual). The latter kind of syllogism will be discussed in Chapter V.

In most of these examples the articles occur in connection with the extreme terms, but in the example "Les consulz..." and in the example "A wyseman..." it is the middle term which is provided with an article.

Bartholomeus Keckermann (1573–1609) also discusses a syllogism with premisses of a kind that are normally said to be indefinite, but, though rather confused on this point, he is far more critical and scholarly than Ramus. In his *Systema logicae* he offers the following example of an invalid argument in the first figure (1614/241):

Aulici sunt divites:	Courtiers are rich;
Ego ero aulicus.	I'll become a courtier,
Ergo ero dives.	Ergo, I'll become rich.

Such arguments may be rejected, says Keckermann, by reference to a rule which was also maintained by Melanchton: from merely particular premisses no conclusion will necessarily follow. From his appeal to this rule in connection with this example it may seem to follow that he regards all indefinite premisses, and individual premisses as well, as particular propositions. This, however, is too hasty a conclusion. Turning to his classification of propositions, we learn that simple propositions are either universal or particular or singular, a *universal* proposition being either *definita* or *indefinita* (1614/215)!

We are simply assumed to *know*, in this special case, that the major premiss is true only when it is understood as a particular proposition.

An abbreviated translation of this work into Dutch was published in 1614 (1614a). The example above, which is still said to be an invalid argument, has now become one with a major premiss in the singular:

Den Hovelingh is Rycke:	The courtier is rich;
Dese is een Hovelingh:	This is a courtier,
Ergo: Dese is Rycke	Ergo, this [person] is rich

(1614a/149).

16. New foundations: hospinianus

In 1560 another Renaissance logician, Johannes Hospinianus Steinanus
from Stein am Rhein in Switzerland, attempted to give the theory of inference a new basis by taking, in addition to the four usual universal and
particular categorical forms, not only singular propositions but also
propositions which he called "indefinitae" as independent non-compound
propositional forms which might occur as premisses and as conclusions of
valid arguments. By thus broadening the supply of propositional forms that
could be employed in the premisses, Hospinianus produced no less than
512 "syllogistic" forms, as against 256 in Aristotle's syllogistic when the
fourth figure is included. Of these forms 36 are logically valid (cp.
Couturat 1961/3, Risse 1964/557) as against 19 in Aristotelian syllogistic
(assuming the major term P and the middle term M not to be empty).

17. Leibniz

At the age of nineteen, G. W. Leibniz (1646–1716) took an interest in the
work of Hospinianus. In his *Dissertatio de Arte combinatoria* (published
in 1666, a hundred years after Hospinianus' booklet) he states that an
argument in which both premisses and conclusion are indefinite can never
be valid (1960/36). This shows that Leibniz did not treat indefinite propositions as if they were universal propositions and probably did not consider them to entail universal propositions. Hospinianus and Leibniz
clearly tried to answer questions of the kind we have repeatedly posed:
what logical conditions have to be satisfied in order that an argument
with at least one definite premiss be valid? Our standard argument: "The
Bantu is Primitive, Saul is a Bantu, hence Saul is Primitive" is such an
argument with an indefinite premiss.

   The rules developed by Hospinianus and Leibniz did not become generally accepted and probably were not adequate for those uses of language
which were dearest to philosophical system-builders.

18. FROM WOLFF TO ÜBERWEG: ἀόριστοι AND THE THEORY OF
NEGATION

In his *Philosophia rationalis sive logica* of 1728, Chr. Wolff describes the
class of indefinite propositions in the following terms: "... propositio
indefinita appellatur, cujus subjectum est terminus communis sive absolute positus sive cum certa determinatione, sed absque signo quantitatis".
In his opinion the proposition "quidam homines non sunt sinceri" is a
propositio indefinita (cp. Ziehen 1920/666). Propositions whose subject
terms have *suppositio simplex* – which is certainly the case for "homo est
doctrinae capax" on William of Sherwood's theory of *suppositio* – are
now explicitly called indefinite and used to exemplify an indefinite proposition,[9] while the example of an indefinite proposition offered by Peter of
Spain was, as will be remembered, a proposition with a subject term
having *suppositio personalis*.

Proceeding now to the classification of judgments according to quantity
in the *Logik* of Immanuel Kant, we observe that the word "in*definit*" does
not occur at all in that work, nor do we find it in the Table of Judgments
in his *Critique of Pure Reason* (1968/140). This is quite remarkable and
also, as it turns out, of considerable historical significance. Kant distinguishes only *allgemeine* (general), *besondere* (particular), and *einzelne*
(singular) judgments. As an example of the first kind he offers the following proposition: "Alle Menschen sind sterblich", "All human beings are
mortal". He then goes on to remark: "With respect to the generality of a
perception a real difference takes place between g e n e r a l and u n i v e r s a l
propositions, *which however does not concern logic*. For general propositions are propositions which contain *only something* of *the* general
[character] of certain entities and which accordingly do not contain sufficient conditions of subsumption, *e.g.* the proposition: one should carry
out proofs carefully; – u n i v e r s a l propositions are those, which state
something *of an object* generally" (1801/157f.; all italics mine).

This description of *generale Sätze* suggests indefinite propositions in
Aristotle's sense. Kant, however, makes the surprising statement that the
classification of the *allgemeine Urtheile* (general judgments) into *generale*
and *universale* does not concern *Logik* at all. Of course such a view is not
likely to promote further clarification.

In his important recent study of Kant's transcendental Logic, Swing concludes that for Kant an *einzelnes Urtheil* (a singular judgment) "is a judgment whose subject is a universal term in the singular form, such as "The horse is an animal" or "A horse is an animal"'" (1969/7f.). This would mean that a proposition like "homo est doctrinae capax", which was classified as *indefinita* by the schoolmen and which was still so called by Chr. Wolff, would have been classified by Kant as *einzelnes Urtheil* (*judicium singulare*). This explains why the expression "indefinit" is not found in Kant's expositions of theoretical logic: the ἀδιόριστοι are called by him "einzelne Urtheile" and classified together with individual judgments (cp. III-4 and Section 1 of the present chapter).

Einzelne judgments are treated by Kant as universal judgments (see further Chapter V; also Swing 1969/8f.). The difference between *generale* and *einzelne Urtheile* and their inferential interrelationships are left untreated.

(ii) In Section 2 of the present chapter it has already been said that there is a close connection between the problem of the articles and the problem of negation. Aristotle (especially in *De interpretatione*) and later the Roman philosophers, especially Boethius, spoke of "infinite" terms to mean terms containing a negation-particle: *non-S, non-P*. Subject terms as well as predicate terms could be infinite in this sense. Boethius called the resulting propositions "infinitae", translating Aristotle's "ἀόριστοι".

Peirce observes that Wolff, for some reason or other, rejected those of Boethius' *propositiones infinitae* which have an "infinite" subject term and retained only those with an "infinite" predicate term (Peirce II/228, Section 2.381; Swing 1969/12). Immanuel Kant, in his classification of judgments according to "quality", distinguishes *Bejahende* (affirmative), *Verneinende* (negative or privative), and *Unendliche* (infinite or limitative) judgments, the latter category indeed being Boethius' *propositiones infinitae* with an "infinite" predicate, not his *propositiones indefinitae* (cp. Ziehen 1920/639, Swing 1969/12). In the German tradition, such propositions are called either "unendlich" or "limitative" (cp. von Freytag 1961/27f.). In a limitative judgment '... *M is not-P*' a certain limitation of "the subject" was said to take place; it is placed within an infinite, albeit limited "sphere" (in as much as *P*, or its extension, does not belong to it), but without any implication as to what the subject positively is like. Thus such a subject term *M* was nevertheless left *indefinite*. This was said to be different when the negation particle was understood to negate the copula; to negate the copula was held to be something different from negating the predicate. Von Freytag informs us that this distinction cannot be made for individual propositions (*loc. cit.*).

One necessary condition for this view is, I believe, to be found in the absence of the theoretical notion of a universe or domain of discourse, which was introduced into theoretical logic by Boole and De Morgan. If such a not all-inclusive domain for a discussion or argument is given, then the term *non-P* is just as well defined as the term P.[10] There may be some doubts in the case when the extension of P is infinite in the mathematical sense; in that case P may designate a recursive set while *non-P* does not. So it is into recursion theory defenders of the Kantian theory of negation should dip in the hope of coming up with some good arguments. But it should also be realized that without a well-defined domain of discourse even a simple universal categorical proposition will not be interpretable whether the extensions of S and P are finite or not; this holds for the simple universal categorical form *every S is P*, as well as for *every S is non-P*.

The rest of the explanation for this weird theory is to be found in the pre-Fregean logic of the quantifiers "every" ("all") and "some", one variant of which was discussed in Section 8; this will be considered further in XII-6.

As might be expected, the original meanings of "ἀόριστοι" and "ἀδιόριστοι", of "infinita" and "indefinita", have in the course of time influenced each other. Thus Trendelenburg writes: "Under the category of Quality the *infinite*, or rather the indefinite, is assigned a place next to the affirmative and the negative judgment" (1862, II/255). And about the exposition of the forms of judgment given by W. Fr. Krug in his *System der theoretischen Philosophie* of 1825, Ziehen writes: "Judgments with no stipulation of quantity he [*i.e.*, Krug] calls *judicia indesignata* and he rightly emphasizes that the term "indefinitus" is already expended in connection with the classification according to quality" (1920/666; my italics). In the following comment by Überweg it is implied that Krug's use of "indefinitus" had become quite normal in the German philosophy of his time: "The rule that a judgment which is undesignated ["unbezeichnet"] with respect to quantity is general if affirmative but particular if negative is grammatical rather than logical and does not hold unconditionally" (1882/216). The German "unbezeichnet" is also clearly a translation of "indesignatum". Not only has the name of an indefinite proposition been changed from "indefinita" into "indesignata", but Überweg refers to a doctrine, which must have been common at that time, implying that an affirmative judgment – for example 'the Bantu is primitive' – is to be understood as universal but a negative indefinite judgment – for example, 'the Bantu is not primitive', as particular. When used as a premiss in a syllogism, the former example would therefore be treated as "every Bantu is primitive", and generally: *every M is P*; the latter example, however, ought to be treated as if the proposition stated were "some Bantus are not Primitive", in general: *some M's are not P*, that is to say: *not every M is P* (or: *not all of M is P*).

The quotation from Überweg makes it clear that since Kant the logic of the "quantitatively" indefinite propositions is often described by formulating "qualitative" conditions, the decisive condition now being whether the indefinite proposition in question is affirmative or negative. This immediately raises the problem of the connection between this doctrine and one which will be discussed in Section 23, according to which the "syllogistic value" of indefinite propositions, *i.e.* the proper treatment of such propositions in syllogistic arguments as either A-, E-, I-, or O-propositions, depends on whether they are uttered in a *matière nécessaire* or in a *matière contingente*.

The Kantian classification is rejected by Sigwart (1904/158).

Due to the fact that, since Kant, the expressions "indefinit" and "unbestimmt" have been used alongside with "infinit" to designate ἀόριστοι, attention has been deflected from the ἀδιόριστοι. This must certainly be the reason why, recently, Lenk, in his *Kritik der logischen Konstanten*, entirely bypasses the ἀδιόριστοι, and why he pays no attention at all to the constants "der", "die", and "das" (1968/55, 387, 410).

19. WUNDT ON LOGIC AS THE STUDY OF INDEFINITE MAGNITUDES

In the logic of W. Wundt the judgments which are called "unbestimmte Urteile" are judgments with an impersonal subject "es" (in German), such as "es regnet", "es blitzt", etc. In the German logical tradition these judgments were called "Impersonalien". This is not to say that Wundt takes no interest in the so-called quantitatively indefinite propositions with a general subject term. Quite the contrary: he is interested in nothing else.

Wundt rejects the usual traditional view that "quantities" are treated only in mathematics and "qualities" only in logic. In his opinion, "qualitative" considerations too are of great importance in mathematics, while "quantitative" conditions are sometimes discussed in logic, contrary to what is commonly supposed: "The most conspicuous refutation of this view is found in the quantification of concepts. This consists in a division of the concept, that is to say in a quantitative operation, and is possible only if the concept is thought as a whole containing the individuals which we ascribe to it as its parts, also in cases where such a division does not directly come into question" (1893/259). Wundt's characterization of logical quantification may certainly be questioned, but this problem need not detain us at the moment. He subsequently introduces an intriguing distinction between definite and indefinite magnitudes ("bestimmte" and "unbestimmte Grössen"): "There is, to be sure, an essential difference between the quantitative aspect of logic and that of mathematics. The latter has generally to do with definite and the former with indefinite magnitudes only" (260).

For systematic reasons, a discussion of the views of Hermann Lotze and Christoph Sigwart, two other important German authors from the same period, will be postponed until Chapter VI.

20. ARE LOGOPHORIC JUDGMENTS ἀδιόριστοι?

According to Th. Ziehen, the theory of the copula which holds the so-called subject-predicate relation to be a *partial identity of contents* (or: *of intensions*) became common in Germany especially in the version of B. Erdmann (Ziehen 1920/612). The reader will remember that one of the characteristics of von Freytag's logic is precisely this theory of the copula as a "partial identity" between *the* contents of concepts. It is therefore likely to be a worthwhile undertaking to enquire how Erdmann treats propositions of the form *the/an M is P*.

Erdmann, whose work von Freytag consulted (von Freytag 1961/37), discusses in some detail a judgment form which he calls "generelle Urteile" (Erdmann 1907/483–7).[11] It is not without importance how this expression is translated into English. Bosanquet uses the term "generic judgment" in a way which suggests that it translates Erdmann's "generell" (Bosan-

quet 1911, I/210f.). It will therefore be excusable if in the following exposition of Erdmann's ideas we translate "generell" by "generic", especially
as a *generelles Urteil* on Erdmann's definition is an elementary statement
("Aussage") the subject of which is a *genus* ("Gattung") "of some level
or other" (483).[12] He takes a genus to be an "allgemeiner Gegenstand"
(482). If "der Mensch" designates a genus, then according to this definition 'der Mensch ist sterblich' will be a generic judgment ("generelles Urteil"). Similarly, if "the Bantu" designates a genus, then 'the Bantu is
Primitive' is another generic judgment.

   Erdmann's own examples of generic judgments contain, without exception, a prenex definite article in the singular: "Das positive Recht ist...;
Das Gehirn der Wirbeltiere wird...; Das Träumen ist...; Der Mensch
hat...; Die antike Welt kannte...; Der süsse Geschmack mancher
Pflanzen und Früchte entstammt..." (484). Our attention is focussed
especially on propositions like the fourth of these examples.[13] Discussing
the difference between '*Alle* Affen sind Säugetiere' ('*All* apes are mammals')
and '*Der* Affe ist ein Säugetier' '*The* ape is a mammal'), Erdmann holds
that in the former the subject is "thought according to its extension" and
in the latter "according to its intension" (content; "Inhalt"; *loc. cit.*, my
italics). The content is more fundamental than the extension, he maintains:
"The content is precisely the logical *prius* of the extension". And then:
"The extension *of an object* is dependent on its content" (203; my italics).
Given these remarks of Erdmann's we are entitled to assume that his
generic judgments are regarded by von Freytag as judgments with concrete universal subjects.

   As a special sub-class of the generic judgments Erdmann mentions real
definitions (486).

   We have already noted in Section 1 of the present chapter that Erdmann
takes his *generelle Urteile* to be the same as Aristotle's indefinite propositions: "They essentially coincide with the judgments to which Aristotle
occasionally assigns a place next to the general and particular propositions under the name of "indefinite (ἀδιόριστοι, indefinitae)" " (*op. cit.*/
484). Here the phrase "essentially" ("im wesentlichen") attracts our
attention; by these words Erdmann enters a reservation so far as Aristotle's own theory of ἀδιόριστοι is concerned.

   The burning question whether generic judgments should be treated as
universal or as particular is only very generally broached by Erdmann:

"Like the singular judgment the general judgment is, moreover, not only related to the universal but may also be related to the particular judgment" (486, my italics). Here too, then, he has reservations, but he does not go on to say what conditions he has in mind. There is not the slightest attempt in his book to clarify this serious issue. A hearer of the spoken proposition or a reader of the written proposition is nowhere mentioned. Unlike Aristotle, he seems to take no interest in the dialogical aspect of logic nor in the possibility of refuting somebody's assertion of a judgment.

On this issue Aristotle was more explicit. To refute an indefinite judgment, it will be necessary to show that the predicate in question cannot be ascribed to any individual thing falling under the subject term, he says in the *Topics* (III 6 120a–5–10). This is also in complete accord with what he says in the *Prior Analytics*, to which we referred in Section 2 of the present chapter (*An. Pr.* I 4 26a 28–30). A proposition (*the, an*) *M is P*, having no more demonstrative force than a particular proposition *some M is P*, is false only if no M whatsoever is P.

This rule has an important exception: if '*P*' should be a genus of (a genus) '*M*', the situation is different, since *every* M is then P. For Aristotle held that a genus may be predicated truly of every member of each of its species, and he explained clearly enough – assuming for the moment and for the sake of the present argument that the framework of genera and species is acceptable – that if there is as much as one exception, one M, S, which is not P, then the statement that '*P*' is a genus of '*M*' is untenable (*Topics* IV I 120b 15–20). From this it follows that a speaker or author who considers himself an Aristotelian in matters of logic is obliged to state explicitly whether he claims that '*P*' is to be regarded as a genus of '*M*', or only that there is *some* connection between '*M*' and '*P*'.

Aristotelian logic thus seems clear enough on this point. But confusion may yet have arisen among his many readers, for Aristotle used precisely the same example in both cases. The proposition "ἡδονὴν ἀγαθὸν εἶναι" is used in *Topics* III 6 120a 5–10 as an example of an indefinite proposition which can be refuted only by showing that no pleasure whatsoever is good, while in *Topics* IV I 120b 15–20 the question is how to refute a claim that τἀγαθὸν is a genus of ἡδονή.[14]

There are still more complications in the Organon. Although in the first part of *De interpretatione* indefinite propositions occur without prenex articles, in the last chapter of that work Aristotle offers examples of a kind rendered in, say, German by articles in the first grammatical case and in which the subject term is quite clearly in the singular: "τοῦ ἀγαθοῦ ὅτι ἐστὶν ἀγαθὸν" (14.23b 8, 23b 32). Both Maier (1896, I/161) and Rolfes translate this by "das Gute ist gut"; Edghill however says: "that *which is* good is good". Assuming that Aristotle meant this sentence as a general but indefinite proposition (whatever that may be), Maier says that in the course of the book indefinite propositions have been brought closer to *universal* propositions. For Aristotle says: "For the judgment that that which is good, is good, *if* the subject be understood in the universal sense, is equivalent to the judgment that whatever is good is good, and this is identical with the judgment that everything that is good is good" (24a 5–10). Add to this, as did Maier, the irregular use in the *Organon* of the articles in indefinite propositions, and we will have to conclude with him (and now Ackrill, cp. Section 2), that although explicit up to a point, Aristotle's logic of indefinite propositions is "not uniformly developed" (*op. cit.*/162).

Erdmann seems to accept that the propositional form *der (die, das) M is P* is not unequivocal. He makes no attempt at separating the two senses, the universal and the particular, and he certainly does not give the impression that he wants to avoid this form of language. Since he does not go into the inferential relationship between generic judgments, universal judgments, and particular judgments I think we may conclude that in his work the *generelle Urteile* have no logic at all.[15]

Neither Wundt nor Erdmann, nor von Freytag, discusses any kinds of *suppositio*.

21. "ABSOLUTE" AND "RELATIVE" JUDGMENTS WITH "LOGICAL" *suppositio* IN PFÄNDER'S LOGIC

We proceed to another influential work on logic in which partial identities are held to be of some systematic importance, the work of the phenomenologist Pfänder which appeared in 1921. This author turns out to deal more extensively with the problems we have formulated than von Freytag does.[16]

Pfänder devotes a good page, no more, to an exposition of "the *suppositio* of propositions" (36f.). But that is more than in any of the works in the German idealist tradition we have met with earlier. A surprise is that

Pfänder ascribes suppositio to propositions (Sätze), not to terms. He only distinguishes three kinds of *suppositio* in which sentences may be "taken" ("genommen"), and he calls these

1. linguistic (material) *suppositio*
2. logical (formal) *suppositio*
3. real *suppositio*.

The second of these is the most important one *for logic*, according to Pfänder. It is, he says, usually indicated by double inverted commas around the sentence in question. It seems that in the present work this function is taken over by single inverted commas (cp. our survey of the use of symbols, p. XVII f.). A sentence with this *suppositio* stands for "the" judgment which is expressed by it. By this employment of quotation marks, then, the author requires that the understanding carries out a re-ification of the judgment (*loc. cit.*).

Pfänder goes on to enlarge upon this. A sentence expressing a judgment is, when taken in logical *suppositio*, no longer the expression of a judgment but of a concept. Pfänder characterizes this concept as "the" concept which has "the" meaning of the sentence as its *Gegenstand*. He illustrates this by means of an example. If the sentence "Schwefel ist gelb" ("Sulphur is yellow") is "taken" in logical *suppositio*, then this sentence means the same as the word sequence "the judgment/Sulphur is yellow" means when the latter is "taken" in real *suppositio*. Hence he implies that the prefix "the judgment" is an operator which assigns an object, a *Gegenstand*, to every categorical sentence in an unequivocal manner. All this is so far little better than a play with words. But from the point of view of cultural history and general philosophical influence it is of the greatest interest that we may conclude that in Pfänder's *Logik* we find a revival of the notion of a philosophical, or virtual, or essential *suppositio*, a *suppositio naturalis* more or less in Ferrer's sense, though not indeed of a *terminus* but rather of a *propositio per se sumpta* (cp. Section 13). A proposition having this *suppositio* cannot very well be anything else than a proposition which expresses something – or better: by which something may be expressed – about a *terminus per se sumptus*.

It is therefore extremely telling that this page is also the last page in his book on which logical or any other kind of *suppositio* is mentioned. *No use is made of the technical term "logische Supposition" in Pfänder's theory of inference.* It is probably assumed in that theory that all sentences to which it is applied are already "taken" in logical *suppositio*. For Pfänder explicitly states: "since logic does not have the task of investigating the sentences themselves nor the projected[17] facts, but only sentences as used as support for arriving at the judgments which it wants to take as objects of investigation, it is clear that logic to a large extent has to take the sentences which it puts forward as examples, in logical *suppositio*". That will clearly have to do.

As usual in the German tradition since Kant, Pfänder distinguishes, with respect to "the so-called relation of the judgment", between categorical, hypothetical, and disjunctive judgments. Disjunctive judgments have the *Formel*: "S is either P or Q" (106), hypothetical judgments have the *Formel*: "S is P if Q is R" (102), and if the statement is made without reference to any condition, then we have "the so-called categorical judgment, whose formula traditionally is "S is P"" (105). It will clearly be enough for us to concentrate on the logic of indefinite categorical judgments.

In his classification by "quantity" of judgments Pfänder, like Kant, makes no use of expressions like "unbestimmt" or "indefinit". He distinguishes judgments "by quantity" into *Singularurteile* and *Pluralurteile*, and prefers (114f.) this classification to the more usual Kantian triple classification into *Einzelurteile, Partikularurteile* und *Universalurteile* (or *Allgemeine Urteile* as some authors say). Having learned this we must expect his theory of inference to be quite unorthodox.

"Universalurteile" is still taken to mean judgments which are expressed by propositions of the form *alle S sind P*, while "Partikularurteile" means judgments expressed by propositions of the form *einige S sind P*. By "Einzelurteile" he understands judgments expressed by propositions of the (ambiguous) form *ein S ist P*.

Now as to the preferred distinction *Singularurteile-Pluralurteile* it turns out that the former are to include all *Einzelurteile*, but by no means only these. Such judgments as "Schwefel ist gelb" and "Der Adler ist ein Raubvogel", which certainly have some kind of general import, are also said to be *Singularurteile* (114).

Apart from the classification into *Singularurteile* and *Pluralurteile* Pfänder also erects another dichotomy, namely that between *Individualurteile* and *Arturteile* (generic judgments). This is a cross-classification of the *Singular/Pluralurteile* distinction. The *Arturteil* "All species of vulture are cowardly" is a *Pluralurteil* and not a *Singularurteil*. *Pluralurteile* of both kinds may have complex subject terms, like "Leibniz and Newton are the creators of the differential calculus" and "The eagle and the vulture are birds of prey". This type of judgment was still counted by Chr. Wolff as belonging to a special sort of propositions called "copulativae", but Kant rejects the class of copulative judgments from his Table of Judgments in order to arrive at the magic number of precisely three in his classification of judgments "according to relation" (Swing 1969/14f.). Nevertheless, we will meet these judgments later on in the logic of Maritain, which harks back to that of John of St. Thomas (cp. XI-3). It is remarkable that Pfänder describes the difference between S *is* P and (S_1 *and* S_2) *is/are* P as one of quantity and not as another kind of judgment expressing a relation between terms; the propositions he chooses as examples are quite clearly used to express a relation of similarity between S_1 and S_2. The explanation must be that he cannot count S_1 and S_2 as two terms since that would force him to conclude that the proposition as a whole had three terms, which is quite inconceivable to any traditional logician.

In Pfänder's terminology, all *Arturteile* belong together with the *universale Urteile* to the *allgemeine Urteile*. We already know some *Arturteile* are *Pluralurteile* ("All species of vulture are cowardly"). But of those *Arturteile* which he expresses by means of sentences of the form *der* (*die, das*) *M is P*, he states more than once that they are *Singularurteile* (119, 122). "Not only the linguistic sentences, but the judgments which are expressed by them as well, are of a singular character", he says about propositions with "sulphur" or with "the eagle" as grammatical subject (119). The concepts 'sulphur' and 'the eagle' are called "singular general concepts" ("singulare allgemeine Begriffe", 143). These concepts are, then, singular and general at the same time.

Our example "Saul is a Bantu" is clearly no *Einzelurteil*, but it is a *Singularurteil* as well as an *Individualurteil*.

Pfänder's rather confusing classification of quantifier-free categorical judgments may be illustrated as in Figure 1 (next page):

From now on, I shall translate Pfänder's expression "Arturteile", too, by "generic judgment".

The subject-concept (as he calls it) of a generic judgment may, in the logic of Pfänder, may mean ("meinen") the species or kind (1) in every case

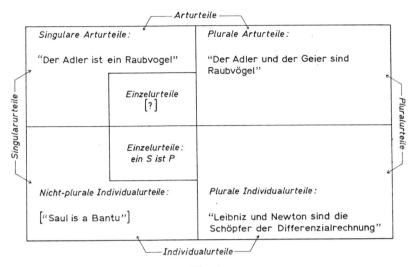

Fig. 1.

or instance ("in jedem Falle"), (2) in the normal case ("im Normalfalle"), (3) in the average case ("im Durchschnittsfalle"), (4) in the typical case ("im typischen Fall"), or (5) in the ideal case ("im Idealfall"; 121f.). The last two cases are especially interesting for the present discussion. How does Pfänder characterize the fourth notion, that of a typical case? We read that "the typical case is the case in which *the* differential essential character of the kind is especially strongly brought out" (122, my italics). The dubiety of this definition from a logical and methodological point of view is not removed by his attempt to make it clearer with an example: "If it is stated, for instance, that Woman is egocentrically narrow-minded, then it is not necessarily meant that Woman is so in every case, nor that *she* is so in every normal case, and just as little that *she* is so in the average case, but what is meant may only be that *where* Woman's character as different from that of *the* Male is brought out in a particularly clear way, *she* confines *her* field of interest completely to herself and to that which seems directly to belong to *her*" (*loc. cit.*, italics mine). The reader will remember that fifteen years after the publication of Pfänder's logic, Hempel and Oppenheim published their original study called *Der Typusbegriff im Lichte der neuen Logik* (1936).

Proceeding now to (5), the notion of an ideal case, we are told that "the ideal case is *the* case in which *the* advantages of the species have reached a particularly perfect expression" (122, my italics). "If, for example, it is stated that Woman is Love, or the German is the Soul, then it is evident [sic] that both should be thought neither in every case nor both in the normal case nor both in the average case nor both in the typical case, but that both kinds of man here must be *taken* in *their* ideal case" (*loc. cit.*, my italics).

Generic judgments falling under (1) are called "absolute" judgments by Pfänder. For the other classes he chooses the common name "relative Arturteile", "relative specific judgments" (259f.), but he does not explain what it is that is relative and to what it is relative, so that consequently the relation at issue is not described. Elsewhere in his book Pfänder enumerates the "four different *species of judgment of relation* ... namely *judgments of comparison, judgments of membership* ["Zugehörigkeit"], *judgments of dependence*, and *judgments of intention*" (49). But he hastens to add that this classification, too, is "not a purely logical classification", but rests upon "real differences within the facts posited by the judgments" (*loc. cit.*).

Pfänder, like von Freytag later, tries to evade all difficult systematic problems by speaking, again and again, of "ontological" or "sachliche" (real) distinctions which presumably do not belong in a study of logic. At any rate, no systematic connection between *relative Arturteile* (relative generic judgments) and *Relationsurteile* (judgments of relation) is to be found in his work.

Both the absolute and the relative generic categorical judgments which are expressed by means of a proposition *the/an M is P* with a grammatical subject in the singular and without complex terms are said by Pfänder to be *Singularurteile* (122). In all probability they are the result of statements which are "taken" *in logischer Supposition*.

22. HERING AND THE PROBLEM OF THE GENERAL LION

J. Hering, another phenomenologist, realized that the semantics and the logic of the article pose a problem. He expressed his opinions on this matter in a contribution (Hering 1930) to the same issue of the *Jahrbuch für Philosophie und Phänomenologische Forschung* in which

Pfänder's logic also appeared, nine years after it was first published. His views are noticeably more modern than Pfänder's.

Hering discusses the meanings of various kinds of propositions with "the lion" as the grammatical subject term. He rejects the idea that there exists, "apart from the lion *hic et nunc*", a new kind of object which might be called "every lion" or "an arbitrary lion" (cp. III-2). His dialogical characterization of the meaning of an expression *an arbitrary M* is very modern: "Instead of "an arbitrary lion will be able to carry out this leap", I can say: "Take this one, it will be able to do it; that one too, etc.""

But he still sees problems. He is especially alert to the problem of the semantics of those judgments which Pfänder classified as relative generic judgments: "The greatest difficulty is the interpretation of a sentence like this: "The lion is often dangerous". One might ask the question whether this is not rather an incorrect formulation of the following thought: "S is p" holds on the whole (as a rule, now and then etc.). We must content ourselves here with drawing attention to this question..." (*loc. cit.*).

Hering is tilting against windmills. His own examples of judgments which he considers as problematic are in any case easily cast in refutable propositional forms by means of the phrase "most" or by means of statistical quantifiers.

It is remarkable that Hering, contrary to Pfänder, rejects the existential import of (the subject term of) universal categorical judgments *every S is P* (542f.) This is certainly not unconnected with his dialogical conception of "an arbitrary lion". He ascribes no systematic role in logic to the *indefiniteness* which he recognizes in the idea '*Lampe überhaupt*': "But to speak about a general entity in the sense of an indefiniteness or of a privation ... seems to me to be completely senseless" (532; cp. Section 18).

23. *Propositions indéfinies* IN THE *Logique de Port-Royal*.
CONNECTION WITH '*definito rei*' IN CASE OF A *universalité métaphysique*

As far as modern philosophy is concerned, that is to say post-Renaissance philosophy, we have paid attention in this chapter only to German philosophers. In the following sections the opinions of some important French and British authors with respect to indefinite propositions will be treated in chronological order.

The dangerous implications of the confused situation regarding the logic of indefinite propositions were recognized by Arnauld and Nicole in their French *Logique de Port-Royal* of 1662 (1965/154). Among their examples we find the following: "L'homme est raisonnable", "L'homme est juste",[18] but also propositions with a grammatical subject in the plural: "Les François sont vaillans". They report that contemporary philosophers eagerly discussed the problem whether those propositions which are called indefinite are to be counted as universal or as particular propositions. Most philosophers accepted the doctrine, so they tell us, that an indefinite proposition should be understood as a universal proposition *dans une matière nécessaire*, but as a particular proposition *dans une matière contingente*. That is to say: if the speaker intends to make an apodeictic statement, then the predicate *P* of the indefinite proposition *M is P* (*the/an M is P, the M's are P, M's are P*) may be ascribed to every S which is M. If the indefinite proposition is not meant to be apodeictic, then *M* (*the/an M, the M's, M's*) refers only to some M's. This is clearly a different position from that explicitly held by Aristotle, Theophrastus and Alexander.

"I find that this maxim is accepted by very able persons", one of the two authors of the *Logique de Port-Royal* writes, "and nevertheless it is quite false" (154). He holds that if some quality or other is ascribed to a general term, then the ensuing indefinite proposition ought, contrary to the current doctrine, to be treated as a universal proposition in any context (subject-matter; "matière") whatsoever. In a contingent context, too, one should not regard such a proposition as a particular but as a universal proposition, which will therefore usually be a false proposition.

Arnauld and Nicole here show a remarkable fear of the ambiguity of such propositional forms as *le M est P* and *les M sont P*. They cannot rest content with the view of so many other philosophers, that propositions of one of these grammatical forms should be considered as universal in some circumstances, but as particular in others. They might have cut the knot by asserting that such propositions ought to be treated as particular in *all* contexts, like Aristotle in the *Prior Analytics* and later on Alexander of Aphrodisias. Very many indefinite statements would then have to be judged true, in fact practically all those which have been used as examples throughout the centuries.

Arnauld and Nicole adopted, however, the position that indefinite propositions have universal import and, consequently, are usually false. An explanation for this un-Aristotelian choice may be found in another classification which was current at the time. It was often said that indefinite propositions are universal propositions *en matière de doctrine*, but that they are particular *dans les faits & dans les narrations*. The authors give the following example of an indefinite proposition of the former kind: "Les anges n'ont point de corps". Now one would be hard put to it to point out any difference between propositions used *en matière de doctrine* and propositions characterized as *real definitions*. But in their discussion of the latter, Arnauld and Nicole do not even once mention indefinite propositions or the property of indefiniteness. Among their examples of indefinite propositions, however, we have found the proposition "L'homme est raisonnable"; and in their exposition of a *définition de chose* (real definition), we read: "Thus one defines man as a rational animal" (164). The intimate connection between the general problem of the logic of indefinite propositions and the problem of the use and the logic of real definitions cannot, therefore, be denied. And indeed, among the conditions which they give for a good definition, we find: "that it be universal" (165). Assuming that they give a *definitio rei* of 'Man' by means of a proposition having a subject term in the singular, and considering that they characterize "l'homme est raisonnable" ("man is rational") as an indefinite proposition, we can see that they must adopt the position that indefinite propositions should be regarded as having universal import, at least in a metaphysical context or subject-matter.

In order to eliminate the dangerous ambiguity of such propositions in unclear contexts and situations, one will, of course, then have to extend this account to all possible uses of indefinite propositional forms, whether with a subject in the singular or with a subject in the plural. Otherwise, the only acceptable solution is to distinguish two kinds of indefinite proposition, using some formal criterion. Arnauld and Nicole do not take this line, however. They do not discuss the "mobility" of terms at all.

As far as I can see, no traditional logician has ever tried to connect systematically the distinction between indefinite propositions as used in a *matière nécessaire* and indefinite propositions as used in a *matière contingente* with Kant's distinction in the theory of negation between *verneinende* (negative) and *unendliche* (infinite) *Urteile*.

24. EXCEPTIONS POSSIBLE IN CASE OF A *universalité morale*

Did this most famous of all logical textbooks until the rise of modern logic have the effect on its readers that the ambiguity, with which other philosophers had burdened the indefinite propositional forms, was eliminated and replaced by absolute univocity? In the light of certain remarks made by Arnauld and Nicole in their book, which we shall quote below, this seems improbable; the confusion in the minds of their readers about the logic of indefinite propositions more likely increased rather than decreased from a study of this book.

For in the section in which indefinite propositions are dealt with we find the following passage: "It is therefore clear that in any context whatsoever, propositions of this kind ["les ours sont blancs", "les hommes sont noirs", "les Parisiens sont Gentilshommes"] are taken to be universal; but that in connection with a contingent subject-matter one contents oneself with a moral universality ["universalité morale"]. As a result one may very well say: Les François sont vaillans: Les Italiens sont soupçonneux: Les Allemans sont grands: Les Orientaux sont voluptueux, although that may not be true of all the particulars, because one is satisfied it if is true for the greater part" (1965/155).

In order to discover whether Arnauld and Nicole consider the use of this kind of sentence compatible with decent logical standards, we must examine how they define the expression "universalité morale": "It will be necessary to distinguish two kinds of universality, one which may be called Metaphysical, & the other Moral.

I call universality metaphysical when a universality is perfect & without exception, like in *all man* ["toute homme"] *is* [*a*] *living* [*being*], that admits of no exception.

And I call universality moral when it admits of some exception, for in practical [moral] matters one is content if things ordinarily are like that, *ut plurimùm*, like what St. Paul reports & accepts;

Cretenses semper mendaces…" (1965/149).

So the proposition "Bantus are primitive" is in general false, just like the proposition "Les François sont vaillans", *unless* one is content with a *universalité morale*. It will be clear that Arnauld and Nicole have thereby opened the door again to illogical thinking and use of language: the proposition "Les Français sont vaillans" is false in its *universalité méta-*

physique, but in its *universalité morale* is acceptable. By *"universalité métaphysique"* the authors understand, however, according to their own words, simply "a universality [which] is perfect & without exception"! (*loc. cit*). We are therefore forced to conclude that their assertions, taken together, betray a completely circular mode of thought. They do not even begin to formulate a criterion by means of which readers or listeners may distinguish those uses of an indefinite proposition which make it universal and hence, in most cases, false.

Notwithstanding the obscurity of the whole discussion I think we may conclude that Arnauld and Nicole would allow a critical listener in a very serious discussion to treat an indefinite proposition as universal. This is a very different attitude from that of members of the school of Fichte and Hegel and indeed of German idealists in general, with the sole exception, perhaps, of Kant. They hold, further, that anyone who considers an indefinite proposition as true although it permits of exceptions, has thereby set it in a less theoretical context, in a *matière morale*, as it might be called.

It is also surprising that on the basis of the Port-Royal logic one cannot make out whether the authors and other contemporary philosophers thought it rationally and ethically acceptable to use the singular form "the/a Bantu is primitive", which has the form of an incomplete real definition, as well as "Bantus are primitive", with a *universalité morale* allowing for exceptions.

25. ARCHBISHOP WHATELY'S PERPLEXITY

Ziehen says (1920/666) that Richard Whately, and following him John Stuart Mill, rejected indefinite propositions. We find in Whately's *Logic*, to start with, a very clear though short discussion of the usual views about indefinite propositions and the dangers that are involved in their use (1859/43). Whately, it turns out, doubts the appropriateness of the distinction between nominal and real definitions of mathematical entities, so perhaps he was a logicist in the philosophy of mathematics; a circle or a square cannot have any *other* properties, he says, than those which are "implied in the *definitions* of those terms" (96), that is to say, in the definitions of the terms "circle" and "square". A nominal definition of "circle" may therefore also be construed as a real definition of certain mathematical entities (circles). This does not hold, he says, for other terms, *i.e.* presumably non-mathematical terms. Their referents may have properties that are not implied by the (lexical?) definition of the general term. To an individual Bantu, who in Whately's opinion goes to make up, together with the other Bantus, the semantical correlate of the word "Bantu", we may or may not ascribe the property of being primitive, whether or not this property or its complement does or does not form a part of the (lexical) definition of the term "Bantu". The names of these properties which are no part of the (lexical) nominal definition of the general linguistic term at issue must therefore be *added to* the nominal definition. One might, of course, say that to carry out this addition is to produce a real definition but, says Whately, "the very word "Definition" however is *not* usually employed in this sense; but rather, "Description"" (*loc. cit.*). To determine such characteristics as are not mentioned in the nominal definition is, he maintains, no task for the logician: "Logic is concerned with *nominal*-definition alone; with a view to guard against *ambiguity* in the use of terms" (96).

In his discussion of indefinite propositions the real definitions are not, however, explicitly mentioned. We do find there an exposition which is closely related to that in the Port-Royal logic. Speaking about propositions with a "Common-term" as subject, sometimes preceded by a universal or by a particular sign, Whately says: "Should there be *no sign* at all to the common term [NB! cp. IV-8], the quantity of the proposition (which is called an *Indefinite* proposition) is ascertained by the *matter*; *i.e.* the nature of the connexion between the extremes: which is either Necessary, Impossible, or Contingent. In necessary and in impossible Matter, an Indefinite is *understood as a universal*: *e.g.* "birds have wings"; *i.e. all*: "birds are not quadrupeds"; *i.e. none*: in contingent matter, (*i.e.* when the terms partly [sometimes] agree and partly not) an indefinite is understood as a Particular; *e.g.* "food is necessary to life"; *i.e. some* food; "birds sing:" *i.e. some* do; "birds are not carnivorous"; *i.e. some* are not, or, all are not" (43).

By adding "or, all are not" to the interpretation of this last example Whately demonstrates, perhaps unwittingly, the unclarity of this doctrine, or what is worse, its circularity: are we to understand that the sentence "birds are *not* carnivorous" is to *mean* "all..." just if all birds happen not to be carnivorous – as a matter of necessity, perhaps? And that if this happens not to be so, then the sentence is to *mean* "some..."? Is there a connection between Whately's strange comment on this sentence and the Kantian distinction between "negative" and "infinite" judgments?

Whately, like the authors of the Port-Royal-logic, presents this (see quotation) as current doctrine, but, again like his French colleagues, he does not present it without some very sceptical remarks, which show clearly enough that he, too, regards the situation as logically unsatisfactory. "It is very perplexing to the learner", he writes, "and needlessly so, to reckon *indefinites* as one class of propositions in respect of quantity. They must *be* either universal or particular, though it is not *declared* which". He does not follow Aristotle (in the *Analytics*), Theophrastus and Alexander in declaring all indefinite propositions to be particular propositions, nor does he plead for the introduction of a new convention according to which the listener may understand all indefinite propositions as universal and therefore mostly as false propositions. But he does display an unmistakable concern with the communicative function of language, for he goes on to say: "The person, indeed, who utters the indefinite proposition, may be mistaken as to this point [*i.e.* the question of whether the proposition is or is not true in its universal interpretation, E.M.B.], and may mean to speak universally in a case where the proposition is not universally true. And the hearer may be in doubt *which* was meant, or *ought* to be meant; but the speaker must mean either the one or the other" (*loc. cit.*). *One cannot mean an indefinite judgment at all.* This account is strongly at variance with that of practically all the idealists.

Whately, alas, does not recommend any rule of behaviour, either, by which the process of communication could be positively affected. But by discussing the ambiguity of indefinite propositions in dialogical terms of speaker and hearer rather than in terms of the ways in which the meaning ought to be "taken", he does at least draw his readers' attention to the danger inherent in the use of these propositional forms.

Like the Port-Royal logic, Whately's logic contains no theory of *suppositio*.

26. MILL AS THE HEIR OF WHATELY: REJECTION OF REAL DEFINITIONS

John Stuart Mill gives as an example of an indefinite proposition: "Man is mortal". He does not even consider the possibility that "man" in this sentence designates a *res*,[19] and rejects the doctrine that (in the theory of judgments, if not in the theory of inference) indefinite propositions belong to a class of their own: "When the form of expression does not clearly show whether the general name which is the subject of the proposition is meant to stand for all the individuals denoted by it, or only for some of them, the proposition is, by some logicians, called Indefinite; but this, as Archbishop Whately observes, is a solecism, of the same nature as that committed by some grammarians when in their list of genders they enumerate the *doubtful gender*" (1965/54).

The notion which Mill, following Whately, rejects on the page in his *System of Logic* to which Ziehen refers turns out to be the notion of definitions of things, *definitio rei* (92). Mill also rejects real definitions of mathematical entities (94) and is, therefore, more radical than Whately. But outside the field of mathematics both authors agree that definitions can be given only of terms.

It is of course no mere coincidence that Mill was also one of the first to attempt the formulation of an inductive logic in the modern sense. There is a close connection between a belief in real definitions and the old notions of deduction and induction (cp. IX-12).

27. "THE MOST PERFECT EXPOSITION OF CLASSICAL LOGIC": KEYNES

John Neville Keynes defines "an indefinite proposition" as a proposition in which nothing is said explicitly about "the quantity" by means of one of the words "all", "some", "many", or the like. He offers the example "Cretans are liars", which we have already met with in our discussion of the Port-Royal logic (Section 23), and also mentions the formula *S is P*. Keynes says: "At any rate the so-called *indefinite proposition* is not the expression of a distinct form of judgment" (1928/105). We may safely infer from this statement that according to Keynes, too, *rein logische Urteile* in the sense of Pfänder do not exist. Whately, Mill, and Keynes all agree upon this crucial point of logical theory.

This contrasts sharply with the following words of von Freytag: "In pure logic it has always been emphasized that every judgment is of the form *S is P*" (1961/61). The explanation of his view is to be found in what he says about particular judgments: in the *reine Logik*, the word "some" means the same as "a species of...". Now if a pure logician, as he ought to do, thinks about this species or kind, then in the case of a particular proposition, too, he will "in reality" be concerned with a judgment about a concept, *Gegenstand* (object), *res* or *Sache*, although with an *indefinite* concept, *Gegenstand, res*, or *Sache*. That is to say, the quantifiers are completely removed from the philosopher's range of vision by the "pure logicians"; every kind of proposition with a general term in the subject position is, "in a sense", of the "form" *S is P*. And this is what matters, not the difference between the quantifiers in the propositions from which these "judgments" are derived. In the terminology of the schoolmen, we might say that the "pure logic" which von Freytag defends today is a logic in which only those propositions are considered in which the subject term has *suppositio simplex* or *suppositio naturalis*. An individual, too, is a species or kind, von Freytag maintains. His logic can therefore be said to be a taxonomic (classifying) logic, a logic of species and genera.

Keynes holds, on the contrary, that an indefinite proposition is only an imperfect means of expressing a judgment, imperfect precisely because the quantifier is undetermined. He thinks, optimistically, that from the knowledge which the listener already possesses about the topic under discussion, or from the context, he will as a rule be able to guess whether an indefinite proposition is meant to be universal or particular. But he immediately adds: "if we are really in doubt with regard to the quantity of the proposition, it must logically be regarded as particular" (*loc. cit*). Although the dialogical workability of a logic in which indefinite propositions are regarded as well-formed is not really increased by this remark, it is nevertheless important that he thereby rejects the "moral universals" of the Port-Royal logic which we encountered in Section 24. When summing up his position, it turns out to be close to that of Aristotle in the *Analytics*: "Logically they [all the indefinite propositions] ought not to be treated as more than particulars, or at any rate pluratives" (*loc. cit.*).

Four decades earlier, Whately had written about negative terms: "A Privative or Negative term is also called *Indefinite* [infinitum] in respect of its not defining and marking out an object; in contradistinction to this, the Positive term is called Definite [finitum] because it does thus define or mark out" (1859/82f.; the brackets are his). Keynes is inclined to agree with Hamilton – and, as we have seen, with Krug (Section 18) – that indefinite propositions (*the/an*) *S is P* ought not to be called "indefinite"

128   HISTORICAL SURVEY

but rather "indesignate propositions". Like De Morgan, he thinks that
particular propositions of the form *some S is P* or *some S's are P* have a
greater natural right to this name, "indefinite", than propositions with
no quantifier whatsoever. Note that this corresponds to the terminology of Ramus in his last period (cp. Section 15). We shall discuss the systematic relationship between particular and indefinite propositions in
traditional logic in Chapters X and XI.

Keynes also discusses the form *a certain S is P* which he calls "singular
indefinite", in contradistinction to *this S is P* which he calls "singular
definite". Propositions of the form *a certain S is P* "possess also the indefinite character which belongs to the *particular* proposition" (103). It is
no accident that Keynes inter-relates the indefiniteness of both kinds of
propositions, *some S are P* and *a certain S is P*. It is an indication that the
logic of existential instantiation has been at issue in the traditional problem
of indefinite propositions.

In the opinion of Heinrich Scholz, Keynes' logic is "altogether the most
perfect exposition of classical formal logic and within the Anglo-Saxon
culture its effect has been as considerable as it has been beneficial" (1967/
46). It seems clear that Keynes should share this praise with both Mill and
Whately. It is very remarkable, therefore, that neither Whately nor
Keynes is mentioned by von Freytag and Jacoby in their surveys of
important traditional logicians. I have already pointed out that they do
not mention De Morgan either (I-3). Keynes himself acknowledges in the
preface to the fourth edition of his work that he is much indebted to
Chr. Sigwart. Von Freytag, on the other hand, is clearly an heir of
Erdmann and Pfänder.

28. A TWENTIETH CENTURY "PHILOSOPHICAL" LOGIC ON INDEF-
INITE PROPOSITIONS AND *suppositio naturalis:* MARITAIN

(i) Maritain's examples of *propositions indéfinies* are: "L'homme est
mortel", "L'homme est injuste". His commentary begins with the remark
that every indefinite proposition "ought to be taken by the logician for
what it is in reality, albeit in a concealed manner" (144). That is to say,
or so it seems, that an indefinite proposition with some kind of general
import ought to be understood either as a universal or as a particular
proposition. This reminds us of Whately, Mill, and Keynes, who regarded

an indefinite proposition as one which in reality is universal or particular. The difference between Maritain's further discussion of the matter and Whately's is, however, quite remarkable. To start with, Maritain points out to his readers that some indefinite propositions, namely those whose subject is a universal nature – "une nature universelle non restreinte" – may be treated as singular propositions. This reminds one of Pfänder's generic *Singularurteile*. His examples of this kind of proposition are: "Man is a species of animal", "Man is the noblest of creatures", "The circumference is the place of points situated at equal distance from the centre" (144).

Secondly, Whately explains that the "real" significance of a proposition is to be found in the meaning given to it by the speaker; Maritain mentions neither speaker nor listener. As an explanation of what it means to "take" a proposition for what it is in reality, he only says that the universal subject is taken *précisement en tant qu'un*, (in as far) as one, and not according to its being in the things, but according to the unity which it enjoys in the mind ("l'esprit"). In *which* mind[20] he fails to report; for Maritain is thinking in terms of the medieval *esse objectivum*. The fact that two or more minds may be engaged in the pursuit of truth is completely overlooked in Maritain's logic.

(ii)   In his discussion of the *suppositiones* of terms Maritain has a surprise in store for us. Like Peter of Spain and the other scholastic supporters of a "philosophical" logic as discussed in Section 14, to which group Vincent Ferrer belonged, he recognizes a *suppositio naturalis* or *suppléance essentielle* (1933/87). But while Peter subsumed both *suppositio simplex* and *suppositio personalis* under *suppositio accidentalis* and contrasted the latter *suppositio* in all its variants with *suppositio naturalis*, we find the distinction between *naturalis* and *accidentalis* in Maritain's logic as a distinction drawn inside the category of *suppositio personalis*, which he also calls "suppléance réelle":

$$
\begin{array}{ll}
s.\ personalis & \left\{\begin{array}{l} 1.\ suppositio\ naturalis \\ \quad (suppléance\ essentielle) \\ 2.\ suppositio\ accidentalis \end{array}\right.
\end{array}
$$

The conception of a "philosophical", essential *suppositio* turns out to have invaded the field of *suppositio personalis*, with which it had nothing to do in Peter's logic. It is therefore not without interest for us to see that Vincent Ferrer's theory of *suppositio* is mentioned in positive terms in Maritain's book (76n.).

The example used by Maritain in order to illustrate *suppositio naturalis* is the proposition "l'HOMME est capable de raisonner". Here the subject term stands for ("supplée pour") a thing ("une chose") to which the predicate belongs "intrinsèquement et essentiellement" (88), and by consequence it has, so he says, *universal suppositio*. We might say: the subject term has some kind of universal import. In the case of a *suppositio personalis* which is only *accidentalis*, the subject term has a definite ("determinée") *particular suppositio* (the same as *suppositio disjunctiva*), provided we are not dealing with a collective subject term (with *suppositio disjuncta*, cp. XI-3). We therefore arrive, so he explicitly states, at the following rule: the subject of an indefinite proposition "has a universal *suppositio* in a necessary context [subject-matter], and a particular determinate *suppositio* in a contingent context" (88). This is precisely the rule that was rejected by Arnauld and Nicole.

Suppositio naturalis is mentioned once more in Maritain's logic. It then turns out that "a proposition in a necessary context" simply *means* ("c'est-à-dire") a proposition in which the predicate belongs to the subject essentially; "one says that the *S* has *suppositio naturalis*" (269). The rule which we just formulated is therefore a tautological rule of no use whatsoever; we might also say that Maritain has formulated a circular definition of "a proposition in a necessary context" and of *suppositio naturalis* or *essentialis*. He does not even realize that the listener, if furnished with his rule, is none the wiser, unless the speaker *tells* him whether he means to make an essential or an accidental predication, for instance by means of a systematic use of article-like operators as indicators of the intended *suppositiones*.

Surprisingly enough, this classification of *suppositio personalis* into natural and accidental *suppositio* is not visible in his schematic survey of his theory of *suppositio* (83).

The *naturalis-accidentalis*-distinction is, we read, a classification with respect to the verb or with respect to the copula (87). For if the subject has *suppositio naturalis*, then "by itself,[21] the copula only says [something about] the relation of predicate to subject in a possible [mode of] existence" (88), in the sense that the subject need not actually exist in order that the proposition be true.[22] Comparison should be made with quantified modal logic and with "free" logics (cp. Lambert 1970).

It seems reasonable to conclude that in Maritain's opinion the subject terms in the examples under (i) all have *suppositio naturalis* (*essentialis*), hence also the subject term of "man is the noblest of creatures", which in the logics of the thirteenth century logicians Peter of Spain and William of Sherwood was said to have *suppositio simplex*. This would mean that as far as Maritain's influence goes,[23] the problem of the logic of logophoric value judgments should be solved not by an inferential theory of *suppositio simplex*, but by an inferential theory of *suppositio naturalis*. He seems to hold that an indefinite proposition whose subject term has *suppositio naturalis* entails the corresponding universal proposition, for he holds that "every man whatsoever is more noble than any irrational creature", and "every individual circumference is the [geometrical] place of points situated at equal distances from the centre". Apart from the fact that an indefinite proposition contains no sign by which the intended *suppositio* of the term can be communicated, the problem is still further complicated by the existence of so-called reduplicative propositions (Section 30).[24]

29. *Suppositio naturalis* THROUGHOUT THE CENTURIES

The following chart is adapted from de Rijk 1973 (Figure 2 below). I have added the last item on the left side and the last two on the right. The stress on the logical primacy of the *Begriff* (concept) in German idealism becomes comprehensible if seen in the light of Peter of Spain's non-contextual *suppositio naturalis*. When Peter speaks of "natural", he refers to the nature of the term; Vincent Ferrer, however, speaks of the nature or essence of the thing or *res* (de Rijk, *op. cit.*). The German notion of a *Begriff* is usually something in between the notion of an interpreted linguistic term and the notion of an essence of a non-linguistic entity, which points to an influence from both schools. Further historical research may lead to the insertion of lines of influence from the upper right part of the chart to the lower left.

The importance of the distinction between contextual and non-contextual definitions of "suppositio naturalis" may be questioned. Maybe we can say that the *suppositio naturalis* of a term sums up the ranges (left

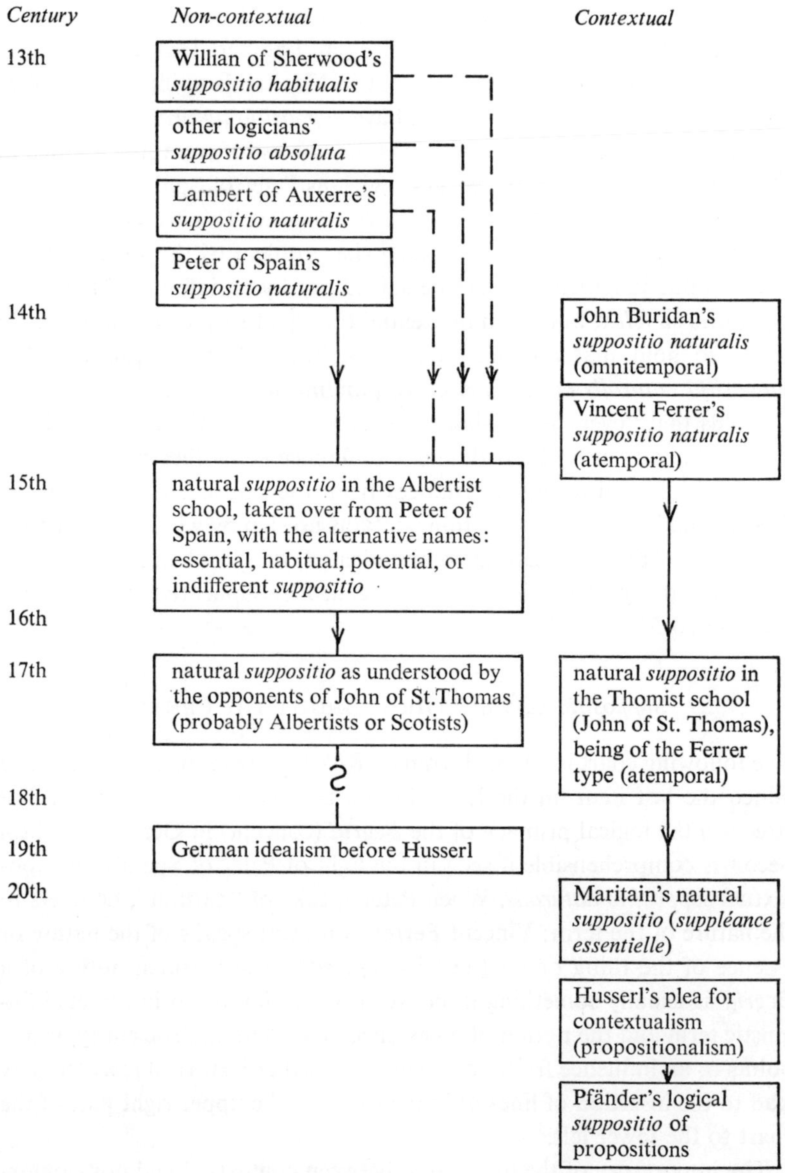
Century
Non-contextual
Contextual
13th
Willian of Sherwood's *suppositio habitualis*
other logicians' *suppositio absoluta*
Lambert of Auxerre's *suppositio naturalis*
Peter of Spain's *suppositio naturalis*
14th
John Buridan's *suppositio naturalis* (omnitemporal)
Vincent Ferrer's *suppositio naturalis* (atemporal)
15th
natural *suppositio* in the Albertist school, taken over from Peter of Spain, with the alternative names: essential, habitual, potential, or indifferent *suppositio*
16th
17th
natural *suppositio* as understood by the opponents of John of St.Thomas (probably Albertists or Scotists)
natural *suppositio* in the Thomist school (John of St. Thomas), being of the Ferrer type (atemporal)
?
18th
19th
German idealism before Husserl
20th
Maritain's natural *suppositio* (*suppléance essentielle*)
Husserl's plea for contextualism (propositionalism)
Pfänder's logical *suppositio* of propositions

Fig. 2.

fields) of all its accidental *suppositiones*, regarded as functions (cp. Section 14), with respect to every proposition in which the term can *possibly* occur. Put in this way the definition is clearly contextual, but in such a manner that mention of the propositional context may well be left out: the natural *suppositio* of a term is independent of any one proposition in particular. Peter of Spain and Vincent Ferrer (though probably not Maritain) might conceivably both agree upon this definition.

That somebody's thought was influenced by somebody else's is no argument against it, not even if the latter is proved to have been wrong. I have discussed *suppositio naturalis* and *suppositio simplex* in some detail in order to be sure that no historical material that might possibly throw light upon the riddle of the logic of the articles should remain unexploited. The illumination, apart from historical, has so far been meagre, but our work is not yet done. One may nevertheless say already that to explain the meaning of propositions like "homo est dignissima creaturarum" and "homo est iustus" by invoking *suppositio simplex* or *suppositio naturalis* is no justification for this form of locution from a dialogical point of view and cannot be seen as an attempt to describe the truth-conditions of propositions of this linguistic form.

30. REDUPLICATIVE PROPOSITIONS:
"IN SO FAR AS" AND "AS SUCH"

(i) *Reduplicative Propositions.* William of Sherwood's fourth type of *suppositio*, illustrated by "homo est dignissima creaturarum", was a kind of *suppositio simplex*, namely *pro significato comparato ad res* and *solum cum reduplicatione speciei*, only by "reduplication" of the species. The second part of the qualification, the expression "cum reduplicatione speciei", has survived into our time in certain philosophical schools. We find it, *e.g.*, in Maritain's "formal" logic, though not in his theory of the *suppositio* of terms but in his classification of propositions. Maritain uses the Latin expressions "ut" and "quatenus" and the French "en tant que", but not the corresponding Latin terms "in quantum" and "qua".

Maritain classifies propositions as *simple* or *complex*; the latter may be either *openly* or *opaquely* complex ("occultement composées"). *Copulative* propositions are what we should call conjunctions (of propositions). Furthermore he mentions disjunctive and hypothetical propositions, all of which are openly complex. The opaquely complex propositions are either *exclusive* or *exceptive* or *reduplicative*. From his examples one may infer that these include propositions of the following verbal forms:

(a)   *only M is P* (exclusive),
(b)   *every M except [those which are] S, is P* (exceptive),
(c)   *the/an M, in so far as [it is an] M, is P* (reduplicative).

At another point in his book (135) we read that propositions of type (c) should be called "specificative", and that they may be paraphrased as follows:

(c′)   *the/an M as such* ["comme tel"] *is P.*

The real reduplicative propositions, on the other hand, are propositions like: "[a] man, as a rational being, is gifted with the faculty[25] of laughter". He clearly has the following grammatical form in mind:

(r)   *the/an S qua M is P,*

where *S* and *M* are different and non-synonymous expressions.

Now all opaquely complex propositions are *exponibiles*, hence also the reduplicative ones. According to Maritain the correct analysis of the form (r) is (in my symbolization):

(r1)   *the/an S is M,*
(r2)   *every* ["tout"] *M is P,*
(r3)   *the M is the reason* ["raison"] *why the S is P.*

Concerning (r1), Maritain's own explanation is this:

il y a en lui [le S] le M,

that is to say that the S carries the M in itself or himself. In his example *le M* is the term "la rationalité".

Some reduplicative propositions demand the following analysis:

(r2′)   *every* ["tout"] *S which is M, is also P,*
(r3′)   *the M is the cause of S's being P.*

Still other ones mean

(r2″)   *the/an S, if M, is P,*
(r3″)   *the M is the condition for* ["moyennant laquelle"] *the S being P.*

Maritain warns us that a great many erroneous results are produced because propositions where a reduplicative meaning is intended ("sous-entendu") are mistaken for simple or categorical propositions. This is often the fate of "formal philosophical propositions" (137). Reduplicative propositions are, in his opinion, quite indispensable. Among the examples of such propositions offered in his *Introduction générale à la philosophie* are:

> everything which is, is good;
> the public good always comes before the individual good;
> one should always obey one's superiors;
> there are human beings who are naturally slaves

(1939/181). All these propositions are, in his logic, true *formaliter loquendo*, *formellement parlant*, but false *materialiter loquendo*, that is if they are "entendu", understood, materially. Some of them are reduplicative; thus in the first of these examples the hearer has somehow to insert "in as far as it *is*", and in the third "in as far as they are superior, that is to say in as far as they do not demand anything which is contrary to the commands of a superior of a still higher rank" (*loc. cit.*).

(ii) *Specificatory propositions.* If the same term is substituted for "*S*" as for "*M*", then we obtain a proposition which is "(seulement) spécificative", merely specificatory: in

(spec)   *the/an M qua M is P*,

or, in other words, in

(spec′)   *the/an M as such is P*,

the subject is once more "specified" before it "receives" the predicate or *forma*.[26] The subject brings its own concept, species or *forma*, being the "formal reason" by which its kind is constituted, to bear upon itself before *P* is ascribed to it (135).

Although Maritain does not explicitly say so, it seems reasonable to assume that a subject term *le M comme tel*, in English: *the/an M as such*, of a specificatory proposition is a term which has *suppositio naturalis* or, in other words, a term which should somehow be "taken" *essentialiter*.

There is an obvious and close affinity between Maritain's logic and Pfänder's. The latter ascribes a "logical" or "formal" *suppositio* to sentences which he considers as philosophically important. The *formale* or *logische Supposition* of a sentence is also indistinguishable from its interpretation as a specificatory proposition whose subject term has *suppositio naturalis*.

(iii)   *Analysis of specificatory propositions as reduplicative propositions.* Since (spec) is obtained by substituting "*M*" for "*S*" in (r), it ought to be possible to analyse (spec) in the same way as reduplicative propositions, (r1)–(r3):

(spec 1)   *the/an M is M* (Plato's Principle),
(spec 2)   *every M is P,*
(spec 3)   *the M is the reason,* or: *the cause,* or: *the instrumental condition for the M's being P.*

The first of these components corroborates the hypothesis I have called "the H-thesis". The third component is completely incomprehensible on any account.

As far as (r) is concerned, Maritain's analysis consists in a conjunction of three propositions, one of which is either *every M is P* or *every S which is M, is P* or else *if an S is M, then it is P.* This would seem to make

> *the/an S qua M is P*
> *this S is M*
> ——————
> ∴ *this S is P*

a valid argument form. Therefore, if (spec) is interpreted in the same way as (r), then this argument form, too, must be valid:

> *the/an M as such is P*
> *S is an M*
> ——————
> ∴ *S is P.*

For example:

(I)
> the Bantu *as such* is Primitive
> Saul is a Bantu
> ——————
> ∴ Saul is Primitive

– *pace* the truth of the premisses.

Now Maritain rejects the interpretation of a specificatory proposition as a simple border-line case of a reduplicative proposition, and he criticizes Goudin for accepting it (136). The right interpretation of the proposition "The evil [person], *in so far as* [he is] evil, should be hated" consists, he informs us, rather of two propositions only: "The evil person is thus specified by a certain determination (evilness)", and "That [ingredient] in him which falls under this determination should be hated" (135). In general:

(spec 1′) *the/an M is specified by the determination M-ness*
(spec 2′) *that [ingredient] in the M, which falls under the M-ness, is P.*

The latter component can presumably be read as follows: for every M-individual it is the case that that [ingredient] in him which falls under the M-ness, is P. For example: that part of, or ingredient in, Saul, which falls under the (concept of) Bantu-ness, is primitive. If this reading is accepted, (I) is *not* a valid argument, since this analysis of (spec) contains no universal component. But perhaps we may understand (spec 2′) in such a way that we obtain a valid argument provided we accept a reduplicative individual proposition (cp. VII-15) as conclusion:

	the/a Bantu *as such* is Primitive
(II)	Saul is a Bantu

Saul, *in so far as* [he is a] Bantu, is Primitive.

The major premiss of (II) cannot be refuted, of course, simply by pointing to non-primitive Bantus as counter-instances.

Let us close this section with a quotation from Frege's *Grundgesetze*: "Dieses "als solches" ist eine vortreffliche Erfindung für unklare Schriftsteller, die weder ja noch nein sagen wollen. Aber dieses Schweben zwischen beiden lasse ich mir nicht gefallen..." (1962/XXIIIf.).[27]

NOTES

[1] Scholz characterizes the English translation of the Organon under the editorship of Sir David Ross as very reliable (Scholz 1967/72). In the present study I have therefore used and quoted from that translation. In addition I have made use of Ackrill's translation of *De interpretatione*.
[2] H. Maier translates the first of these two examples as follows: "Die Glieder der conträren Gegensätze fallen in dieselbe Wissenschaft" (Maier 1896/160). The plural

"die Glieder" is undoubtedly a more correct translation than Rolfes's singular. But for our purposes it is of interest to notice that the linguistic feeling of some – in fact, of many – Germans leads to a preference for the grammatical singular, as in the translation by Rolfes. Aristotle in any case used grammatical subjects in the plural as well as grammatical subjects in the singular in his examples of indefinite propositions. Edghill's translation comes closer to the meaning of the forms *any M is P* and *all which is M, is P* than the German translation by Rolfes. Peter of Spain also discusses this example, in Section 6.07 of his *Summulae logicales*. He says that the predicate in the proposition "Omnium contrariorum eadem est disciplina" has *suppositio simplex*, but without saying what the predicate of this proposition is; the (logical) subject is not mentioned. In Rolfes's translation the singular form of the grammatical subject "das Konträre" gives the impression that the logical subject term has *suppositio simplex*. Aristotle does not use Peter of Spain's quantifier "omnium" in his formulation, but it is nevertheless likely that Peter took his example from him.

[3] "Cursus enim inest homini gratia alicuius singularis", says William of Sherwood in explanation of the fact that he contributes a *suppositio personalis* to the subject term of this proposition (quoted from Kneale and Kneale 1964/249). "The postulated identity between a thing and its essence creates an atmosphere in which "man runs" is as meaningful as "this man runs"" (Angelelli 1967/120). Cp. XI-4.

[4] Cp. Angelelli 1967/96.

[5] Attempts to construct a theory of types of the kind of which Peter of Spain's theory of *suppositio simplex* is an example were not continued after the decline of scholastic logic; by and large they were forgotten in the centuries which followed them. The logic of Maritain is an exception to this rule. Von Freytag's logic on the other hand shows no trace of such a stratification.

[6] Angelelli (1967/132) holds that in the logic of Thomas Aquinas, "homo ambulat" implies a proposition about a universal, albeit a universal-in-the-thing. In that case, what is meant is probably a universal in an individual person. This corresponds to von Freytag's notion of an *individual logophore*. Von Freytag's logophores are clearly a kind of logical nuclei, or rather, they are the subject-part of such nuclei, considered as a bundle of predicates *of* something. Exactly the same nucleus occurs in 'homo' and in 'Socrates', or rather, nuclei with identical predicate-bundles. Their subject or carrier – the logophore – is called "determinate, individual" with respect to 'Socrates' and "indeterminate" or "indefinite" with respect to 'homo' (von Freytag 1961/39). The interpretation of "homo ambulat" which is rejected by Angelelli is that corresponding to the interpretation (1) of "homo currit".

[7] Cp. II-4 above. Cp. Fearnside and Holther 1963/60.

[8] The ethical content of this statement can also be found in Spinoza; cp. Hampshire (Strawson (ed.) 1968/56). Hampshire connects this with Spinoza's "nominalistic logic" (*op. cit.*/66). "Nominalistic" here clearly means "non-essentialist", not "non-Platonist" in the sense of the logical "Platonism" of Russell, Church and others.

[9] Angelelli mentions (1967/102) an example from the year 1600: "Indefinita est quae habet subiectum universale sine ulla nota, ut homo est iustus". The author is Julius Pacius, and the book was called *Institutiones Logicae in usum scholarum Bernensium* (Bern 1600).

[10] The logic of limitative propositions is useless, since the classification of propositions containing a negation element into limitative and others is not viable. Take for example the proposition "every Bantu is not civilized". If we define "civilized" as "not primitive", then we are faced with the problem of where to place this proposition in

Kant's classification. According to the nominal definition we just introduced, it is synonymous with "every Bantu is not not primitive". Kant's explanation of his classification does not allow for an unequivocal answer to the question whether this proposition, which contains two consecutive "not"s, is an infinite – limitative – proposition or not (cp. Prior 1967). This is not made any clearer by dropping the quantifier: the logical implications of "the Bantu is not not primitive" in Kant's logic are unclear and open unlimited possibilities for fruitless reflection. Cp. von Freytag 1961/27f., 1967/113; Maritain 1933/137. Above all, see Swing 1969/3–27. Wilcox (1971) points out that without "a doctrine of the distribution of the complement of terms" the influential theory of distribution of terms will always remain incomplete and unsatisfactory. "Knowledge of all of a class gives us, in itself, no knowledge of its complement" (*loc. cit.*).

[11] Cp. Section 18, about Kant's *generale Urtheile* and about his opinion that the difference between *generale* and *universale* judgments is no concern of logic.

[12] Cp. Jespersen's terminology, VI–30.

[13] In full his example is: "Der Mensch hat reflektirendes Wahlbewusstsein". In the terminology of the schoolmen, Erdmann seems to have in mind a judgment whose subject term has some kind of *suppositio simplex* or *naturalis* rather than *personalis*.

[14] Cp. V-3.

[15] Frege mentions Erdmann as a good example of a logician who does not distinguish between different levels of concepts and who consequently does not distinguish between a characteristic ("Merkmal") of, or *in*, a concept and a property of the concept itself (Frege 1962/XIV). This distinction, its variants and its history form the main topic in Angelelli's work (1967). Frege regards the lack of this distinction as typical of psychologism in logic; where everything is "idea", this and many other important logical distinctions are likely to be overlooked. Cp. VII–9.

[16] Cp. Angelelli 1967/122.

[17] According to Pfänder, *the judgment designs* ("entwirft") the factual situation: "As the judgment designs the fact, which differs from it, it determines the latter, taking its starting-point in itself; hence in so far the *judgment* is *primary*, the *fact secondary*. Being entirely construed from the judgment, the fact is quite dependent upon the judgment. It is designed by the judgment as a projected image by the projection lamp" (36). I used the same metaphor, not knowing Pfänder then, in an essay on "egocentric" logic (Barth 1968b/286). Cp. notes 20 and 21 below.

[18] Cp. note 9. See also Angelelli 1967/118.

[19] Cp. Angelelli 1967/121.

[20] "Ich frage: in wessen Geiste?" ("I ask: in whose mind?") (Frege 1962, I/2).

[21] "The writers, and editors, of the Logical text-books which run in the ordinary grooves – to whom I shall hereafter refer by the (I hope inoffensive) title "The Logicians" – take, on this subject, what seems to me to be a more humble position than is at all necessary. They speak of the Copula of a Proposition "with bated breath"; almost as if it were a living conscious Entity, capable of declaring for itself what it chose to mean, and that we, poor human creatures, had nothing to do but ascertain what was its sovereign will and pleasure, and submit to it" (Lewis Carroll in *Symbolic Logic*; quoted from Carroll 1970/268). Cp. note 20.

[22] De Rijk writes about John of St. Thomas: "John starts with a definition of supposition taken as a term's standing for something in the proposition and rejects a modern view of supposition which seems to identify supposition with signification. The many adherents of this view, he says, do not admit the old and approved rule that affirmative

propositions the subject term of which refer to nothing (*de subiecto non supponente*) are false" (1973/74). The view referred to here coincides with the result of Russell's analysis of "The present King of France is bald". Nevertheless "his natural supposition is that according to which the subject term of a proposition is taken to stand for all its supposits regardless of (the time of) their actual existence" (de Rijk, *op. cit.*/76) Maritain may perhaps be forgiven for being able to give a consistent interpretation of John's views (cp. Doyle 1953). Cp. our discussion of Jacoby's views in X-12 below.

[23] At the death of Maritain in 1973, Pope Paul VI acknowledged him as his philosophical teacher.

[24] For systematic reasons a discussion of Maritain's treatment of *suppositio copulata* and *suppositio disjuncta* will be postponed to XI-3.

[25] This should be seen in connection with the earlier faculty psychology or psychology of "what's". Cp., *e.g.*, Woodworth 1952/12ff.

[26] The vocabulary of traditional theoretical logic betrays its origin in a metaphysics of sex: there is a *subjected* entity, or subject, which *receives a potential* or *dynamic form*. The aversion to polyadic predication probably derives to no small extent from the impossibility of one form's *copulating* with more than one *subject* at a time. The word "copula" as a logical technical term was introduced by no other than Abelard.

[27] "This "as such" is an invention excellently suited to the needs of unclear authors who are unwilling to say yes or no. But this wavering between the two answers I will not put up with…".

CHAPTER V

FROM THE HISTORY OF THE LOGIC
OF INDIVIDUAL PROPOSITIONS

1. INTRODUCTORY REMARKS ABOUT THE USE OF THE WORDS "SINGULAR" AND "INDIVIDUAL"

In English the expression "propositio singularis" is translated as "singular proposition", in French as "proposition singulière", in German sometimes as "einzelnes Urteil" (Kant, Herbart), sometimes as "singuläres Urteil" (Hegel). There are authors who classify propositions in another way than the usual one and they often introduce a special terminology. Ramus speaks in terms of "enonciation propre" and of "axioma proprium" (IV-15) and Pfänder ascribes different meanings to "Singularurteile", "Einzelurteile", and "Individualurteile" (IV-21).

In German, "singular" is a grammatical and "singulär" a logical term. Kant used "Einzelurteil" as a translation of "propositio singularis". Pfänder's definition of "Einzelurteil" is, however, much narrower than the meaning usually ascribed to this term. In Pfänder's logic an *Einzelurteil* has the verbal form *ein S ist P*, where *S* is a general term. If the subject term is a proper name, as in "Saul is a Bantu", the proposition in Pfänder's terminology does not express an *Einzelurteil*, although it does express a *Singularurteil*.

For these reasons I shall in general employ the exact terminology of the author in question when describing theories and views. I will make one exception to this rule: when the author under discussion makes no distinctions between propositions other than those common among the schoolmen, I shall translate "propositio singularis" and "proposition singulière" as "singular proposition". But I shall use the expression "singular proposition" merely as a technical term of the logical systems under discussion without attributing any meaning to it in advance. If on occasion I want to refer to the grammatical form of a proposition, I shall employ the expressions "in the singular" and "in the plural".

In the preceding chapter I spoke deliberately of "individual propositions" and not of "singular propositions". "Propositio singularis" is a technical term in the tradition and in order to discuss traditional logical networks it is necessary to employ a terminology which is neutral with respect to doctrines of traditional logic. In modern logic, "individual" is used in a much broader sense than in other parts of philosophy, *i.e.* in the sense of an element of the domain of discourse. This domain need by no means consist of individuals in a metaphysical or in an ontological sense. One is free to choose a set of abstract, or of fictitious, or of complex entities as the domain of discourse. For this reason and also because propositions with more than one subject are admitted, modern logicians do not usually talk in terms of "individual propositions" and often avoid the expression "singular propositions" as well, in which case they resort to the expression "atomic propositions".

The schoolmen often used the expression "terminus discretus" for a term which was a name or a description of an individual. It might be thought a good idea to speak of discrete terms and to call a proposition about an individual "a discrete proposition". But except when reporting their theories I shall not follow the usage of the schoolmen. For although Bantus are discrete entities, and triangles as well, modern quantification theory does not depend on the elements of the domain of discourse being *discrete* elements. The domain of discourse may very well be not only densely but even continuously ordered in one or more dimensions; an important example is the set of real numbers, which may be named or described "individually" but which are not separable and which cannot therefore be called "discrete". The modern use of "individual" does not carry the traditional connotations of "discretus".

The expression "individual proposition" would therefore seem to be more suitable, as well as being fairly neutral with respect to both old and new logical theories. The following domains are of particular interest to us in the present study: (a) the domain of all human beings (sometimes: animals, sometimes: plants), and (b) the domain of all figures in a plane. An element of one of these domains will be called "an individual", and a proposition of the form *S is P* where *S* is a proper name or a definite description of an individual in our sense (or has the grammatical form of such a description, although it may fail to refer) will be called "an individual proposition". The individual propositions of the old metaphysics will therefore also be individual propositions in our sense, while the converse does not always hold. There is no doubt whatsoever that the traditional logic counted propositions like "Saul is a Bantu" and "Socrates is mortal" as well as propositions like "the philosopher who drank the hemlock was Plato's teacher" as singular propositions, expressing singular judgments. It is therefore quite clear that all our individual propositions belong to the singular propositions of the tradition, but it does not follow that the converse also holds.

Three points should be called to mind in this connection.

(i) We have posed the problem of the validity or invalidity of the inference form: *the/an M is P, S is (an) M, therefore S is P*, in which the major premiss is assumed to be an indefinite proposition, *e.g.*, "the Bantu is Primitive", and the minor premiss an individual proposition, *e.g.*, "Saul is a Bantu". In order to solve this problem we shall not only have to know how an indefinite premiss is treated but also how individual premisses were dealt with in the tradition.

(ii) If the Locke-Berkeley problem is (or contains) a logical problem and if propositions about an entity *the* triangle (or: *a* triangle) play a part in it, then the Locke-Berkeley problem is (contains) the problem how to reach an indefinite conclusion "*the/a* triangle..." starting out from one or more individual propositions as premisses.

(iii) Individual propositions are obtained as conclusions when the *dictum de omni et nullo* is applied to one universal together with one individual premiss: *every M is P, S is an M, therefore, S is P*. To this *dictum* corresponds the rule of universal instantiation of modern logic. This is not one of the rules in which the choice of an individual constant must be made "critically" (cp. III-2), which explains why this rule never gave occasion to serious philosophical disagreement.

Summing up: our special interest is in the logic of individual *premisses*, separately or in connection with an indefinite premiss, as explained in (i) above, whether or not the conclusion is an indefinite proposition as discussed in (ii).

2. ARISTOTLE ON INDIVIDUAL PROPOSITIONS AND THE DOCTRINE OF EKTHESIS

In *De interpretatione* Aristotle distinguishes, as we saw in IV-2, between two classes of propositions; they can be characterized as (i) propositions with a general import and (ii) propositions with an individual import.

In the *Prior Analytics*, however, the latter kind of propositions, being propositions with an individual subject, are not mentioned. Aristotle confines himself in this work to a distinction between (a) universal, (b) particular and (c) indefinite propositions ("ἀδιόριστοι"). We discussed the latter classification in Section 2 of the preceding chapter.

Aristotle never formulated a logic of individual propositions. His logical interest is directed toward such terms – Aristotle himself says: "things" – which can both be the subject of a proposition and be predicated of other things (43a 42f). Such is the case with the term "Man", for we can say "(a) Man is a Vertebrate" but also "(a) Greek is a Man". On the other hand, the term "Socrates" cannot be used freely as a predicate. Łukasiewicz explains Aristotle's lack of interest in individual terms by pointing out a fundamental requirement of Aristotle's syllogistic: "that the same term may be used as a subject and as a predicate without any restriction. In all three syllogistic figures known to Aristotle there exists one term which occurs once as a subject and then again as a predicate... Syllogistic as conceived by Aristotle requires terms to be homogeneous with respect to their possible positions as subjects and predicates. This seems to be the true reason why singular terms were omitted by Aristotle" (Łukasiewicz 1957/7; cp. Kneale and Kneale 1964/60, 69). Maier is of the same opinion (1896/164). Albrecht, however, maintains that Łukasiewicz's explanation misses a quite fundamental point. According to Albrecht, Łukasiewicz did not succeed in bringing out the connection between the *lack of individual terms* in Aristotelian syllogistic on the one hand and the doctrine of *ekthesis* on the other. This doctrine is credited by Albrecht with an immense logical importance, while Łukasiewicz thinks that the method of ekthesis "is of little importance for the system" (59; cp. Albrecht 1954/56) and regards it as based upon a logical procedure which falls outside of the boundaries of syllogistic.

When an individual is occasionally mentioned in the *Prior* or *Posterior Analytics*, Aristotle seems to let the individual proposition in question fall into the class of particular propositions. According to Maier, this is so at least when a special diagram of a triangle, Γ, is being discussed (*Pr. An.* 21.67a; cp. Maier *loc. cit.*). There is the Pittacus-syllogism in the third figure, at the end of the *Prior Analytics*, but it is difficult to conclude from Aristotle's discussion of that syllogism whether he regarded an individual proposition as a particular or as a universal proposition. Prior does not seem to be able to infer anything from that discussion either (Prior 1962/157).

3. COMPARISON WITH INDEFINITE PROPOSITIONS IN THE ORGANON

It seems that we may recapitulate our findings as follows: Of the indefinite and of the individual propositions in the logic of Aristotle it may be said: (1) that neither kind of proposition is of systematic importance in his logic, and (2) that propositions of both kinds are treated in the syllogistic as if they were particular propositions.

There are certain facts, however, which greatly obscure the issue. To start with, Aristotle formulates all propositions, the universal and the particular propositions as well, in the singular. We have already discussed this in IV-8. Secondly, his not very happy choice of examples, which we discussed in IV-20, should be mentioned. Furthermore, propositions lacking an "all" ("every"), "some", or "no" are sometimes considered in Aristotle's works to have a universal significance while in the theory of inference in the *Prior Analytics* he usually treats them as particular propositions. This is confusing enough, and, to make things worse, the subject terms sometimes occur with and sometimes without an article. Where they occur with a article in the singular or at least are translated that way ("die Lust..."), one can imagine that the reader is tempted to regard them as having the same logic as individual propositions. The logic of indefinite propositions will then depend on the logic of the so-called singular propositions.

Add to all this that Aristotle's use of variables is a very peculiar one from the point of view of modern logic.

William Kneale, heaving a sigh, says that Aristotle "started a queer fashion in the use of general terms" (274). He probably has in mind, among other things, that feature of Aristotle's syllogistic which Łukasiewicz used to explain why Aristotle neglected individual propositions and which has also been discussed extensively by Martha Kneale: every term must, without any restrictions, be able to function as a predicate in a categorical proposition. Or, to put it in another way: *the variables as introduced by Aristotle were all of the same kind.* This Aristotelian use of variables, together with the special features of the

Greek language, led, as Martha Kneale says, to "a philosophical obscurity which affected logic for many centuries" (1964/63).

More specifically, this is one of the reasons for the confusion of individual with general propositions in later philosophies. Aristotle's use of variables should be compared with his metaphysical doctrine of οὐσία, the "nature" or "essence" of things. For the sake of the doctrine of the logical-ontological nuclei or essences, "*ousia* had to become a sort of kaleidoscopic entity, neither universal nor singular, or... both" (Angelelli 1967/119).

4. EXPOSITORY SYLLOGISMS

(i) E. W. Beth's article *Über Lockes 'Allgemeines Dreieck'* (1956) was first delivered as a lecture under the title *Het expositorisch syllogisme (The expository syllogism)*. As was mentioned already in the Introductory chapter above, this article connects the traditional problem of proof by ekthesis with the mediaeval doctrine of expository syllogisms.

What precisely is an expository syllogism? Duns Scotus (1266–1308) offers the following definition: "Syllogismus expositorius est ille, cuius medium est terminus discretus." He clearly saw that it would be necessary to distinguish several kinds of discrete terms, for he goes on to say: "Sed notandum est, quod terminus discretus est triplex" (quoted from Prantl 1955, III/142 n. 624; cp. Section 1 of the present chapter).

Unlike his younger contemporary, William of Ockham, Duns Scotus held that expository syllogisms may be formulated in each of the three figures: "De syllogismo expositorio affirmativo dico, quod in qualibet trium figurarum potest fieri bonus" (Prantl *op. cit.*/231 n. 206). One of his examples of a valid expository syllogism in the first figure runs as follows:

> Socrates est musicus,
> album est Socrates,
> igitur album est musicum

(Prantl, *loc. cit.*). This conclusion is an indefinite proposition. Bocheński ascribes the same doctrine to the fifteenth-century logician Stephanus de Monte (Bocheński 1956/271).

(ii) But the expression "expository syllogism" was often given a much narrower meaning. William of Ockham (1285–1349) defines it as a syllogism with two singular premisses *in the third Aristotelian figure*. By this he means that it is a third figure syllogism with a "discrete" ("discretus"), hence individual, middle term:

(1)   *m is P*
             m is S.

According to Ockham, from these premisses one may draw "not only a singular but also a particular or an indefinite conclusion, however not a universal conclusion, just as one cannot draw a universal conclusion from two universal premisses in the third figure" (*Summa Logicae III*, part 1, Ch. 16 1–5; my translation). If I understand Ockham correctly, the following conclusions may all be validly inferred from these premisses:

a certain S is P, or: *this S is P, some S are P*, [*the/an*] *S is P* [indefinite].

This is to say that "indefinite" propositions may be *introduced* given two singular premisses with the same subject term, for they may occur as conclusions of logically valid expository syllogisms.

There must have been people in Ockham's time who criticized this logical procedure with the argument that it (sometimes) leads to a *fallacia accidentis*, but Ockham does not agree with them: "Et universaliter in talibus non potest assignari fallacia accidentis sicut aliqui assignant, non plus quam hic: Sortes est homo, Sortes est animal, igitur animal est homo. Et ideo multum errant qui in talibus assignant fallaciam accidentis et destruunt omnem modum arguendi et omnem disputationem" (Ch. 16 64–68; cp. IV–11 above).

In *Summa logicae II*, Ch. 27 90–95, Ockham writes: "Similiter hic non est syllogismus expositorius: 'Sortes est albus; Sortes est animal; igitur animal est albus', sed hic est fallacia figuræ dictionis". The reason why he regards this argument as invalid is that "animal" is neuter and "albus" masculine. We may probably conclude that he would regard the argument: "Sortes est albus, Sortes est homo; ergo homo est albus" as valid, although 'albus' is usually not considered as an essential characteristic of 'man'.

(iii) In the syllogisms which Ockham calls expository, indefinite propositions do not occur as premisses. He emphasizes in an unmistakeable manner that the subject term *m* should in no way have *suppositio* for different things: "Si autem subiectum, sive sit pronomen demonstrativum sive nomen proprium sive pronomen demonstrativum sumptum cum aliquo termino addito, supponat pro aliquo quod, quamvis sit *unicum et simplex* et unum numero et singularissimum et *tamen est plures res*, non tenet consequentia per rationem syllogismi expositorii" (Ch. 16 12–15; my italics). Therefore Ockham cannot admit terms with *suppositio (personalis) communis* as middle terms in expository syllogisms, but he can admit terms with *suppositio (personalis) discreta*. This is also Moody's interpretation of the passage just quoted (Moody 1965/217). Now the reason for this is, Ockham says, that an argument with two premisses having the same subject term is invalid when these subject terms are *particular (some M)*. This would seem to imply that he takes an indefinite proposition, which in English would be expressed as *the/an M is P*, to have the logical force of a particular proposition. For such an indefinite proposition is assumed to have some kind of general import, hence the schoolmen would not consider its subject to have only *suppositio discreta*. This implies that two indefinite premisses

> *the/an M is P* [*M's are P*]
> *the/an M is S* [*M's are S*]

do not permit of any conclusion, not even an indefinite conclusion. For the first premiss could be true due to one M-thing (this M, or m_1, having the property P), and the second could be true due to another M-thing (that M, or m_2, with the property S), so Ockham's argument runs.

(iv) Ockham also acknowledges the existence of valid syllogisms with individual middle terms in the second figure:

> Homo est Sortes
(2a) animal est Sortes
> igitur animal est homo.

But he justifies them by reducing them to syllogisms with singular middle terms in the first figure. He regards the major premiss as convertible: "Ista: 'Homo est Sortes', convertitur in istam: 'Sortes est homo'" (Ch. 16 74–75). He has now reduced (2a) to

> Sortes est homo
(2b) animal est Sortes
> igitur animal est homo.

Ockham holds that such a syllogism (2b) is valid, "quia dictum est prius, quod in prima figura non refert an maior sit universalis an singularis'. (77–78). Presumably this implies a reference to his third and his eighth chapter, in which he has explained that a premiss of a syllogism in the first figure may be singular. The third chapter contains this example of a valid syllogism:

(3)
Omnis homo est animal
Sortes est homo
igitur Sortes est animal.

Ockham considers the validity of this argument to follow directly from the *dictum de omni* which together with the *dictum de nullo* was assumed to explain the validity of syllogisms in the first figure.

In the eighth chapter we find the following example of a valid syllogism in the first figure:

(4)
Sortes est albus
omnis homo est Sortes
igitur omnis homo est albus.

This is an argument leading from a singular premiss *m is P* to a universal conclusion *every S is P* by means of a minor premiss connecting the terms *S* and *m*. Nowadays we ascribe the following (valid) logical form to this argument (using the letter "I" for "is identical with"):

(5)
$$\frac{Pm \qquad\quad (x)[Sx \to I(x, m)]}{\therefore\ (x)[Sx \to Px].}$$

But (5) is not a syllogistic mood, for we have employed two different translations of the word "est".

(v) Bocheński characterizes Ockham's justification of the validity of (4) as revolutionary when seen from the point of view of the logic of the *Analytics*. This is perhaps slightly exaggerated since there are, as we have seen, several features in the logic of Aristotle which point towards a confusion of individual with indefinite general propositions. But individual propositions do not seem to have had a different logical force from that of particular propositions when they occurred as premisses. Now Ockham holds that just as the subject term, *every M* or *all M*, of a universal proposition has *suppositio* for everything it signifies, so the subject term of an individual proposition also has *suppositio* for everything it signifies because it signifies one thing only. These are his own words: "... quia sicut subiectum universalis supponit actualiter pro omni suo significato, ita etiam subiectum singularis supponit actualiter pro omni suo significato, cum non habeat nisi unum" (*op. cit.*, Ch. 8 13–15).

Here a singular proposition is compared to a universal proposition, and the comparison is used in the theory of inference. In the light of the last but one quotation from the *Summa Logicae* (ch. 16 77–78) we can express this as follows: Ockham treats singular premisses as universal premisses, provided the subject term refers uniquely to one individual. The validity of (4) is thereby reduced to the validity of the mood Barbara.

Suppose that the same procedure is applied to the premisses (1) of those syllogisms which, in the terminology of Ockham, were the real expository syllogisms. Then those syllogisms turn out to be in the mood Darapti:

$$\text{every } M \text{ is } P$$
$$\text{every } M \text{ is } S$$
$$\overline{\text{ergo, some } S \text{ is } P,}$$

or else: *S's are P* (indefinite) or (*a certain*) *S is P* (individual). Not surprisingly, Darapti is one of the contested moods. It was said in I-3 above that the traditional logicians usually regard this mood as logically valid but that modern logicians are only willing to do so on condition that the extension of *M* is not empty.

It is now clear that there is a connection between Ockham's view about the logical form of singular premisses and the opinion that Darapti is a valid mood. The validity of an expository syllogism in the third figure can be defended by an appeal to Darapti if and only if (1) Darapti is unconditionally valid and (2) singular premisses can be treated as universal premisses. This is to say that the analysis of singular propositions as universal propositions seems to be a necessary condition if one wants to explain the validity of expository syllogisms in the context of syllogistic, instead of introducing an independent inference form corresponding to the modern rule of existential generalization (cp. Section 13 below).

It would have been un-Aristotelian to allow premisses with singular terms into syllogistic, just as it is probably un-Aristotelian to treat them as universal. Ockham does the latter, but he is not so un-Aristotelian as to admit a wholly new and independent class of well-formed propositions which can occur as premisses of syllogisms. By Ockham's reduction of individual propositions to universal ones, the obscurities already noted in Aristotle's own logic were given a theoretical justification. Ockham's reduction was therefore an important further step in confusing individual with general propositions.

(vi) Bocheński (1956/20) does not venture to offer a final verdict on the following question: was "Ockham's reduction", this *grundsätzliche Veränderung der Syllogistik* by which quantifier-free propositions were not only admitted as premisses but were also reduced to universal propositions, wholly his own invention or should it be blamed on one of his predecessors? Peter of Spain still treated singular propositions as particular propositions (Ashworth 1970). But it is in any case wrong to ascribe this novelty to the logicians of the sixteenth and seventeenth centuries, as was for a long time common. Those logicians contributed still further to the decay of Aristotle's theory of the quantifiers, but that is another matter.

At the beginning of the present section I used a small letter, "*m*", as a variable for the middle term of an expository syllogism, because modern logic always uses small letters as individual variables. The schoolmen used variables very sparingly, if at all. But if the notational apparatus of Aristotle is being used, then the logical form of the individual premisses of an expository syllogism will as likely as not be construed in the following manner:

this M is P
this M is S.

Bocheński emphasizes that the schoolmen considered all three terms in an expository syllogism to be constructed from class names (270).

At least from Duns Scotus and Ockham onwards the schoolmen often supplied propositions with discrete subject terms with the prefix "omne quod est". In Duns Scotus' own words: "Ideo dicendum est aliter, quod syllogismus expositorius tenet gratia formae in omnibus terminis, dum tamen premissae regulentur debite per "dici de omni" vel "dici de nullo", ita quod terminus discretus distribuatur mediantibus istis dictionibus "quod est", ut ista "Socrates currit" debet resolvi in istam "Omne quod est Socrates, currit", et tunc syllogismus expositorius est consequentia formalis" (quoted from Prantl 1955, III/143 n. 624; cp. Kneale and Kneale 1964/273).

(vii) The reduction of expository syllogisms in the third figure to the mood Darapti is not a theory of the logical procedure which Aristotle called "ἔκθεσις". What ekthesis is, how it was conceived in the later tradition and to which contemporary techniques it corresponds, will be discussed in the last chapter of this book. On the basis of certain historical indications which were given to him by Bocheński, Beth nevertheless held that there is a systematic connection between the Locke-Berkeley problem of the logical function of individual propositions in deductions and the medieval discussion about the validity of expository syllogisms. In one sense this is most certainly the case, since in the theory of expository syllogisms a theory is offered, for the first time in history, of a number of valid argument forms containing individual propositions, be it as premisses or as conclusions. It is immediately clear that the interpretation of all individual premisses as universal premisses did not contribute to the solution of what was later called "the Locke-Berkeley problem".

Especially dangerous is the combination of the *omne quod est*-reduction of singular to universal premisses with (i) the Platonic version of *suppositio simplex* in the logic of Peter of Spain, or with (ii) the doctrine of a philosophical, virtual, essential *suppositio naturalis* of a *terminus per se sumptus*. In Ockham's own logic this danger is not present, since he accepted neither (i) nor (ii).

5. EXPOSITION AND *exemplum* IN THE LOGIC OF THE RENAISSANCE

(i) *Influence from rhetoric.* The logic of the Renaissance bears witness to influence from rhetoric, especially from the rhetoric of Cicero. This influence can be studied in the doctrine of the *exemplum*, which rated a place of its own in a large number of works on logic in the sixteenth century. In the *Rhetorica ad Herennium (De ratione dicendi)*, which is usually ascribed to Cicero, "*exemplum*" is defined as a recording of something which is done or said in the past, reporting also the name of the person who said or did it.

This kind of argument is not, of course, conclusive. In the sixteenth century, however, it is often treated in books on logic and put on a level with arguments which are, or which were supposed to be, conclusive. This is the case for instance in the works of the anti-Aristotelians Grammarus (*Legalis dialectica*, 1524) and J. L. Vives (*De disciplinis libri XX*, 1531).[1] But the Spanish logician Fonseca, who considered himself to be a pure Aristotelian, also allotted an important place to *exemplum* in part IV of his *Institutionum dialecticarum libri octo* which was published in 1567. In his theory of the *consequentiae*, arguments are grouped into four classes. The fourth of these contains the following kinds of argument: syllogism, enthymeme, and finally *inductio* or *exemplum*.

It is really not very surprising that the persuasively rhetorical kind of argument called "exemplum" was also regarded as of great importance once expository syllogisms had become current in works on logic. For an expository syllogism is also an inference from an example which is given in the premises. A third-figure expository syllogism contains no information other than the description of an example. Words like "example", "Beispiel" are also used by present-day authors when describing such syllogisms. Thus Risse reports that Snellius, in his *De ratione dicendi et excercendi logicam* of 1595, defines the third syllogistic figure in general (called by him "syllogismus contractus") as "argument from an example, employed as a middle term" (Risse 1964/190). Similarly, William Kneale describes the *syllogismi contracti* of the 1572 edition of the *Dialecticae libri duo* by Petrus Ramus as "arguments from examples which Aristotle would have presented in his third figure" (Kneale and Kneale 1964/304).[2]

(ii) *Exemplum and similitudo.* The author of *Ad Herennium* holds that an *exemplum* is used for the same reasons ("causae") as a comparison ("similitudo") (*Ad Herennium*, IV xlix 62; 1954/382f.). An example permits us to draw a particular conclusion. This may therefore indicate a willingness on the part of the author of *Ad Herennium* to construe particular propositions as expressions of similarity. In any case they were so construed in the later tradition, as I shall show in later chapters (VII, X-7). In all likelihood the impact of neo-Platonism prompted this interpretation of particular propositions. For as Risse reports:

"In the sixteenth century Plato's dialectic is often combined with Cicero's doctrines" (1964/72).

(iii) *When* is an argument from an *exemplum* logically valid, and when is it merely persuasive? At least an expository syllogism with a particular proposition as conclusion does not deserve this derogatory predicate, "merely persuasive", whether or not Ockham's reduction of individual premisses is accepted. This is to say that a third figure expository syllogism is an impeccable argument form provided the word "aliquis" or the word "quidam" occurs in the conclusion, so as to avoid the interpretation that its subject term has *suppositio simplex* or *suppositio naturalis*.

Now William of Ockham as well as Duns Scotus also drew an *indefinite* conclusion from two singular premisses. Suppose that such a conclusion, containing neither "quidam" nor "aliquis", is given an interpretation which in some way or other differs from the interpretation of the corresponding particular proposition. If we still want to distinguish this expository syllogism from a rhetorical *exemplum*, we shall be in need of a very accurate logic of various kinds of indefinite proposition.

As long as no such logic is available, we must conclude that whether or not the words "merely persuasive" are used with justification is a question that does not depend on the particular example chosen, but on the form of the conclusion.

(iv) *The third figure* occasioned heated discussions among the logicians of the fifteenth and sixteenth centuries. Listen to Lorenzo Valla (1407–1457), one of the earliest authors to be counted as belonging to the Renaissance: "Tertia, quae ab istis constituitur figura, nihil in se habet sanitatis, sed tota plane insana est, ut pudeat me vicem eorum, qui vel invenerunt eam vel probandam putaverunt..." (quoted from Prantl III/166 n. 79). Another less important Renaissance logician, Gislenus, also rejected the third figure. Melanchton, too, rejects it in the first (1520) edition of his dialectics, the *Compendiaria dialectices ratio*.

In the re-written version of this text, published under the title of *Dialectices libri IV* in 1528, he reconsidered the matter. The third figure and expository syllogisms in this figure are now given a place of honour: "Magna laus est syllogismi expositorii, et usus eius aliquando esse in subtili disputatione potest... Est autem syllogismus expositorius, in quo medium est singulare. Et fit in tertia figura commodissime" (quoted from Risse, *op. cit.*/88). Writing at about the same time, Johannes Sturmius, the pupil of Agricola who formed the link between him and Ramus, regarded the third syllogistic figure as the most important form of proof. "Ordine igitur directo non nisi tertiae figurae fit conclusio," is his opinion as expressed in the third part of his *Partitionum dialecticarum libri IV*, published in 1549 (quoted from Risse 1964/46). In his later years Ramus, too, held the third Aristotelian figure to be the most important. The Dutch scientist and logician Simon Stevin published his *Dialectike* in 1585. There we find the following example of an expository syllogism:

> John is poor,
> John is a Dutchman,
> [hence] some Dutchmen are poor

(quoted from Beth 1948/42, my translation).

The battle of the expository syllogisms was, however, not yet decided.

(v) *Why* would many authors not accept the third figure and expository syllogisms in this figure? An answer to this question could be that they feared that an indefinite conclusion, while formally correct according to William of Ockham and Duns Scotus, might be given an essentialistic interpretation of the kind that Peter of Spain would describe in terms of *suppositio simplex* or *suppositio naturalis*.

I doubt that this fear was ever very prominent among the adversaries of this argument form. It is more likely that they all felt what the seventeenth century author Cornelis Martini expressed in his *Commentatiorum logicorum adversus Ramistas* (1623): this argument form is not a real syllogism, for it pertains to the senses rather than to reason ("cum ad sensum potius ad rationem referatur"; quoted from Ashworth 1970). This is precisely Alexander's old argument for excluding the procedure of *ekthesis* from logic, which to him meant syllogistic.

6. MELANCHTON'S EXAMPLE

The fear of promoting paradigmatic thought by permitting indefinite conclusions with a general import may not have been the cause of the dispute, but it is certainly a realistic fear.

Melanchton's example of an expository syllogism turns out to be the following:

> Hoc est animal
> Et hoc est homo,
> Ergo homo est animal.

Ockham and Duns Scotus[3] regarded the form of this argument as valid. But the example Melanchton chose has the peculiar property that its ambiguous conclusion is true even when it is interpreted as a *universal* proposition or as a proposition which entails a universal proposition. Keckermann quotes (1614/252) a similar example, referring to Melanchton:

> Haec res calefacit
> Haec res est zinziber
> Ergo zinziber calefacit.

This conclusion, he says, is *universalis indefinita*, the indefinite propositions being a *sub-class* of the class of universal propositions (1614/215). Nevertheless he states quite explicitly, as one out of three rules on validity in third-figure arguments, that "in the third figure there can be no [definite?] universal conclusion" (Keckermann *op. cit.*/252).

Keckermann writes: "Conclusio quae est universalis indefinita, quasi definitur & explicatur ad sensum monstrato zinzibere; Tales Syllogismi, inquit idem [Melanchton], omnium primi se humane menti offerunt; Primum enim apprehendunt sensus singularia, post mens abstrahit species, componit & dividit; Ita in syllogismo expositorio dum oculi cernunt aliquid inesse subiecto singulari, postea mens componit, & Syllogismum expositorium extruit" (*loc. cit.*). Notice that Keckermann uses the expression "the mind *abstracts* the species". These examples may be read as reports on particularly successful pieces of *induction*, in the modern sense of "induction" as well as in the traditional sense (cp. IX-12). But when the indefinite conclusion is given a universal interpretation, the argument form is clearly *not* logically valid and does not belong to "formal" logic, even today.

In several modern languages the conclusion would most certainly be formulated with a prenex article.

It would seem that unless Melanchton held a theory of *suppositio simplex* like that of William of Ockham, such expository arguments with an indefinite conclusion (*the/an*) *M is P*, which may be interpreted by the listener as particular *or* as universal, deserve to be characterized as rhetorical, inconclusive and merely persuasive *exemplum*-arguments, rather than as syllogisms in an extension of the theoretical logic of Aristotle. In general, insight into the nature of logical generalization and into induction cannot be expected to improve when the rhetorical doctrine of *exemplum* is conjoined to the theory of expository syllogisms – especially not when the theory of *suppositio* is either rejected *in toto* as a piece of hair-splitting, or else dominated by the "philosophical" version of the *suppositio naturalis* of terms *the/an M as such*. An adversary of the third figure expository syllogisms may be expected to be *a fortiori* against admitting the rhetorical *exemplum* into logic ("dialectic"), provided he saw any difference between these two argument forms at all.

The third version of Melanchton's dialectic was published in 1548 under the title *Erotemata dialectices*. It set the fashion in logic for the German protestants for several decades. In this work, Melanchton lists *syllogismus*, *enthymema*, *inductio*, *exemplum* and *sorites* as the *bonae consequentiae*; the *sorites* is an addition to the 1520 edition. That *exemplum* is a special kind of argument which may be used as part of a proof seems to have been a common opinion in the sixteenth century, current from Wittenberg to Coimbra.

7. THE RAMISTIC MOODS

Petrus Ramus is held to have been the most fervent anti-Aristotelian of the whole sixteenth century. To the valid moods in each of the three original Aristotelian figures he adds two new valid moods, obtained by choosing (affirmative or negative) singular propositions in both premisses. Ramus called the middle term of such a syllogism "terminus discretus sive singularis". It will be recalled that the logicians of the fourteenth century ascribed *suppositio discreta* to individual terms. Ramus talks of *syllogismi proprii* and not of *syllogismi expositorii*. He writes in his *Dialectique* of 1555: "Aristote use aussi de ces syllogismes propres, sans penser toutefois que se soyent syllogismes et les appelle expositions" (1964b/129). This implies that Ramus consciously identified expository syllogisms with the Aristotelian method of *exposition* or *ekthesis*.

Dassonville observes that this passage was omitted from the 1576 edition of the *Dialectique (loc. cit.)*.

Piscator, a follower of Ramus, constructed special mnemonics for the Ramistic moods: "Burburu", "Cylurynt", "Durupti", and so forth, copying the style of the names constructed by Peter of Spain for the moods of the Aristotelian system: "Barbara", "Darapti" and so on. The "u" and the "y" signify that the proposition is a singular one, with "u" for an affirmative and "y" for a negative singular proposition (cp. Risse 1964/133).

In addition, and unlike William of Ockham, Petrus Ramus gave examples of syllogisms, which he considered valid, with only one individual premiss and one premiss which may be called indefinite, as we saw in IV-15. In most of those examples the subject term in the indefinite premiss, and the subject term of the individual premiss as well, occurs as one of the extreme terms. But in the works of Ramus we also found two examples in which *the subject term of the indefinite premiss occurs as the middle term of the syllogism*. This middle term is formulated with an article ("Les consulz...", "A wyseman..."). It seems almost impossible that a fusion of this kind of first figure arguments with the form which Piscator calls "Burburu" should not have taken place in the minds of logicians and other philosophers who were influenced by their works. In fact there is little reason to assume that these authors themselves distinguished between these forms of argument.

Ockham's *Summa logicae* contains no examples of expository syllogisms with indefinite premisses and he gives a warning that the subject term of a premiss may not have other than discrete *suppositio* (Section 4 above). We therefore conclude that the introduction of expository syllogisms or *syllogismi proprii* where the subject term of an indefinite proposition occurs as the middle term is really a contribution of the logicians of the Renaissance. But when he says that Duns Scotus (Pseudo-Scotus?) and William of Ockham paved the way for the Ramistic moods with *discrete*, *i.e.* individual, middle terms, Scholz was certainly right (1967/39). Scholz also mentions Melanchton (1497–1560). Melanchton was practically a contemporary of Ramus (1515–1572), but Dassonville (Ramus 1964b/18) is also of the opinon that Ramus was influenced by Melanchton. Scholz, writing in 1931, does not mention Stephanus de Monte (cp. Section 4 above), but he was mentioned by Bocheński in 1956.

In the logic of Ramus, the use of discrete and other singular middle terms may be explained by the fact that he identified the concept of an *individuum* with the concept of a *species*. Risse (1964/134) does not see anything remarkable in this identification, but Ong is of a different opinion. He holds that this identification implied a doctrine which was to have disastrous consequences (Ong 1958/203, 354, notes 14, 16). The result was complete chaos, since Talon, Ramus' faithful collaborator, did not always distinguish the meanings of "genus" and "species", notwithstanding the fact that Ramus himself identified an *individuum* as a *species*.

I find it hard to judge precisely how far Ramus deviated here from his contemporaries or from the schoolmen. But it is at least clear that his two examples of valid syllogisms with an *indefinite* premiss which we met in IV-15 could not be ascribed to William of Ockham.

Ramus' first logic, the *Dialecticae institutiones*, dates from 1543 and was re-issued by Risse in 1964. In it we find the following first-figure *syllogismi proprii* (fol.24):

> Iunius suo, & suorum sanguine patrię libertatem peperit:
> Brutus erat Iunius:
> Brutus igitur suo, & suorũ sanguine patriae libertatem peperit.

> Sylla non miseretur ciuium:
> Cornellius est Sylla:
> Cornellius igitur non miseretur ciuium.

As with the examples he offers us in the two other Aristotelian figures, the middle term is here a *proper name*. A French or German or English translation will therefore not contain any articles. But in the *Dialectique* of 1555 we do find definite articles in the middle terms of the Ramistic modes (Ramus 1964b/133):

> La magnanimité de Thémistocle est louable.
> La magnanimité de Thémistocle est confiance.
> Quelque confiance donques est louable.

> La timidité de Xerxe n'est poinct louable.
> La timidité de Xerxe est défiance.
> Quelque défiance donques n'est poinct louable.

These examples are in the *third figure*. The first is a syllogism in the

mood Durupti; the second is in the mood Fylupton. They have the following
form:

> *The M of x is P (not P).*
> *The M of x is Q.*
> ───────────────
> *Ergo, some Q is P (not P).*

When these premisses are taken at their face-value, the *terminus medius*,
the M of x, is a definite description. The entity described is *the* value of a
function *y is M of x*. Clearly the rules of quantification are, in the logic
of Ramus, not separated from the theory of identity.

Scholz characterized the Ramistic terms as follows: "These are, by
approximation, expressions which modern exact logic characterizes as
descriptions of individuals, expressions of the form "the so-and-so""
(1967/39). He bases this statement on the Latin translation of the
Dialectique, the *Dialecticae libri duo* of 1556. The examples quoted here
from the original French edition and the examples in IV-15 above
support this characterization. Scholz is far from content with Ramus'
treatment of singular propositions, which are "himmelweit davon
entfernt..., die zu diesen Syllogismen gehörigen Schlussregeln explizit zu
formulieren." He shakes his head in horror at Ramus' logical nonchalance:
"What would Aristotle have said of this kind of modesty in logic!"

8. HOSPINIANUS

The logic of Hospinianus was studied by K. Dürr (1949a). In order to
characterize the basic tenets of this logic Dürr makes use of the technical
term "einfacher Ausdruck", taken from Łukasiewicz ("simple proposition"). He uses this expression for the irreducible propositional forms on
which a theory of inference is based. Aristotle made use of only four such
simple forms, A-, I-, E- and O-propositions. No changes in this syntactic
doctrine were brought about by the schoolmen, who allowed individual
propositions as premises of their syllogisms but who considered them as
reducible to the A-form. And because of his identification of an individual
with a species one cannot really say of Ramus either that he consciously
introduced another syntactic form into his logic. At any rate, the number
of forms he recognized was not greater than six.

Hospinianus, however, both consciously and explicitly breaks with the classical set of basic syntactic forms. He constructs his syllogistic on the basis of eight kinds of simple proposition, *i.e.* four affirmative and four negative forms, each group containing a universal, a particular, an indefinite, and a singular form. Both the negative and the affirmative singular forms are treated by him as irreducible.

Dürr reports an opinion held by Hospinianus which throws some light on the problems in the present work. Hospinianus assumed that a universal affirmative proposition *all S are P* or *every S is P* is *false* when *P* is a proper name. He offers an example which Dürr translates as follows: "Jeder Dreiruderer war Centaurus", where "Centaurus" is a proper name. As we have seen, Ockham used a proposition of this kind in his example of a valid first-figure syllogism with an individual middle term: "Every man is Socrates". Such propositions were regarded as false by Hospinianus. [4]

9. THE *vis universalis* OF SINGULAR PROPOSITIONS

About a hundred years later, the English mathematician John Wallis (1616–1703) pointed out that the six Ramistic modes with singular premisses could be reduced to Aristotelian modes, provided that every singular proposition be treated as a universal proposition.

This is indeed true. And in fact it was, as we have seen, an old truth, well known by the schoolmen, who also knew valid arguments with singular premisses. Wallis' discovery cannot be said to have been anything more than a rediscovery.

Wallis did not stop at this rediscovery, however. He drew the further dubious conclusion that singular propositions *ought* to be treated in this manner, and he expressed his point of view in highly suggestive terms: "propositio singularis, in dispositione syllogistica, semper habet vim universalem" (quoted from Scholz 1967/40, who says "... vim universalis").

What more speculative minds may have made of this "force" attributed to singular propositions can only be guessed. It would be most interesting to know more about the manner and the degree of influence of this technical treatment of singular propositions (i) on the various conceptions of the first *proprietas terminorum*, the *significatio*, which was mentioned in IV-3, and also (ii) on the doctrine of a "philosophical" or "virtual" or "natural" *suppositio* of a *terminus per se sumptus* which we discussed in IV-14.

Wallis published his interpretation of singular propositions in 1643.

In 1662, the same doctrine was advanced by Arnauld and Nicole in their influential work on logic, to be known as the Port-Royal logic. The authors write that "les propositions singulières tiennent lieu d'universelles dans l'argumentation" (1965/115).

10. LEIBNIZ

Leibniz studied Hospinianus's investigations carefully, as well as his extension of the set of basic propositional forms. But he did not adopt Hospinianus's account of singular propositions.

Hospinianus seems to have treated individual propositions as well as indefinite propositions as particular. Leibniz accepts the (Aristotelian?) opinion that indefinite propositions are to be treated as particular propositions. But as to singular propositions, Leibniz settles for the view that they should be treated as universal.

His justification is taken from Johann Raue. In the opinion of Dürr (1949/17), Raue's view can be rendered as follows: a proposition of the form

the individual a has the property E

is equivalent to a proposition of the form

for every x, if x is identical with a, x has the property E.

This interpretation of *a is E* differs from the explanation given by Wallis, who held that a singular proposition has the "force" of a universal proposition because the *whole extension* of the individual is at issue. Dürr remarks that "the extension of an individual" was not defined.[5] At least as problematic is the view that in a singular proposition we are concerned with the total *content* of "the" individual concept behind the singular subject term.

Throughout the centuries, the *infimae* (or: *ultimae*) *species* in the conceptual pyramids of *genera* and *species* were considered especially problematic. It is well known that the content of an individual concept is, if anything, still more indefinite than the content of a general concept. This is also admitted by von Freytag (1967/36).

Leibniz contrasted the propositions he called theorems with singular propositions and propositions describing observations. The former he conceived of as propositions having eternal truth, a characteristic that does not in his opinion belong to singular propositions. It is a short step from here to the metaphysical position, so common in later German philosophy, that singular propositions should be, and in fact are, "derived" from propositions which are of the nature of theorems (cp. VII-4).

Certainly Leibniz took an interest in individuals. But it should by now be clear that the difference in point of logical principle between Wallis, Leibniz and the schoolmen is much smaller than the difference between Leibniz and those philosophers who either treat both singular and indefinite propositions as if they were particular propositions *some S is P* (with Aristotle), or else count them as belonging to two entirely independent classes (with Hospinianus). Hospinianus's praiseworthy attempt was not to have any influence on European thought. He was not allowed to carry out the logical revolution whose seeds were sown in the expository syllogisms of the schoolmen. By their acceptance of the reduction of singular propositions to universal propositions, the prestige of writers like Leibniz and Euler (1707–1783) led others to rely upon it, so that before the rise of modern logic no one ever again questioned the correctness of this interpretation (cp. Scholz 1967/40).

11. WOLFF: DEFINITE DESCRIPTIONS AND THE *dictum de omni*

The very influential German philosopher Christian Wolff (1679–1754) had very close relations with Leibniz. It is therefore no surprise that the Wallis-Leibniz account of singular propositions is to be found also in his *Philosophia rationalis sive logica*, published in 1728. This work contains a syllogism with an individual premiss which has a definite description as its subject term. About this proposition, Wolff writes: "Propositio maior: *Inventor calculi differentialis est autor Theodiceae*, equivalet universali, unde & per modum universalis effere potest: *Quicunque est inventor calculi differentialis, ille etiam est autor Theodiceae* (1728/298, Section 355). This supports Dürr's interpretation of Johann Raue.

Now the singular subject term in Wolff's example is a definite description: "Inventor calculi differentialis". But Wolff lacked the means for expressing the difference between predication and strong (strict) identity. The result is that in his analysis the difference between "every one who *is* (an) Inventor of the calculus" and "everyone who *is identical with the* Inventor of the calculus" is entirely lost. No doubt the absence of a definite article in Latin makes it easier to overlook this distinction.

Wolff held that the whole theory of inference can be reduced to one principle: "Dicto de omni et nullo nituntur omnes syllogismi" (297). At the same time he holds that syllogisms express the way in which we reason. It follows that he considers every human argument to be based on the *dictum de omni et nullo*.

Wolff calls arguments with individual premisses not by their scholastic name, but by the name given them by Ramus: "syllogismi proprii". He explains their validity by saying that individual propositions may be considered equivalent to universal propositions: "propositiones propriae universalibus aequipollere possunt" (322).

This he takes to have the following consequence: "Erunt itaque affirmativi syllogismi proprii tertiae figurae in modo Darapti & negativi proprii in modo Felapton" (*loc. cit.*). He does not use the names introduced by Piscator, "Durupti" and "Fylupton", but reduces them to Aristotelian moods which he holds to be based upon the *dictum de omni et nullo*.

Of third-figure *syllogismi proprii* he offers four examples, two in *Darapti* and two in *Felapton*. I quote only one example in each mood.

Isaacus Nevvtonus est inventor methodi fluxionum.
Isaacus Nevvtonus est autor Principiorum Philosophiae naturalis mathematicorum.
Ergo Autor Principiorum est inventor methodi fluxionum.

In this syllogism, said to be in the mood Darapti, the word "quidam" or "aliquid" ("some") is missing from the conclusion. We remember that Ockham accepted third-figure expository syllogisms with *indefinite* propositions as conclusion. This example of Wolff's is of a different kind, for the conclusion has a *definite description* as a subject. I assume here that Wolff would approve of the following translation:

Ergo, the author of the *Principia* is the inventor of the *methodus fluxionum*.

This syllogism is one in the mood Duruptu, using the nomenclature of Piscator.

In the following example of a *syllogismus proprius*, which Wolff holds to be in the mood Felapton, the word "quidam" does appear:

> *Aristoteles* non habuit distinctam rerum cognitionem.
> *Aristoteles* fuit magni nominis Philosophus,
> Ergo Quidam magni nominis Philosophus non habuit distinctam rerum cognitionem.

So we may say that Wolff was just as careless in matters of logical theory as Petrus Ramus was before him.

Furthermore, Wolff held that a *propositio propria* is always convertible. He illustrates this by means of an example in which the copula "est" expresses strict identity and therefore is symmetric: "Leibnitius est Autor Theodiceae. Sed eadem converti potest: Autor Theodiceae est Leibnitius" (296). This he holds to be applicable to the minor in the second of the two examples above ("Aristotle was a famous philosopher"), which in his opinion is equivalent to the proposition: "Quidam magni nominis philosophus fuit Aristoteles" ("some famous philosopher was Aristotle"; 324). Like Ramus, Wolff is unable to formulate the theory of "all" and "some" without reference to the notion of identity.

Wolff considers the first figure as the most perfect of the four syllogistic figures (Patzig 1963/87f.).

12. KANT

I. Kant (1724–1804) holds practically the same opinion on the interpretation of individual propositions as Leibniz and Wolff. Both in his *Kritik der reinen Vernunft* and in the *Logik*, which was published by Jäsche, Kant distinguishes three kinds of judgment in respect of "quantity", *i.e.* universal ("allgemeine"), particular ("besondere"), and individual ("einzelne") judgments.

The *Logik* contains the following passage on individual judgments: "Individual judgments should according to form be rated in practice equal to universal judgments; for in both cases the predicate holds of the subject without any exception. For instance in the individual proposition *Cajus is mortal* there can no more be an exception than in the universal proposition: *All human beings* are *mortal*. For there is only one Cajus" (1801/158).

The same view is expressed in the *Kritik der reinen Vernunft* (1968/141), but Kant justifies it with the remark that *einzelne Urteile* – he probably has their subject terms in mind – have no extension at all. From this he concludes to the usual view in the tradition, that in an individual proposition something is predicated of something without exception, just as in universal propositions.

He then goes on to make a remark of some interest. The amount of knowledge contained in an individual judgment ("einzelnes Urteil (judicium singulare)") is, he says, infinitely smaller than the knowledge contained in judgments with a general validity ("gemeingültige Urteile (judicia communia)"). Like Leibniz before him, he concludes that individual judgments do deserve a place of their own in a "Tafel der Momente des denkens überhaupt (*obzwar freilich nicht in der bloss auf den Gebrauch der Urteile untereinander eingeschränkten Logik*)" (*loc. cit.*, my italics). But this can only mean that in the theory of inference individual propositions are to be treated in the same manner as the *gemeingültige Urteile*. In Kant's view the difference between these two kinds of propositions falls entirely outside theoretical logic.

In his early essay *Die falsche Spitzfindigkeit der vier syllogistischen Schlüsse*, Kant champions another theory which, conjoined to the usual interpretation of individual propositions, was to have tremendous consequences for German philosophy. He tells us that "In der sogenannten ersten Figur sind einzig und allein Vernunftschlüsse möglich, in den drei übrigen lediglich vermischte" (1905/51). We have just seen that Wolff already held this view. Kant expands on it a few pages later, saying that "the highest rules of all arguments of Reason ["Vernunftschlüsse"] immediately lead to that arrangement of concepts which is called the first figure, so that all other displacements of the middle concept only will yield a correct argument when they lead, by simple immediate inferences, to propositions which are combined in the simple arrangement of the first figure" (1905/57f.).

We shall meet this view again in the logic of Hegel (Chapter VIII) and shall return to it once more in IX-9 and IX-12.

THE LOGIC OF INDIVIDUAL PROPOSITIONS

Kant seems uninterested in the problem of expository syllogisms. In his opinion, the tenets *nota notae est etiam nota rei ipsius* (the mark of the mark is also a mark of the thing itself; that is to say, the transitivity of the copula) and *repugnans notae repugnat rei ipsi* (1905/49) are fundamental logical principles.

13. LAMBERT AND THE *dictum de exemplo*. COMPARISON WITH MELANCHTON

The philosopher-scientist J. H. Lambert (1728–1777) was Kant's contemporary. His work on logic, published in 1764, was called *Neues Organon*. Quite unlike Kant, he distinguished three independent principles of inference in addition to the principle upon which the validity of the first figure was commonly said to be based, the *dictum de omni* ("What holds for all A, holds also for every A"). These three other fundamental principles were: the *dictum de diverso*, the *dictum de exemplo*, and the *dictum de reciproco*.

Lambert formulates his *dictum de exemplo* in the following manner: "Wenn man Dinge A findet, die B sind, so gibt es A die B sind" ("When one finds things, A, which are B, then there are A which are B"; 1965, I/142). All arguments in the third figure, the figure of the *Beispiele und Ausnahmen* (examples and exceptions), are said to be based upon this *dictum*. He explicitly denies that the first figure is more perfect than the other syllogistic figures (141), and is therefore no less radical than Hospinianus.

Lambert distinguishes two ways of justifying existential statements:

1.   *Es giebt B, die C sind. Denn M ist B und C.*
2.   *Es giebt B, die nicht C sind. Denn M ist B und nicht C*

(138, my italics). He offers a concrete application of the *dictum de exemplo* (147):

        Die Erde ist bewohnt
(1)   Die Erde ist ein Planet
        Folglich ist wenigstens ein Planet bewohnt

("The earth is inhabited; the earth is a planet; hence at least one planet is inhabited"). Lambert's *dictum de exemplo* clearly is a principle which explains the validity of a *syllogismus expositorius* in the third figure, in which a *particular* conclusion is drawn from two individual premisses. Lambert considers this *dictum* as independent, which is to say that, like Hospinianus, he does not regard (1) as an argument in Darapti.

It is important to note that his conclusion contains the existential operator which we missed in the example of Melanchton that we discussed in Section 6.

In modern logic the validity of (1) is proved by an appeal to the introduction rule for the existential quantifier "there is" or "some", called "existential generalization". The premisses are of the form Bm, Cm. This yields the intermediate conclusion Bm & Cm. By applying the rule of existential generalization to this conjunction one derives the conclusion of (1): (Ex) $[Bx$ & $Cx]$, read: *some B is C*, or in Lambert's words: *at least one B is C*.

Summing up: just as the old *dictum de omni* corresponds to the *universal instantiation* of modern logic, so Lambert's *dictum de exemplo* corresponds to *existential generalization*. Kant did not even recognize the latter rule; neither he nor Lambert formulated a rule for universal generalization or for existential instantiation.

Until Frege it was commonly held in the European philosophical tradition after Descartes that logical rules ought to be provided with an epistemological foundation (cp. Dummett 1967/225). This had the result that logical rules themselves were often formulated in epistemological terms. Lambert's formulation of his *dictum de exemplo* shows that, rightly or wrongly, he did not emancipate himself from this usage either: "Wenn man Dinge A findet...". Most twentieth-century logicians would probably say that the schoolmen showed, in their theories of the *consequentiae*, a still better insight into semantical and logical questions than Lambert, whose *dictum de exemplo* has an epistemologically formulated antecedent. Quite recently, however, Hintikka has advocated an interpretation of the (Fregean) quantifiers in terms of *seeking and finding* (1973/58–121 *passim*). It is certainly true that modern logic is usually too exclusively tied to the needs and to the terminology of mathematicians. On the other hand, the demand that logical rules be formulated literally in epistemological terms does not seem likely to promote understanding of the difference between *generalization* (existential and universal) and *induction*. But Lambert in any case deserves our praise for having tried to supplement the current logic with an independent rule for the introduction of an existential operator ("es giebt", "wenigstens"). The interpretation of individual as universal propositions is necessary if expository third-figure syllogisms are to be reduced to Darapti; in the logic of Lambert this interpretation can be missed.

14. HEGEL

G. W. F. Hegel (1770–1831) takes the view of Leibniz, Wolff, and Kant on the logic of individual propositions as his starting-point. In a section on *Das singuläre Urteil* in his *Wissenschaft der Logik* (1813) he begins with the statement "Das Einzelne ist allgemein" ("The individual is universal"; 1966, II/288). He approaches this traditional doctrine not without considerable scepticism: "But a This is n o t essentially a universal [a general] entity" (288), "but such an i m m e d i a t e individuality is n o t general" (277), and "The singular judgment thus has its nearest truth in the particular judgment" (288). This gives us reason to hope that he will go on to produce a better theory. But in the end he does not reject the old doctrine and, as soon as he starts amending it, he goes from bad to worse. The statement "Das Einzelne ist allgemein" is, he says, an expression of "the positive judgment" (274). The word "not" which occurs in two of his own statements quoted above makes him conclude: "this judgment, which in respect of its universal form is p o s i t i v e, must be taken negatively" (288). This negation has, in his opinion, bearing upon the subject, and that means: "N i c h t e i n D i e s e s ist ein Allgemeines der Reflexion" ("N o t a T h i s is a universal of Reflexion"; 288). He already said earlier: "The positive judgment has its truth above all in the negative: T h e i n d i v i d u a l i s n o t abstractly u n i v e r s a l", or "The individual [the singular] is a particular" (279). This implies that, for Hegel, the particular judgment form is a *negatives Urteil*. And, indeed, by "some" ("einige") he understands "only some". He does not recognize a pure existential operator but mixes it up with a notion of negation. This has not made the logic of individual propositions any clearer (cp. VIII-20).

Hegel certainly knew Lambert's work. He calls him "the dry, sensible Lambert" (257). In Hegel's logic there is no discussion of expository syllogisms, to say nothing of the *dictum de exemplo*.

Like Kant, Hegel also discusses the problem which Beth called "the

Locke-Berkeley problem", but without referring to any other writers (470). He adopts the view of Alexander of Aphrodisias, C. Martini (Section 5) and others, that proofs based upon ekthesis are *äusserlich* (external) and do not belong to philosophy, although they may be valid in geometry, the science of "outer" space (471). The only exposition he approves of is "die Exposition des reinen Begriffes" ("the exposition of the pure concept"; 356).

Hegel's use of the first syllogistic figure will be discussed in Chapter VIII.

15. HERBART

Hegel's contemporary J. F. Herbart (1776–1841) shows an interest in the logical role of the articles which is remarkable for his time. In his *Lehrbuch zur Einleitung in die Philosophie (Introductory Textbook in Philosophy)* we find this passage about the logic of individual ("einzelne") propositions: "Of individual propositions logicians say that they should be regarded as equal to universal propositions, since they do not admit of any indefinite limitation of quantity. Yet here one ought probably to distinguish more carefully. What was said holds for a definite subject, *e.g.* (the) Vesuvius spouts fire; but it does not hold when the meaning of a general expression is limited to some individual or other by means of the indefinite article; *e.g.*, "a human being has invented that" (1834/82). He is clearly aware of the fact that the indefinite article may have a logical function, but he overlooks the logical function of definite articles in forming definite, *i.e.* uniquely referring (or non-referring) definite descriptions. This may be partly due to the fact that in German a definite article may also precede a non-descriptive proper name ("der Vesuv"). Concerning subject-predicate propositions the subject term of which is either a proper name or a definite ("bestimmte", see quotation) description, Herbart also accepts the theory that they should be construed as universal propositions.

16. COMPARISON BETWEEN THE TRADITIONAL THEORY OF DEFINITE DESCRIPTIONS AND RUSSELL'S THEORY: ANALOGY INSTEAD OF IDENTITY

I shall now undertake a detailed comparison of the analysis of *the/an M is P* given by Ockham, Wallis, Leibniz, Wolff, Kant, to mention only some, with the analysis given by Russell in his theory of descriptions of 1905. As Wolff says, the traditional understanding of the proposition

(1)   (the) inventor of the calculus is (the) author of the *Theodicy*

is that it is equivalent to a universal proposition. This can be no other than the proposition which he himself goes on to formulate:

(2)   everyone who is inventor of the calculus, is also author of the *Theodicy*.

In present-day notation this is a proposition of the form

(3) $(y)\,[My \rightarrow Ay]$ & $(Ex)\,Mx$.

In traditional logic one does not allow empty terms so that the term M in itself has existential import. This proposition (3) is equivalent to

(4) $(Ex)\,[Mx$ & $(y)(My \rightarrow Ay)]$

and to

(5) $(Ex)\,[Mx$ & Ax & $(y)(My \rightarrow Ay)]$.

Following Russell, most present-day logicians, however, ascribe this form to (1):

(6) $(Ex)\,[Mx$ & Ax & $(y)(My \rightarrow y = x)]$,

where the expression $y = x$ takes the place of the expression Ay. In order to make clear that $y = x$ here means: y is absolutely identical with x (*hic et nunc*, if the domain of discourse consists of actually existing things), one often prefers the notation $I(y, x)$:

(6) $(Ex)\,[Mx$ & Ax & $(y)\,(My \rightarrow I(y, x))]$.

In order to make the analysis less confusing than a complete analysis of Wolff's example would be, I have treated the predicate "author of the *Theodicy*" as an ordinary predicate and not as a uniquely referring expression. For the purposes of this analysis we might, instead of "is (an) author of the *Theodicy*", have chosen the predicate "is (a) genius" (cp. Scholz 1967/39).

The important thing is that one cannot expect to be able to prove (6) from (5). For (5) entails:

(7) $Ma \; \& \; Aa \; \& \; (y)(My \rightarrow Ay)$,

to be read as

(7′)   *a is (an) M and a is (an) A and every M is A,*

where the name "Leibniz" may be substituted for "a". On the strength of such a statement or judgment one cannot in general infer that

(8) $(y)(My \rightarrow y = x)$,

which is to be read as

(8′)   *every M is identical with a*, or: *all M's are identical with a,*

is a true proposition. This can be inferred only if we can somehow demonstrate the truth of

(9) $Mb \rightarrow b = a$,

(9′)   *if b has the property M, then b is identical with a,*

for *an arbitrarily chosen constant b* (a name for an arbitrarily chosen individual, in the sense that any opponent of the person who holds (8) may have his choice). But on the strength of the given analogy

(10) $\begin{cases} Ma \; \& \; Aa \\ Mb \; \& \; Ab \end{cases}$

one cannot draw the conclusion that

(11) $a = b$

is true in the strong sense of

(11′) $I(a, b)$.

This would be required in order to ensure the truth of (6), the modern analysis of a proposition "the inventor of the calculus is...". But of course we cannot assume that (11) follows from (10). For this is the same as accepting an inference form according to which one may always proceed from a given analogy between a and b to the conclusion that a is strictly identical with b (*hic et nunc*, if a and b are actually existing things). If that is rejected as illogical, (6) is seen to be stronger than (5). For there can be no doubt that (6) entails (5), while *condition (5) can be satisfied even if there were more than one inventor of the calculus.* And Wolff treats the judgment (1), whose grammatical subject may be rendered by the incomplete expression $\imath x M x$, precisely like a judgment expressed by $(x) [Mx...]$.

Now Newton, too, invented the calculus. Suppose that he was also co-author of the *Theodicy*.[3] The relevant interpretation of (5) would then be true, but that of (6) would be false. Or choose for "A" the predicate "was a Genius"; under this interpretation, too, (5) is true and (6) is false.

We are now able to formulate a conclusion of great importance. It is quite clear that *the traditional theory of definite descriptions as exemplified by Wolff will predispose those who adhere to it to content themselves with analogies where strict identity is required.*

This conclusion was reached independently of any knowledge of the properties ascribed to "est" in the analysis of (1) (cp. Section 11). But if we assume *predication* to be a symmetrical and also transitive relation of equality between a logical subject and a logical predicate then we come to the same result even if we start out from the modern analysis (6) of definite description statements. For the addition $x=y$ in (6) can then be omitted, since in this case the truth of $x=y$ is already implied by the analogy

(10′)   *x is M, x is A*
         y is M, y is A.

The "identity" of Newton and Leibniz is guaranteed by (10′): Newton is Leibniz, and if only one kind of "is" is recognized, then it is of course quite unnecessary to add "Newton=Leibniz", or "I (Newton, Leibniz)", to (3). So we see that *the identity theory of the predicative copula helps to make the traditional theory of definite descriptions plausible,* since these two theories have the same implications.

17. SINGULAR PROPOSITIONS IN THE LOGIC OF MARITAIN

Maritain does not quite know what to do with the doctrine that singular propositions can be reduced to universal propositions. At one place in his book he explicitly says that the singular propositions "nevertheless constitute a distinct category of propositions, and one is much mistaken

if one thinks that one always can assimilate them to universal ones"
(1933/145). In order to justify this conclusion he points to the fact that the
expository syllogism "this apostle is Judas, now this traitor is Judas,
hence this traitor is apostle", or in general

> *this P is [the individual] M*
> *this S is [the individual] M*
> _____
> *hence this S is P,*

is a valid argument. No conclusion can be drawn, however, if these
premisses are interpreted as universal propositions (*tout P est M, tout
S est M*). For from two universal premisses in the second figure, nothing
follows in Aristotelian logic. This fact "shows that here the singular
premisses are something other than simple equivalents of universal
propositions" (*loc. cit.*).

Later on we nevertheless read: "I remark that in the syllogism, every
singular term is equivalent to a *universal* term, in the sense that since its
extension is reduced to a sole definite individual, it clearly contains the
whole of the extension of the subject which it signifies, while this reduced
extension is not delimited in advance" (229f.). This is the same explanation which was also given by Wallis. Maritain concludes: "In the theory
of the Syllogism, therefore, one only has to take account of universal
propositions (to which singular propositions are assimilated) and particular propositions" (230).

In a footnote to this page, his common sense shows trough: used as the
major premiss, a singular proposition is by no means equivalent to a
universal one. Maritain's example is this:

> *cet homme est menteur,*
> or Pierre est homme,
> donc Pierre est menteur

(*loc. cit.*). Far from being a valid syllogism in the mood Barbara, this
argument sins against the rule: "Aut semel aut iterum medius generaliter
esto" (221), or so Maritain thinks. In fact it commits the fallacy of four
terms, since "homme" and "cet homme" are different terms.

Maritain's scepticism about the Ockham-Wallis doctrine is of course
sound, but he is clearly incapable of replacing it by any coherent theory
about the logical status of singular propositions. In spite of his argument

about the validity of the second-figure expository syllogism ("cet apôtre est Judas..."), he maintains later on that expository syllogisms are really not syllogisms at all. They only have the external appearance ("l'apparence extérieure") of a syllogism, they are not logical *inferences* at all, but only a presentation or *exposition* of the matter under discussion in such a manner that it can be observed by means of our senses ("une présentation sensible", 278). Herein we recognize once again the opinion held by Alexander of Aphrodisias, who characterized proofs in which individual propositions occur as proofs δι' αἰσθήσεως, by Martini and by Hegel (Section 14).[10]

18. THE LOGICAL FORM OF INDIVIDUAL PROPOSITIONS WITH DEFINITE DESCRIPTIONS AS GRAMMATICAL SUBJECT: RUSSELL

As we have seen, the analysis and the treatment of so-called singular propositions has differed greatly throughout the ages. Three positions can be distinguished:

(i) Singular propositions should be treated *as particular* propositions (the Aristotelian I-form). This was probably Aristotle's view. It was also the opinion of Peter of Spain, and was not uncommon at the time of Wallis and of Leibniz.[11] Hegel, too, was inclined toward this view.

(ii) When singular propositions occur as premisses, they should be treated as if they were explicitly universal, of the Aristotelian A-form. This position was taken by Duns Scotus, Ockham, Wallis, Arnauld and Nicole, Leibniz, Wolff, Euler, Kant, and in our time by W. Albrecht; it can probably also be ascribed to Petrus Ramus. It was rejected neither by Hegel nor by Maritain.

(iii) Singular propositions constitute an independent category of basic expressions, in addition to the four simple forms in the logic of Aristotle, A, E, I, and O. This is Hospinianus's standpoint. Maritain divides his allegiance between this position and the previous one.

In comparing these three views with the most well-known account in modern logic, let us first focus our attention on individual propositions in which the grammatical subject is a definite description. The usual (Russellian) modern analysis agrees with Hospinianus in so far as it does not subsume them under any of the four simple Aristotelian forms, but disagrees with him in so far as he regarded such propositions as completely irreducible to any other propositional forms whatsoever. The usual modern (Russellian) analysis is as follows:[12]

(1) $the\ S\ is\ P \leftrightarrow (Ex)\ [Sx\ \&\ Px]\ \&\ (y)\ (z)\ [(Sy\ \&\ Sz) \rightarrow I(y,\ z)].$

In words:

(2)   *the S is P $\leftrightarrow$ some S is P and all S-things are identical.*

The position described under (i) is represented by the left-hand term of the analysis, while that described under (ii) is explained by the necessity of conjoining the right-hand term. If Russell's analysis is accepted, it can be said that all who ever took part in this historical discussion have pointed out something that was correct as far as it went. When Russell's analysis is rendered in form (2), it can be seen to be a combination of these three positions (i)–(iii). But it should be remarked that the modern way of representing the left-hand conjunctive term of the analysis of (2) does not coincide with the traditional one (cp. IV–8).

It is also of importance to observe the following. It is natural to attribute some kind of existential import (relative to the domain of discourse) to the grammatical subject of an individual proposition, be it a proper name, a definite description *the S*, or a demontrative description *this S*. For most people will accept the inference from *this S is P* or from *the S is P* to *some S is P*. Consequently, anyone who adopts the majority view (ii) above *must* attribute existential import to the grammatical subject of the categorial A-form *all S's are P*. If one does, then Darapti is valid. And in fact this is von Freytag's position.

19. REPLACING PROPER NAMES BY DESCRIPTIONS

In principle, proper names may be omitted. It is generally possible to replace a proper name by a definite description without any damage being done to the communication process. What cannot be omitted are the atomic propositional functions, expressed by means of individual variables: *Px*, *Rxy*, etc.[13]

But individual (proper) names may, if one wishes, be replaced by definite descriptions *the (one and only) M*.[14] If this procedure is followed descriptions of entities of all kinds will become very prominent in discourse and reasoning. Today, this is not held to imply that we can resort to the procedure of *describing* things as a method of substantially augmenting our knowledge, provided the descriptions are well chosen. As we shall see below (Chapter VIII), earlier philosophers believed that it was possible to obtain new knowledge in this manner.

20. IDENTIFICATION BY DESCRIPTION

A singular proposition *S is P* with a proper name for its grammatical subject may be regarded as belonging to a simple (irreducible, atomic) propositional form. Today this is very common. The existential import of proper names is expressed in the rule of existential generalization.

It is likely that Russell's theory of descriptions will not be the last word on the subject. Investigations in modal logic make it likely that his theory will be supplemented or amended by setting it within a wider conceptual framework. We cannot go into this contemporary discussion of modalities and descriptions here. I shall restrict myself to a comparison between the views of Ockham and Wolff (and the majority of the other logicians in the tradition), Russell and Hintikka with respect to the form of identifying propositions *the M is s*, or *s is the B*.[15] I shall use the equality sign for strict identity. From (6) in Section 16, Russell's theory of definite descriptions, it follows that:

(1) $a = the\ M \leftrightarrow (Ex)\ [a = x\ \&\ Mx\ \&\ (y)\ (My \rightarrow y = x)]$.

Hintikka[16] offers the following analysis:

(2) $a = the\ M \leftrightarrow$ $Ma\ \&\ (y)\ (My \rightarrow y = a)$,

while a modern symbolization of the analysis of Ockham, Wolff, and others is simply:

(3) $a = the\ M \leftrightarrow$ $(y)\ (My \rightarrow y = a)$.

Hintikka's analysis can be seen to be half way between the Ockham-Wolff analysis and Russell's analysis with respect to propositions of identity.

(2) taken together with the assumption that the descriptive phrase $\imath x Mx$ is referentially non-empty entails the *definiens* of Russell's analysis

of sentences of the form *the unique M is P*. In other words: Hintikka's theory as expressed in (2) entails

(4)   *the M exists and is P* $\leftrightarrow (Ex)\ [Mx\ \&\ Px\ \&\ (y)\ (My \rightarrow y = x)]$.

The proof is given in Hintikka 1958. Hintikka himself expects (2) to fail in fictional contexts.

NOTES

[1] The data in the first part of this section are taken from Risse 1964.

[2] For a comparison between Snellius and Ramus with respect to *syllogismus contractus*, see Risse 1964/190.

[3] I have not been able to trace the role of Duns Scotus (Pseudo-Scotus?) in this matter.

[4] Hospinianus' work was not obtainable in any Dutch library, so I could not investigate whether he gives any analysis or example of a valid syllogism with a premiss in which the subject term is a definite description. The answer to this question is, however, not of too great importance, for Hospinianus did not have much influence, although his work was studied by Leibniz. – Cp. note 12.

[5] The phrase "the extension of an individual concept" has sense if "an individual concept" is defined as a unit class (or: singleton), {a}, where a is that individual to which the individual (subject) term refers. For a, and nothing else, belongs to (the extension of) this class.

[6] Cp. Albrecht 1954/36. Burkamp regards it as "understandable, although regrettable, that the concept of individuality as further developed by Duns Scotus and Leibniz has often become quite alienated from its original meaning" (1932/87). In the opinion of the present author, this alienation should be studied in the light of the interpretation of singular propositions as universal propositions. For Leibniz, see VII-4 below.

[7] "Terminus autem singularis vel indicatur nomine proprio, vel particula demonstrativa, vel determinatione quadam singulari", writes Wolff (*op. cit.*/241). Angelelli reports that Fonseca "counts descriptions (existence and uniqueness) as proper names" (1967/88 n. 186). None of the mediaeval authors I have consulted show any signs of having treated individual propositions with proper names as subject terms differently from individual propositions with definite descriptions as subject terms. Ashworth (1970) arrives at the same conclusion.

[8] These are two genitives: *the I of c is the A of t*. This example is therefore more complicated than examples of the propositional form: *the S is P*, where S and P are non-complex general terms, but Wolff shows no awareness of this fact.

[9] "Ceterum notandum, quamvis fieri soleat, ut plures personae eodem nomine proprio gaudeant, nos tamen nomen proprium hic sumere pro termino singulari, quatenus non nisi unicum individuum denotat... Quamobrem si idem nomen proprium pluribus competat personis, ad tollendum ambiguitatem aliae adhuc determinationes singulares eidem adjiciendae, ne secunda ac quarta propositio perperam applicetur" (Wolff *op. cit.*/296). Wolff uses the classical terminology of "sumere" and "quatenus" in order to prevent propositions with subject terms which are ambiguous (non-unique) descriptions or proper names from being given a universal interpretation.

[10] Cp. I-9, IV-15.

[11] Cp. Parkinson 1965/27 n., Ashworth 1970/20, Risse 1970/452.

12 The analysis may also be contracted into one complex existential statement (cp. (6), Section 14):

(3)   *the S is P* ↔ $(Ey)[Sy \ \& \ Py \ \& \ (x)(Sx \rightarrow I(x, y))]$.

The right-hand term of the conjunction inside the square brackets, the formula: $(x)(Sx \rightarrow I(x, y))$, is precisely the form which Hospinianus held to be always false (Section 8; cp. Dürr 1949a/480). Did Hospinianus want to ensure that non-uniquely referring cases of *the M is P* are false? If so, then the price he had to pay was too high, for it will make *all* propositions of this form false.

13 Cp. Quine 1966/171.

14 Russell replaced proper names *a* by definite descriptions *the M*. Any proposition *a is P* can in principle be replaced by *ιxMx is P*, where *M* is the name of some property by which the individual called *a* can be identified. Quine maintains (1950/223) that we may also replace *a* by *ιxax*, in which expression the proper name *a* is treated as a monadic predicate. Instead of "Saul is black" we then obtain "the individual x, such that x is [predicative "is"!] Saul, is black". The definite description *ιxax*, is subsequently treated as in Russell's theory of definite descriptions.

15 In 1960 Lejewski, building upon the ideas of Leśniewski, suggested a theory of descriptions which has certain affinities with Hintikka's theory. Important differences are (1) Lejewski's return to only one kind of variable, and (2) his use of a weakly reflexive relation of predication (copula): a proposition *a is a* is taken to be true provided *a* denotes something, "is" being the copula in question.

16 Cp. Hintikka 1958; 1969/39.

CHAPTER VI

SINGULAR – GENERAL – INDEFINITE

1. Logophoric Judgments as Singular Judgments

Von Freytag's little book contains scarcely any information on the logic of logophoric terms. We therefore have no choice but to give considerable weight to his brief remarks on this topic. The reader will remember that he says: "…"der (die, das)" heisst das Element" (1961/65; cp. III-4). He is of the opinion that *individuelle* judgments and indefinite generic judgments have the same logical form: "the judgment… treats it [the individual, E.M.B.] like a genus-concept… Individual judgments like "Socrates is [a] man" have, then, logically the same structure as "Europeans are men" or "Sinners are men" (65f.). About the universal categorical judgment form *alle S sind P* he says: "As we saw, this judgment says, from a logical point of view: "S, *taken* as a genus or as an individual, stands in relation of identity to P"" (66; my italics).

From this we may conclude that in von Freytag's logic, a judgment with a logophoric subject term *the/an M*, or *das M* (*überhaupt*), is treated syllogistically no differently from an individual judgment with the same linguistic form, the subject term of which is intended to provide an identifying description of a special individual.

We therefore apply the traditional syllogistic treatment of singular propositions to propositions with concrete universal (general) terms *the/a M* as subject. As our point of departure we shall take the Raue-Leibniz analysis, which is in any case no less precise and no more misleading than that of Wolff (V-10; cp. Rescher 1964/120). We then realize that in the syllogistic theory of inference, a proposition

(1)   *the/an M is P*

or

(1')   *der (die, das) M ist P*

in a logophoric interpretation

(1″)   *ıM is P*

is interpreted as follows:

(2)   *everything which is identical with the/an M, is P.*

Our example of an indefinite proposition, "the Bantu is primitive", becomes

(3)   everything which is identical with the Bantu is primitive.

The analysis (2) brings the relation of identity – or put more correctly: *a* relation of identity – to the fore. It is now quite clear why that relation is so important to von Freytag: it is basic to the logic of concrete universal terms.

I now repeat the question which I have already formulated several times:
do the premisses

(4)   the Bantu is primitive

and

(5)   Saul is a Bantu

entail the conclusion

(6)   Saul is primitive,

or do they not?

It will be clear that this question may be given the following formulation: is the argument

(7)
$$\left\{\begin{array}{l}\text{everything which is identical with the Bantu is primitive} \\ \text{everything which is identical with Saul is (a) Bantu} \\ \hline \text{everything which is identical with Saul is primitive}\end{array}\right.$$

a logically valid argument, or is it not? Is it a valid syllogism, or is it a fallacy?

2. PRIMITIVE PARADIGMATIC LOGIC

The answer to this question clearly is that (7) is a valid syllogism in the mood Barbara if and only if the expressions "is (a) Bantu" and "is identical with the Bantu" are interchangeable. The minor may then be formulated thus:

(8)   everything which is identical with Saul is the Bantu,

that is to say

(9)   Saul is *the Bantu*.[1]

If the argument (7) is logically valid, then this judgment (9) is true if the minor premiss in its original form (5) is true. Now this judgment (9) has alarming consequences. Irrespective of whether "is" is supposed to be a predicative or an identifying "is", or – as in von Freytag's logic – both at the same time, this judgment (9) implies that an observation of the one Bantu Saul is at the same time an observation of *the Bantu*. In other words, the *vis universalis* of the singular subject term *S*, or *this M*, suffices to enable us to determine the content of the quasi-singular concept '*the M*'. When "the Bantu" is understood to mean the same as "the Bantu as such" or "der Bantu überhaupt", we have here a remarkable form of reasoning indeed. In this form of reasoning it suffices to observe, as an *exemplum*, one Bantu only, say Saul, in order to determine the characteristics of *the Bantu* (*as such*), for Saul is *the Bantu*.

This result is of such an overwhelming importance that I shall repeat my argument in general terms, *i.e.*, using variables. Assume that someone regards the argument form

(10)
> *the M is P*
> *S is (an) M*
> ———————
> *ergo, S is P*

as logically valid, on traditional syllogistic grounds. This is to say that he regards the argument form

(11)
> *everything/all which is identical with the M is P*
> *everything/all which is identical with S is (an) M*
> ———————
> *ergo, everything/all which is identical with S is P*

as valid. If he justifies this position by an appeal to Aristotelian syllogistics, he clearly takes (11) to be an application of Barbara, which is to say that he regards *is (an) M* and *is identical with the M* as synonymous and interchangeable expressions, for every expression *M*. Otherwise he would commit the fallacy of four terms, the *quaternio terminorum*. If furthermore "is" and "is identical with" (or: "is identically the same as") are synonymous expressions, which is so often said to be the case in the pure logic of identity, then the expressions *M* and *the M* are interchangeable terms. One then has the opportunity to substitute *the M* for *M*, or *der (die, das) M* for *(ein, eine) M*, whenever that substitution would seem to lead to interesting results.

One can also reason as follows: assume the major of (11) to be given. If it is also maintained, for all terms *M*, that

(12)   *(an) M is identical with the M*

(12')   *(ein(e)) M ist identisch dasselbe wie der (die, das) M*,

then with the *dictum de omni* we may conclude that

(13)   *(an) M is P*,

and provided "is" is transitive, this conclusion together with the minor of (11) yields the conclusion of (11). It is clear that (12) is a necessary and sufficient condition for the universal validity of (11), and hence of (10).

When the condition (12) is conjoined to the assumption of the transitivity of the copula "is", we can infer the validity of (10) without making the detour *via* (11). It is then unnecessary to apply the Ockham-Wallis-Leibniz-Wolff-Kant reduction to the premisses of (11).

A modern symbolization of (11) yields the same result. If we understand the premisses and the conclusion of (11) in the manner of von Freytag, then a modern symbolization of the logical form of this argument will be:

$$(x) [x = the\ M \rightarrow Px] \ \& \ (Ex) [x = the\ M]$$

(14) $(y) [y = s \rightarrow My] \ \& \ (Ey) [y = s]$

$$(z) [z = s \rightarrow Pz] \ \& \ (Ez) [z = s].$$

This argument form is invalid in modern logic because the propositions *Ms* and *s = the M* are not interchangeable, whatever term *the M* is, and irrespective of whether " = " stands for strict identity or for some other kind of similarity of "partial identity". But should anybody adopt a theory of predication implying the interchangeability of *Ms* and *s = the M* in arguments, then and only then will (14) be valid for him. This may be checked by means of a semantic tableau.

Assume hypothetically that someone considers (10) as a valid argument form. He thereby assumes that every property P which we have ascribed to 'the Bantu' as a consequence of our observation of John, may also be truly predicated of the Bantu Saul. This result is absurd. It remains as absurd if we systematically replace "John" and "Saul" in our argument by "the concept John" and "the concept Saul", or by "my concept (of) John" and "my concept (of) Saul", and "observation" by "Anschauung".

This is clearly a completely irrational paradigmatic mode of thought. Any individual Bantu, for instance Saul, may be chosen as a *paradigm* or *exemplum* of '*the Bantu*'. I shall call this mode of thought "primitive paradigmatic logic" ("PPL" for short).

The formal characteristics of PPL are:

(i)   (*an*) *M is identical with the M* is true for every predicate *M*;

(ii)   the copula is a transitive relation *is identical with*

(cp. Veatch 1952/362).

One also arrives at PPL from the following four principles:

(iii) all judgments with the verbal form *the/an M is P* are treated in syllogistic in the same manner as singular propositions;

(iv) the Ockham-Wallis-Leibniz-Wolff-Kant reduction is to be applied to singular propositions;

(v) *the M* may be replaced by (*an*) *M* and *vice versa*;

(vi) no distinction is to be made between "is" and "is identical with".

By means of the above argument we have carried the hypothesis of the validity of (10) *ad absurdum*, provided the major premiss of (10) is assumed to have some kind of logophoric meaning. PPL here served us as an absurdity. A weaker conclusion would be this: if PPL is rejected, *then* (10) is invalid, given that the major premiss has a logophoric meaning.

When *S* is the proper name of an individual, no theoretical logic, traditional or modern, pays much attention to whether the predicate term is or is not preceded by an indefinite article, as is the case in *S is a P* (cp. Quine 1965/118). If someone insists that it be used, then our argument will go through only if the indefinite article is either included in or excluded from all minor premisses in this section and all other judgments '*S is* (*a*) *P*' and '*S is* (*an*) *M*' as well. One can easily imagine that somebody will attempt to refute our derivation of PPL from traditional principles on the ground that in English or in German it is sometimes natural to speak of *an M* (*ein M*), but at other times of plain *M*. In order that this refutation may be successful, it will be necessary to construct a new formal logic in which the difference between *an M* and *M* is marked in such a way that (10) becomes valid without entailing PPL. I do not see how this could be done.

It should be remarked that we have focussed our attention upon such predicates *M* ("man", "Bantu", "language", and "state") as can be preceded either by a definite or by an indefinite article, whatever the position of *M* in the proposition may be. The problem of the logic of mass propositions like "Sugar is sweet", "Copper is red", "Iron can be magnetized", etc., are in any case left open so far as the philosophical purposes of this book are concerned.

No one will be prepared to accept PPL as it stands. No philosopher writing about logic will maintain publicly that it is sufficient to observe one Bantu, John, in order to ascribe the observed properties to another Bantu, Saul, thereby making appeal to the validity of (1). No traditional logician would be prepared to defend this mode of thought, at least not in this simple and direct form.

This leaves the question unanswered what the logic of a logophoric "the" can possibly be. We still do not know how to answer this question; we only know that (10) ought to be rejected as a universally valid argument form. Von Freytag and Veatch give us no information about this problem. All they have to say about the logic of "the" has already been taken into account in our discussion. We are faced with a complete mystery.

3. PRIMITIVE PARADIGMATIC LOGIC REJECTED BY PFÄNDER

Pfänder devotes one section to a discussion of the logical relationship between individual judgments and generic judgments ("Arturteile"; 257–261). This section contains the following statements: (1) an immediate inference ("der unmittelbare Schluss") from an individual to the species ("Art") is never valid, (2) an immediate inference from many individuals to the species is not valid either, (3) an immediate inference

from an *absolute* generic judgment is *always* valid, (4) an immediate
inference from *relative* generic judgments to the relevant individual
judgments is *not always* ("nicht immer") "valid" (259). He exemplifies
the kind of valid inference characterized sub (3) by:

> Der Mensch ist ein seelisches Wesen
> > (Man is an animate being)
>
> ∴ dieser Mensch hier ist ein seelisches Wesen
> > (this man here is an animate being)

One of his examples of an invalid inference of type (4) is the inference from
the premiss: "Man has (in the normal, adult, healthy case) the capacity
for self-determination" to the conclusion: "this particular man has this
capacity" (259).

Pfänder states: (5) "From the truth of a generic judgment, the truth
of the corresponding individual judgments will follow immediately only
when the generic judgment is an absolute one, *i.e.* when it concerns the
kind in every instance ("Fall")" (*loc. cit.*). This brings him to the following
conclusions: (6) from the falsity of an individual judgment one may
immediately ("unmittelbar") validly infer the falsity of the corresponding
absolute generic judgment, but (7) from the falsity of an individual
judgment one can "*not* conclude immediately to the falsity of one of these
relative generic judgments" (260).

Pfänder thus holds that the argument form

> *the/an M is P*
> *S is an M*
> ———————
> *ergo, S is P*

is sometimes valid, and sometimes invalid. If *the/an M is P* expresses an
absolute generic judgment, it is valid, and if *the/an M is P* expresses a
relative generic judgment then this argument form *may* be invalid – at
least this is the conclusion if his statement (4) is to be taken literally. But
in regard to his statements (6) and (7), it would seem to be more in
harmony with his intentions to say: if the major premiss expresses a
logophoric judgment, then this argument form is valid if and only if the
major premiss *the/an M is P* is an *absolute* generic judgment and invalid
if and only if this premiss expresses a *relative* generic judgment.

4. TWO KINDS OF DEFINITE ARTICLE WITH A GENERALIZING FUNCTION. COMPARISON WITH MARITAIN

Our first objection to the account given in Pfänder's logic must be that in a concrete case the opponent as often as not cannot know whether *the/an M is P* is meant as "absolute" or as "relative", especially when he has a written statement in front of him with little or no pragmatic data to support his interpretation. This certainly constitutes a serious objection to Pfänder's theory of the indefinite judgment form. But it is this "theory" of one "absolute" and one "relative" interpretation of the same sentence which saves Pfänder from the accusation of holding a primitive paradigmatic logic.

The objection can be met only by the introduction of two (or more) articles for use in generic statements, for instance "the$_{abs}$" and "the$_{rel}$". The logic of Pfänder's *singulare Arturteile* may then be expressed as follows:

Valid	Invalid
the$_{abs}$ *M is P*	*the*$_{rel}$ *M is P*
S is an M	*S is an M*
———————	———————
ergo, S is P	*ergo, S is P.*

The situation is similar to that in the logic of Maritain. In IV-30 we found with respect to that logic:

Valid	Invalid
the M as such is P	*the M is P*
S is an M	*S is an M*
————————————	———————
ergo, S, in so far as it is M, is P	*ergo, S is P.*

Suppose someone says: "The Bantu is primitive". If the proponent is a follower of Pfänder, the critic who wants to refute this proposition by pointing out one or more non-primitive Bantus may expect the reply: "I only meant *the* Bantu in the typical/normal/ideal case".

Is the opponent obliged to put up with this? If so, it should be realized that the proposition "the Bantu is highly cultured" with respect to the same "case" (the typical, the normal, or the ideal case) is just as tenable, unless a logic is developed for "the$_{rel}$".

A number of mutually dependent questions now come to mind:

(i) what is the logical relationship between the articles "the$_{abs}$" and "the$_{rel}$"?

(ii) how are judgments of the form *the$_{rel}$ M is P* (or sometimes, in English: *a$_{rel}$ M is P*) to be contested (for judgments of this form cannot possibly *all* be true)?

(iii) is there any single elimination rule, or are there perhaps several different ones, for "the$_{rel}$"?

(iv) is there an introduction rule for "the$_{abs}$"?

These questions are formulated upon the (unproved) assumption that in some cases at least, the use of a logophoric article in relative judgments is unavoidable, and that no reduction is possible to "quantified" propositions with a time-indication in what would traditionally be called "the subject term",[2] like propositions of the form *x% of all M's-at-t are P*, or of the form *most M's until now have been P*, or to any other propositional form. Suppose that all generic judgments *the/an M is P* were reducible to judgments of such forms as these, which do allow for critical objections. Pfänder could then have recommended his readers to use prenex articles in generic judgments *only* when they were meant as *absolute* generic judgments. He does not so advise his readers, but prefers the muddle resulting from the use of the same structural ("logical") words "der", "die", and "das" in several different, yet undefined senses.

5. PFÄNDER ON THE LIMITATIONS OF TRADITIONAL SYLLOGISTIC

In his theory of inference, which covers more than a hundred pages, Pfänder discusses Aristotelian syllogistic, and in the accepted manner. He subsequently devotes one section to a number of important shortcomings of this Aristotelian logic – in his words, the "traditional" logic. This is at variance with the attitude of von Freytag, who, as will be remembered, holds that Aristotelian logic is sufficient for all philosophical purposes, although he, too, does seem to put a particular stress on logophoric generic judgments. One of the things for which Pfänder blames Aristotle's theory of inference is that in that theory, "the *quantity* of judgments is simply taken only in the sense of *universal* or *particular* judgments". Pfänder comments: "But we have already seen that there are also judgments which lie completely outside the domain of this quantity, because in these judgments the subject-concept does not primarily delimit a set of objects ["Gegenstände"]. And these judgments, *viz.* the generic ["Art-"] and the individual judgments, do also permit the formation of syllogisms" (319 f.).

6. A THEORY OF ENTAILMENT?

Pfänder thereupon makes the heroic resolution to develop a theory of inference different from and more fundamental than Aristotle's syllogistic, a system that may be used for *all categorical judgments*, including those which are not contained in the traditional classification in respect of "quantity" (320 f.). Not belonging to this classification are, as will be remembered, in the first place the *singulare Arturteile*, the singular generic judgments, which are *indefinite* judgments. We saw in IV-21 that Pfänder prefers the classification into *Singularurteile* and *Pluralurteile* to the traditional classification into *Einzel-*, *Partikular-* and *Universalurteile*. But as soon as he starts his discussion of the new theory of inference, he no longer uses the technical term "Singularurteil". The new theory of inference is said to pertain to *categorical* judgments, to "the so-called *categorical* judgment, whose traditional formula is "S is P"" (320 f., 105; cp. IV-21).

Now if the "formula" for categorical judgments in Pfänder's sense is *S is P*, it is natural to ask whether every categorical judgment is also a *Singularurteil*. This is a question which receives no clear answer in Pfänder's book. I shall try to answer it in Chapter X.

There is no doubt but that Pfänder considers those *Singularurteile* which are expressed by means of propositions of the form *der (die, das) S ist P* or of the form *S ist P* as "categorical". This is to say that his new theory of inference will be applicable to all those *Singularurteile* which are categorical in the well-known sense of being unconditional and non-disjunctive, and thus to *singulare Individualurteile* as well as to *singulare Arturteile*. It should be remembered that the latter class embraces the *relative* as well as the *absolute singulare Arturteile* or absolute singular generic judgments.

Let us take a closer look at Pfänder's new theory of inference. His inference rules (or principles) are formulated in terms of an expression "mit sich mitführen", which may be translated as "to entail" but which he assumes to stand for a logical relation between two *Gegenstände*. By means of this technical term he attempts to develop a fundamental theory of *mittelbare Schlüsse* ("mediate inferences") with two categorical premisses. This theory turns out to consist of the following three *formalontologische* principles; I offer the German text first and a translation into the language of first-order predicate logic afterwards:

> (1)   das M-sein *eines beliebigen Gegenstandes* überhaupt führt das M-sein desselben Gegenstandes mit sich (schliesst es aus), der Gegenstand S ist M,
>
> ――――――――――――――――――――――――――――――――――――
>
> der Gegenstand S ist notwendig P (nicht P);
>
> (2)   das P-sein *eines beliebigen Gegenstandes* überhaupt führt das M-sein desselben Gegenstandes mit sich (schliesst es aus), der Gegenstand S ist nicht M (ist M),
>
> ――――――――――――――――――――――――――――――――――――
>
> der Gegenstand S ist notwendig nicht P;
>
> (3)   der Gegenstand M ist P (nicht P), der Gegenstand M ist S,
>
> ――――――――――――――――――――――――――――――――――――
>
> einige Gegenstände, die S sind, sind zugleich P (nicht P)

(320–326; schematized by me). If we choose "x" as a *Gegenstand*-variable and if we abbreviate "führt mit sich mit" by "ent", then these principles may be expressed more simply as follows, from which it becomes clear that Pfänder's theory of inference is not particularly impressive or elucidatory:

> (1') $(x)[Mx \text{ ent } Px]$
> Ms
> ――――――――
> $\therefore Ps$
>
> (2') $(x)[Px \text{ ent } Mx]$
> $\sim Ms$
> ――――――――
> $\therefore \sim Ps$
>
> (3') Pm
> Sm
> ――――――――
> $\therefore (Ex)[Sx \ \& \ Px]$

(1') and (2') express that the *modus ponens* and the *modus tollens* are valid for this relation of entailment. (3') is more interesting, for it shows that Pfänder recognized third-figure expository syllogisms with respect to any domain of *Gegenstände*, but he does not discuss their importance or use.

If preferred, the quantifiers in (1') and in (2') may be omitted and the variable "x" be replaced by "s". – The alternative cases (in parentheses) in (1) and (2) above need not be formulated separately.

Given the purposes of this new theory of inference as stated by Pfänder himself (320 f. especially), he must have meant the first premiss in (1) and (2) as an explication of the categorical judgment form '*der* (*die, das*) *M ist P*':

$$der\ (die,\ das)_{cat}\ M\ ist\ P \leftrightarrow (x)[Mx\ ent\ Px],$$

where the formula to the right pertains to the domain of all *Gegenstände*. According to Pfänder, individuals as well as individual qualities,[3] species ("Arten") and genera ("Gattungen") are *Gegenstände* (137).

Three remarks should be made:

(i) Pfänder formulates no introduction rule for "das_{cat}". This article, then, is not semantically defined, and the same then also holds for his entailment-relation *mit sich mitführen* which he defines in terms of "das_{cat}". In other words, we do not know what it *means* to say of a certain case that the introduction of a statement of the form $das_{cat}\ M\ ist\ P$ is justified. The structure of that conceptual pattern to which this mode of speech belongs has not been characterized and brought into a systematic connection with this language form. In a dialogue a proposition $das_{cat}\ M\ ist\ P$ cannot therefore be distinguished from a sheer dogma or prejudice. Pfänder's *reine Logik* is, then, anything but a cut and dried system ready for use, even if we were to restrict ourselves to his logic of "das_{cat}".

(ii) It should be noted that Pfänder himself is unclear about his own systematic goals. In a short section on immediate inferences ("Unmittelbare Schlüsse durch Entfaltung der in einem Urteil implizierten Urteile"), the following example is offered of an immediate inference, *i.e.* an inference from one premiss, employing no middle term: "The judgment "This eagle is dark brown" entails the judgments: "This is an eagle",…, "This animal is a bird of prey", etc." (287). Here a conclusion *this S is P* is drawn from a premiss of the form *this* [S] *is M*. We could have followed him if he had added a major premiss: *the*/*an_{cat} M is P*; in this case: "The$_{cat}$ Eagle is a bird of prey". But he does not do that; the inference is characterized as "unmittelbar" (immediate).

It is not hard to replace his example by one wherein "Adler" is replaced by "Bantu", so as to point out the importance of this discussion.

(iii) The third remark will be relegated to a section of its own (Section 7 below).

7. THE PROBLEM OF RELATIVE GENERIC JUDGMENTS AND THE VARIABLE COPULA

Our third observation is even more fatal to Pfänder's logic. In his new theory of inference not one word is said about the difference between relative and absolute categorical judgments. With respect to the semantics of "das$_{rel}$", not a single principle is formulated. The relation between an "absolute" and a "relative" categorical judgment *'the/an M is P'*, and the relation between the latter judgment and individual judgments *'this M is P'*, are not treated by Pfänder at all. Nor, in his theory of inference, does he ever refer to the distinction in his theory of judgment between *Singularurteile* and *Pluralurteile*; similar complaints may be made about the logic of Maritain (cp. VII-14 below).

In Section 5 above I expressed the opinion that the inference form *the/an M is P, S is an M, ergo S is P* is meant to be valid in Pfänder's logic if *the/an M is P* is used to express an *absolute* judgment, but invalid if it is used to express a so-called *relative* judgment. The only definition of the expression "ein absolutes Arturteil" in the whole of Pfänder's logic is that it is a judgment *'the/an M is P'* such that an inference of the above-mentioned kind will be valid, in the sense of completely safe. To say that the inference form under consideration is valid *provided* the major premiss expresses an absolute judgment therefore in reality amounts to saying that this inference form is *not* valid.

How is this all to be understood? What is the conception behind this incoherent and even inconsistent talk about a new theory of judgments and a new theory of inference?

Light is thrown on this riddle by still another classification of categorical judgments which is offered in Pfänder's book, *viz.* his distinction between (1) assertoric, (2) problematic, and (3) apodeictic. This is, he tells us, a division according to *modality* ("die Modalität des Urteils", 106). An assertoric judgment ('*S ist tatsächlich, wirklich P*') is a judgment in which "the judgment is not logically qualified [lit.: damped, "abgedämpft"] in any way" (97). In a so-called problematic judgment ('*S ist vielleicht, möglicherweise P*') the "logical stress" ("Behauptungsgewicht") is more or less reduced (93). Finally, an apodeictic judgment ('*S muss P sein, ist notwendigerweise P*') is, so Pfänder informs us, not only not logically qualified ("logisch abgedämpft"), for this holds for assertoric judgments as well, but "über sein angemessenes Vollgewicht hinaus bekommt dann der Behauptungsschlag einen aus irgendwelchen logischen Quellen zerfliessenden grösseren oder geringeren Überschuss an Wucht, und das Urteil macht nun einen *überhöhten Anspruch* auf Wahrheit" (98).

This is pretty untranslateable, but is part of the only extant book on phenomenological logic.

Pfänder then draws an important conclusion about the meaning of "is": "The "Is" in the formula for the categorical judgment should therefore not be regarded as determinate [definite, "bestimmt"], either in respect of the quality of the relation [of predicate to subject], or in respect of the modality of the statement" (106). *The copula, even in a categorical judgment, is variable*, not only in respect of *Qualität* (which perhaps means no more than that the copula may be affirmative or negative), but also – and for us that is of much greater importance – in respect of the *Modalität* of the judgment.

In the light of this information, *Pfänder's attempt, even his wish, to construct a new, phenomenological, theory of inference may be entirely ignored from a theoretical standpoint, for in that theory he does not even mention the variable copula. His new theory of inference and his logical reflections in general are without any systematic value whatsoever.* He does not offer a theory of entailment as a form of strict implication.[4] He introduces two distinctions with respect to categorical judgments *the/an S is P*, each of which operates as an *escape clause or Way Out*: first a distinction between absolute and relative judgments, and secondly, a distinction between apodeictic, assertoric, and problematic judgments. Are the relative judgments problematic, and vice versa? Nothing is said by Pfänder about any relationship between the absolute-relative distinction and the apodeictic-assertoric-problematic distinction, neither of which plays the slightest role in his theory of entailment.

Is Pfänder's logic the logic of the phenomenological movement?

8. COMPARISON WITH LOTZE, SIGWART AND JERUSALEM

Lotze chooses the sentence "der Mensch ist sterblich" ("Man is mortal") as an example of *das generelle Urtheil* and offers the following analysis: "wenn *irgend ein* S ein Mensch ist, so ist dieses S sterblich" ("if some S or other is a human being, then this S is mortal") (1912/94). He nevertheless distinguishes between this judgment and (that expressed by) "all human beings are mortal". This is to say that he takes "wenn… so…" to be a stronger kind of implication than the implication in a universal judgment.

Lotze's opinion is shared by W. Jerusalem, who in 1905 launched an attack upon the "pure logic" of Husserl and others (Jerusalem 1905/189).

Given the words "irgend ein" ("some… or other") in the analysis and assuming that no one has ever used "if… then…" in such a sense that *modus ponens* does not hold, it follows that Lotze and Jerusalem regard the inference form

> *the/an M is P*
> *S is an M*
> ———————
> *S is P*

as logically valid, contrary to the teachings of Pfänder.

Lotze does not assume any variability or modification of the *copula*, nor do we find anything in his logic about classes of "relative" judgments. But he does hold that a mark ("Merkmal") always has a degree, a certain "degree of that intensity which is characteristic of it" (161), although unfortunately he does not relate this to his theory of inference. He assumes that a predicate P may be analysed into several "species and modifications" which he calls p^1, p^2, p^3, so that from premisses M *ist* P, S *ist ein* M the following conclusion may be drawn: S *ist entweder* p^1 *oder* p^2 *oder* p^3 (*loc. cit.*). There is, then, no question but that in Lotze's logic two premisses *der* (*die, das*) M *ist* P and S *ist* (a species/kind of; a modification of) M cannot both be held to be true unless some kind (or modification?) of P-ness is attributed to S. The possibility that we may have to attribute some "modification" of P to S is not very attractive but seems to hinge upon the presence of a modifying word in the minor premiss. The crucial question now becomes what kind of modification of the predicate he ascribes to the subject term of the conclusion. It seems likely that, using scholastic terminology, Lotze generally assumes the subject term M in *der* (*die*, *das*) M *ist* P to have personal *suppositio*. In his own words: "it is precisely the individual human beings and the individual [physical] objects in: "Physical objects fill up space" which are the real subjects of the general judgment" (94; cp. VII-27 below).

Sigwart discusses "the indefinite relatives (who and what, quisquis) which are unable to do anything else but express that the subjects of which the one predicate [A] holds, also have the other one [B], so that the expression thereby becomes equivalent to a universal judgment" (1904, I/288). By a universal judgment ("allgemeines Urteil"), Sigwart understands a judgment (expressed by a proposition) of the form *alle A sind B*, i.e., an A-proposition (I/216). In his reaction against the metaphysical traditional logic, therefore, he goes further than Lotze, who was prepared to conclude from the generic to the universal judgment, but not conversely. Sigwart boldly maintains: "Man is mortal – *all* men are mortal – *what* is a man, is mortal – certainly all mean the same thing, the *necessary* going together of being a man and being mortal" (I/288, italics mine). He does not discuss more than one kind of indefinite relative judgment *das M ist P*. In Sigwart's logic the difference between *das A*, *alle A*, and *ein A*, which was considered to be of paramount importance by Husserl, Pfänder, and many others, is assumed to be non-existent.

9. THE RELATIVITY OF RELATIVE GENERIC JUDGMENTS

Pfänder maintains: "the copula may... vary according to the modality". His thinking is clearly closely related to the metaphysico-logical tradition, even more so than that of Lotze, to say nothing of Sigwart. Lotze, as we have seen, still ascribed a degree of intensity to *every* mark or property. Even Hering speaks in traditional terms about graded generalities. He sees a gradual transition from a lowest species or singularity, such as carmine-red of a definite shade, to what he takes to be the highest generality within the region to which colouredness belongs, namely the idea of sense quality as such ("sinnliche Qualität schlechthin", 1930/531).[5] Von Freytag, too, talks of degrees in his logic, *viz.* degrees of concreteness or ontological independence. *Teilhabe* or participation, the Platonic μέθεξις, is *abstufbar* (L. Nelson 1962/523), and is said by von Freytag to be an ontological relation (1961/38).

It is interesting to observe that Sigwart assumes "relative identity" to be another expression for a *degree of identity*, a notion he himself rejects (1904/110).

This assumption seems to me to be justified. Our investigations up to this point lead us to the following hypothesis: both the problematic and the relative generic categorical judgments in Pfänder's logic are conceptually connected with this notion of degrees of identity, to be expressed by a variable copula. This hypothesis explains why he speaks about *absolute* and *relative* judgments and why he regards the latter as *Singularurteile* and not as *Pluralurteile*.

Hering uses the expression "Wesenskern" (502), which suggests that the truth-value of relative generic judgments is related in some way or other to the truth-value of judgments about this "nucleus of Being". The latter judgments are those which are called "absolute" generic judgments by Pfänder, and the relative judgments would be the judgments which ensue when the thinking subject has penetrated less deeply into the *Wesen* of the *Gegenstand* or topic under investigation.

I cannot find any other explanation for the fact that Pfänder regards relative generic judgments as *Singularurteile* and not as *Pluralurteile*.

10. *Exemplum* AND THE METHOD OF VARIATION

In Section 2 it was said that when the expressions *is an M* and *is the M* (*überhaupt*) are used interchangeably, we arrive at Primitive Paradigmatic Logic.

Now Husserl and Pfänder do intend to distinguish logically between *ein M* und *der* (*die, das*) *M überhaupt*. Pfänder also assumes a distinction between *ist* as identity and *ist* as partial identity, but this distinction finds no expression in his theory of entailment. Nor is he consistent in his application of the quality-operator (or -operators) "das". He makes no distinction in his new theory of entailment between "das_{abs}" and "das_{rel}".

The price which Pfänder pays, in order to avoid the accusation that he has simply given a formulation of PPL, is a complete lack of any semantics and logic for "das_{rel}", and also the lack of an introduction rule or other kind of semantics for "das_{abs}", hence this logical particle is dialogically undefined.

Instead of an introduction rule or other kind of semantics for "das_{abs}"

and "das$_{rel}$" Pfänder offers an extremely short description of the so-called method of (eidetic) *Variation* (347). The description of this mysterious method, which takes up only a few lines, ends as follows: "The more precise exposition of this way, and also of the *special* procedures which have to be followed especially when we conclude to the species in the normal case or in the average case or in the typical case or in the ideal case, is not the task of logic, but of epistemology in the sense of a methodology" (348). This of course does not answer our demand for introduction rules for "das$_{rel}$" and "das$_{abs}$". Pfänder does not attempt to develop a semantics from certain epistemological tenets or vice versa.

In the pithy phrase of Adorno, the phenomenological doctrine concerning *das Wesen* is "precisely *the* attempt to emancipate the Essential Being ["das Wesen"] from "the example"" (1956/131). If so, then phenomenology is an attempt to realize an important goal in the school of Hegel (cp. VIII-20 below).

The point of departure for the method of variation in the phenomenological *Wesenforschung* or research into Essential Being is nevertheless *the example* (cp. Pfänder *op. cit.*/330). It is therefore completely impossible to understand why Pfänder, who takes as fundamental an inference form which turns out to be the form of an expository syllogism and of Lambert's *dictum de exemplo*, does not make the slightest attempt to bring this inference form into connection with the method of variation.

11. RELATION, ANALOGY, PARADIGM: "A GLIMPSE OF CHAOS"

In older literature there is great confusion with respect to the expressions "παράδειγμα", "ἀναλογία" and "ἀνάλογος", and "ὁμοῖος". According to E. Haenssler, Aristotle's "παράδειγμα" is – wrongly, he thinks – interpreted as "analogical inference" in the whole of modern philosophy. This interpretation is most misleading, since the word "analogy" has, and always had, connotations which may be traced back to the Pythagoreans, namely the kind of similarities or identities between relationships which are called either harmonies, or proportionalities, or isomorphisms (structural identities). In all these cases four or more entities are involved, which are usually of the same ontological kind, for instance all of them individuals. Aristotle, too, used "ἀναλογία" for four-subject propositions,[6] not "παράδειγμα". According to Haenssler, Aristotle's usage does not allow for a clear distinction between the meanings of "παράδειγμα" and "ὁμοῖος". With respect to the meanings of these three terms the situation has been made even worse since Aristotle's time, he holds, because it has become common to translate "παράδειγμα" with "inference by (or from) analogy" ("Analogieschluss"). Anyone who takes a look at the traditional expositions about reasoning from examples, analogies, and similarities will get, according to Haenssler, a "glimpse of chaos" (1927/96).

In this chaos the lack of a workable logic of relational or many-place predicates comes to the fore. It has already become quite clear that this chaos is intimately related to defective theories of the logical relationships between individual propositions, "quantified" propositions, and generic propositions of the form *the/an M is P.*

12. Conclusions

In our search for introduction and elimination rules for logophoric uses of the articles we have obtained the following results:

(i) Duns Scotus and William of Ockham allow indefinite propositions as conclusions in expository syllogisms. In other words, they derive indefinite propositions from two singular premisses (Chapter V).

(ii) Since the days of the schoolmen the most current traditional analysis of singular proposition is the one which reduces them to universal propositions (Chapter V).

(iii) From the Renaissance onwards many philosophers treat some, but not all indefinite propositions, as universal propositions, but without delimiting this sub-class of indefinite propositions by means of a special prefix (Chapter IV). As a result, the propositions called indefinite and those called singular cannot be separated from each other, so that the question we formulated in IV-1 has to be answered in the affirmative.

(iv) Not only the interpretation of singular propositions as a kind of universal propositions, but also the interpretation of some indefinite propositions (*e.g.*, the *singulare Arturteile* in Pfänder's logic) as propositions referring fundamentally to one entity – hence as a kind of universal propositions – was made plausible by and encouraged by the theories of *suppositio simplex* ("homo est digmissima creaturarum") and *suppositio personalis determinata* (IV-3 to IV-12), and especially by the theories of *suppositio naturalis* (IV-5, 14, 21, 28, 29). Von Freytag confirms that singular (or as he calls them: individual) and indefinite propositions are not distinguished in traditional logic (IV-1, VI-1).

(v) The analysis of singular propositions mentioned under (ii) above may be regarded as a consequence of a faulty analysis of universal propositions (IV-8) together with lack of conceptual and notational distinctions between a predicative copula and a copula of identity (V-11). Instead of this distinction between two copulas Pfänder offers the notion of one gradually variable copula (VI-7), *i.e.*, of infinitely many copulas.

(vi) The analysis mentioned under (ii) encourages a mode of thought wherein one contents oneself with analogies, expressed by propositions of the form *a resembles* (*is similar to, is equal in some respect to*) *b*, in cases where strict identity, as expressed by $I(a, b)$, is required if fallacies are to be avoided (V-16).

(vii) In a system of thought without a clear distinction between the predicative "is" and the "is" of identity, the treatment of logophoric indefinite propositions as singular propositions, which are subsequently interpreted as universal propositions, leads to a primitive paradigmatic form of thought. It can only be avoided by means of special devices, such as a careful distinction between two kinds of well-defined articles (VI-2).

(viii) The measures that have actually been taken in order to avoid PPL (theories of *suppositiones*; later: variable copula) do not yield a dialogically acceptable logic. In all the cases we have so far investigated either an elimination rule or an introduction rule for the articles is missing, or both (IV-30, VI-7; see also VIII-21 below).

13. GLIMPSES OF GRAMMAR

(i) *Jespersen on the arbitrary generic person.* In his discussion of "the Generic Singular and Plural", Jespersen (1924) explains that these are linguistic expressions "for a whole species", and that they occur when words like "all" and "ever" and "some" are not used. Jespersen, then, simply assumes the existence of a species for every case which he wants to consider and cannot therefore contribute substantially to the clarification of our problem, but what he has to say confirms the conclusions and impressions which we have noted so far.

He sets out to describe the use of various expressions for "the indefinite person", or, as he prefers to say, for "the generic person". Among the expression he has in mind are "one", "a man" ("What is a man to do?"), and the like. He feels that the difference between the meaning or use of such expressions for "the indefinite person" and the use of nouns in sentences like "man is mortal" is hard to define, and that it is often a matter of feeling rather than a cognitive issue.

He twice discusses the combination of the grammatical singular with the indefinite article, as in the phrase "a dog".

He first characterizes this combination by saying that the article occurs here as a "weaker version of "any" ", and explains this as follows: one (a) dog is here taken to represent a whole class, which is to say that an *arbitrary* individual is regarded as *paradigmatic*.

It is an essential part of my thesis that "arbitrary" and "paradigmatic" should not be regarded as logical synonyms and, in a good logic, not even as closely related notions. The combination of 'arbitrary' with 'paradigmatic' yields precisely that hybrid concept which Beth, in his article on the Locke-Berkeley triangle, opposed as logically redundant and misconceived, and which earlier governed the notion of a variable.

Later on Jespersen refers to the combination of the singular with the indefinite article as a possible expression for "the indefinite person", or "the generic person". His use of "person" here is quite general and covers dogs and other logical individuals as well as human beings.

It may safely be said that in Jespersen's analyses there is no difference whatsoever between the notion of a *generic* person, expressed, *e.g.*, by "a man", "a fellow", "a dog", or "a triangle", and the notion of an *arbitrary* individual person (man, dog, or triangle) understood as a *paradigm*, whatever that may be.

The discovery that this hybrid notion is embedded in the semantics of many "natural" languages ought not to come as a surprise. That older logics and older conceptual structures have influenced the common usage of natural languages seems no less obvious than the converse. If the conceptual situation in, or behind, traditional logic is taken into consideration it is only to be expected that linguists will encounter a number of intractable problems in their efforts to analyse the *syncategoremata* of these languages.

(ii) *Negative indefinite sentences.* One of these is clearly the problem of negative indefinite sentences. Thus Kraak writes that "the traditional differentiation [between definite and indefinite articles][7] is insufficient when it comes to giving a justification for certain phenomena of negation; and there are other facts, too, which necessitate the introduction of further distinctions" (Kraak 1966/122). This much might also be expected from what was said in Sections 2 and 18 of Chapter IV.

Although he does refer extensively to Aristotle's works, including the *De interpretatione*, Kraak[8] surprisingly enough fails to mention its seventh chapter, which seems to be of special importance in connection with notions of negation and their inter-relationship with the logic and grammar of the articles. For in that chapter Aristotle draws, or attempts to draw, a distinction between the notions of (mutually) contradictory and contrary pairs of propositional forms and propositions. In most textbooks on logic the reader is given the impression that Aristotle only applied these expressions, "contradictory" and "contrary", to his "quantified" propositional forms: the A-, the E-, the I-, and the O-forms. Ackrill (1963) reminds us that this is incorrect: Aristotle also spoke of contrary and of contradictory ἀδιόριστοι. As Ackrill points out, this gravely obscures Aristotle's theory of negation, which is easily understood so long as we limit ourselves to quantified propositions. To the task of constructing a clear and useful theory of negation for sequences of words that contain none of the operators "all" ("every"), "some", or "no", even Aristotle's intellectual powers proved inadequate.

Is it an inevitable task for the human intellect? However that may be, Aristotle's theory of contrary ἀδιόριστοι seems to have been influenced by the Greek and, more especially, by the Platonic outlook on comparative relations. "The Greeks tended to consider 'hot' and 'cold' not as relative positions on a single temperature scale, so much as separate and distinct [and contrary, E.M.B.] substances (as we see from Anaxagoras Fr. 8, for example, where he protests that 'the hot' and 'the cold' are not 'cut off from one another with an axe'), and this tendency was, no doubt, encouraged by the common use of the definite article plus the neuter adjective as a substantive" (Lloyd 1966/81 n.).[9] In my opinion the causal connection ought to be reversed.

(iii) *Singularis – pluralis – indifferentialis.* In recent years, linguists of all schools have been taking an increasing interest in the articles, including the zero-article, and of course they now and then touch upon uses which are logophoric in the sense of II-4.[10] The remarks in this section are not meant as a survey of the present state of linguistics in respect of articles; I merely intend to draw the reader's attention to certain salient issues of potential common interest. One of these is the kind of logophoric use of articles that occurs in propositions which some earlier logicians would have discussed by means of the notion of *suppositio naturalis*. This notion, it will be remembered, has survived in Thomist and also in other (idealist) circles. Among the several names which were used for this "philosophical" *suppositio*, Gerard of Harderwijk mentions "(suppositio) indifferens" (de Rijk 1973/73).[11] A recent linguistic publication (Mattens 1970), which, as far as I can see, falls outside the main-stream of contemporary linguistics, cuts through the grammatical knot by introducing a third grammatical category, *indifferentialis*, on a par with singular and plural. The author's intention is to provide for what he calls "the a-numerical use of nouns in common usable Dutch". It is likely that under the "numerical use of nouns" he counts every employment of nouns in sentences which contain "quantifiers", *i.e.* words like "all", "every", and "some" (cp. IV-8 above). It must be left to the grammarians to judge the merits of Mattens' grammatical proposal, but it testifies to the grammatical relevance of the problems discussed in the present investigation.

NOTES

[1] Cp. Angelelli 1967/118. See also the end of V-3.
[2] Cp. III n. 13.
[3] Cp. Angelelli 1967/22.
[4] Cp. Hughes and Cresswell 1968/335f. and the bibliography in that work.
[5] This, too, should be compared with Angelelli 1967/22 (cp. note 4 above).
[6] Cp. IX-6 below.
[7] About the choice of article in indefinite propositions in various natural languages, see Kraak *op. cit.*/125, Vater 1963/13, 60, 112–115, Mattens 1970/148.
[8] As well as Angelelli; cp. XI-4.
[9] See the discussion by Castañeda in *The Journal of Philosophical Logic* (1972). Cp. VII-11, VIII n. 22, n. 23.
[10] See, *e.g.*, Dirven 1971/157ff. Robbins (1970) hardly mentions logophoric uses of articles (6, 238f.; cp. Yotsukura 1970/27). See, however, Bacon 1973.
[11] Gerard writes: "Quarto vocatur *indifferens*, quia terminus habens suppositionem naturalem potentiam habet ut teneatur per indifferentiam pro supposito sue speciei et

pro ipsis individuis, uno vel pluribus, et non contrahitur per comparationem ad diversa predicata cuius est exemplum. Ut cum dico "*homo*" per se, ibi stat pro omnibus potentibus naturam humanam participare; et potest addi quodcumque predicatum, quia potest dici: "*homo est animal*", "*homo currit*", "*homo est species*" " (quoted from de Rijk 1973/73). Gerard defends the non-contextual outlook on "natural" *suppositio*.

CHAPTER VII

THE IDENTITY THEORIES OF THE COPULA

1. INTRODUCTION

It will be recalled that Maritain regards his distinction between *suppositio naturalis* and *suppositio accidentalis* as a division within *suppositio personalis*, and as a division pertaining to the *verb* or to the *copula* (IV-28). If the subject has *suppositio naturalis*, then the copula expresses the relation of subject to predicate only in a *possible* existence.

It is quite clear that the theory of "est", *i.e.* of "the" copula, will also be of the greatest importance in Maritain's version of traditional logic. The same holds for Pfänder's logic, from which I quote still another characteristic passage on the copula, modalities and relations: "The modality of judgments concerns the secondary function of the copula, the *assertive function*, which may be given either diminished ["abgedämpftem"], or full, or intensified logical weight. The *relation* of judgment also concerns the copula, and again the secondary function of *assertion*" (101). Pfänder's position with respect to modalities is the same as Kant's (1968/174).

Finally, when von Freytag argues against a logical distinction between singular ("individuelle") and indefinite propositions, he tries to justify this by referring to his own definition of *reine Logik* as a theory of identity and diversity, the copula being an expression of partial identity (I-5).

All this, and our conclusions (v)–(vii) at the end of the last chapter, make it desirable to go into the traditional theories of the copula in more detail, in the hope that such an investigation will help to clarify the conceptual background of the various traditional schools of thought. We shall clearly have to pay special attention to those theories of the copula in which the "is" of predication (attribution) is brought into close connection with means for expressing *analogies*, in the sense of similarities.

The nature of the relation between 'S' and 'P' in "the (true) judgment", 'S is P', has always been a much debated topic. But all traditional logicians seem to have assumed that predication was closely related to or the same as identification. This assumption is dropped in modern logic. Present-day (modern) logicians hold that the sense of "is" in "Dick is Tom's oldest brother (*is the* oldest brother of Tom)" is radically different from that of the "is" in "Saul is brown", and that both of these differ from the "is", or "are", in covertly universal propositions like "Bantus are Africans" or "the/a Bantu is an African".

In what follows it is important to remember that the expression "the subject term of a proposition" in traditional logic is ambiguous, since the quantifiers were usually regarded as attaching to the first non-logical word, *A*, and the combination of the two, *all A, some A*, or *no A*, was often taken as the subject term of the proposition, its meaning being the *logical* subject (term, concept) of the judgment (cp. IV-8, and Chapter X below). For this reason I shall in this chapter mostly render the forms of quantified propositions by means of the variables "*A*" and "*B*": *all A are B, some A is B*, etc., thereby allowing for a difference between *S* and *A*.

The logical tradition may seem to distinguish five theories as to the nature of "the" copula or, as the Germans often call it, *die Urteilsrelation*. It is doubtful whether these five would-be different theories really amount to more than two, but I shall start out by taking the words of the various authors at their face value, in which case five embryonic theories may be discerned.

In addition to the identity-theories, of which there are four, one sometimes speaks about a theory of inherence or attribution.

The identity theories differ from each other in various respects. In the first place, some authors talk in terms of "identity", others in terms of "partial identity": (a) the subject-predicate relationship is an identity relation; *i.e.*, if *S is P* is true, then the logical subject S or '*S*' *is* the logical predicate P or '*P*', where "is" possesses such properties as are required of an identity relation (in the opinion of the authors in question); (b) the subject-predicate relationship is only one of *partial* identity (between S and P, or between '*S*' and '*P*').

Far more important, however, is the difference between (1) the extensional identity theories on the one side and on the other side (2) the inherence theory and (3) intensional[1] (comprehensional) identity theories of the copula.

2. EXTENSIONAL IDENTITY THEORIES OF THE COPULA

(1) There are two types of extensional identity theories of the copula:

(1a) The "is" in, *e.g.*, *every (all) A is B* expresses the identity of two extensions. By the predicate term *B* is only meant, the supporters of this theory hold, those and only those B-things which are also A. In other words: the expressions *every (all) A* and *B* both refer to the intersection of the class of A-things and the class of B-things.

This interpretation invites the introduction of a second quantifier between the copula and the predicate: *all A is some B's.*[2] W. Hamilton (1788–1856) strongly advocated this "quantification of the predicate", as it was called. One may then write: *all A = some B's.* G. Ploucquet was another adherent of this interpretation, to the annoyance of his student Hegel and, says Hegel, to the annoyance also of Mendelssohn (Hegel 1967, I/148). This approach suggests an algebraic formulation of logic and must be seen as a prelude to the work of Boole and Schröder. If the predicate really is "quantified", the copula will indeed be symmetrical.

An extensional theory became the most common theory in the fourteenth century and counted William of Ockham among its supporters (Boehner 1952, Moody 1953/36f.). It is an intriguing problem to disclose the precise theory adopted by the schoolmen. In some way or other it was linked up with the theory of *suppositio*. Ockham says that a particular affirmative proposition, *some A is B,* is true if and only if the subject and the predicate terms have *suppositio* for the same thing (*Summa logicae* 2.3). Is it possible to describe what Ockham means in the language of modern logicians? I think it is, and find myself in complete agreement with the exposition given by Matthews (1964). He holds that since in modern logic we have access to double quantification, we can describe the intention of Ockham and his colleagues correctly in the following manner:

$$(Ex) [Ax \ \& \ (Ey) (By \ \& \ x = y)].$$

The corresponding interpretation of the universal affirmative propositional form (leaving out existential import) can be rendered in our symbolism as

$$(x) [Ax \rightarrow (Ey) (By \ \& \ x = y)].$$

As Matthews shows, this symbolization is in accordance with what

Ockham and others have to say about *descensus* from subject terms and from predicate terms as well, as long as they have *suppositio personalis*, as Ockham generally assumed them to have.

It is quite clear that, in the absence of internal quantification, some other instrument must be invented by which to account for valid descents (derivations). The theory of *suppositio* was precisely such an instrument.

(1b) "The word "is" fundamentally expresses a partial identity of extensions." This way of speaking corresponds more or less to the analysis of *all A is (are) B* as $A \subset B$, where the extension of the subject term of a universal proposition is a part of the extension of the predicate, although it is hard to see why the *identity* should be said to be partial.[3] This theory, or terminology, is sometimes called the subsumption theory (Erdmann 1907/343). Erdmann and Honecker (1927/123) hold that it comes close to Aristotle's position. They seem to ascribe to Aristotle a theory of the copula corresponding to a logic of classes.[4] Erdmann, who does not himself support this theory of the copula, characterizes it as the oldest, and, in his time, the most common conception of the copula (*loc. cit.*)

If either (1a) or (1b) is supposed to be a theory covering *every* kind of proposition, then it will be difficult to distinguish these "theories" from each other. The a-variant is better adapted to propositions like "Scott *is* the author of Waverley", the b-variant to statements like "(all) cats are mammals" or "the cat *is* a mammal".

With respect to propositions like "some cats have blue eyes" it is hard to give preference to one of these "theories" over the other.

3. THE INHERENCE THEORY OF THE COPULA

(2) The inherence or attribution theory of the copula is the doctrine that the copula expresses a relation between the extension of a (logical) subject term S and the meaning, intension (Leibniz), comprehension, content, or sense of a (logical) predicate P. In the days of the schoolmen this theory was an alternative to the theory of extensional identity (Moody 1953f., de Rijk 1967/569f.). The supporters of the inherence theory seem to have thought, and to think, in terms of *the* meaning or intension of any predicate P of a natural language, as a *forma* or universal nature belonging to the expression P and to all things of which P can be truly predicated.

The inherence theory is particularly plausible with respect to quantifier-free propositions – and with respect to judgments '*S is P*' where the quantifiers have already been absorbed into *S* and *P*.[5]

In a remarkable passage, J. Maritain opposes the *logique de l'inhérence ou de la prédication* to the *logique de la relation*,[6] the latter being the new logic as advocated by Bertrand Russell. Maritain himself takes the side of the former, the logic of inherence. His own description of the function of the copula shows, however, that the inherence theory, too, is some kind of identity theory: "La logique... ramène toute énonciation (catégorique) à l'expression d'une identité par le moyen de la copule" (*loc. cit.*) If, as is sometimes the case, our statements express something other than simple identity between a logical subject and a logical predicate ("la simple identité (*in re*) d'un prédicat et d'un sujet"), then this is because the statement in question is not a simple (categorical) proposition ("simple énonciation"), but a more complex, *e.g.* a hypothetical proposition, containing other connectives in addition to the copula.

In spite of this identification, Maritain maintains that the (logical) predicate is attributed *à titre de forme* to the logical subject S *à titre de matière*. With respect to the proposition "cet homme est blanc" ("this man is white") this means that "the mind... says that whiteness is a (is one of the) mark(s) of this man" (150), and this is equivalent to saying that the mind "makes the S (re)enter into the extension of the Pr, saying that "this man" is a (is one of the) things(s) having whiteness". In this example of an individual proposition it is hard to distinguish his so-called inherence theory of the copula from the subsumption theory, or partial extensional identity theory (16). Now the mind does not straightforwardly identify the (logical?) subject, or in his notation: S, with the predicate, Pr, but carries out the identification "of a concept *which is a function of S* and a concept *which is a function of Pr*, in such a way that the mind, when carrying out this identification, will necessarily have to take either the extensional or the comprehensional point of view" (150). The identifying mind may already choose between an intensional and an extensional point of view at the moment when the judgment is formulated, but Maritain maintains that the notion of comprehension and its use is both more natural and more fundamental than that of extension (148f.).

For more than one reason his theory is most unclear. One is that he does not attempt to characterize those *functions of* S and Pr which yield the concepts that are finally identified. But Maritain is much clearer than most recent traditional logicians as to what he regards as the logical subject, S, and the logical predicate, Pr, of a proposition. S and Pr are themselves functions, in his logic. One cannot say, perhaps, that he really defines these functions, but they turn out to be *expressed* by quantifiers, articles, and demonstrative propositions, all of which may therefore be called functors. The argument of each of *these* functions is (expressed by) the term (expression) following immediately after the functor in the proposition; *e.g.*, in *all A are B, A* is the argument of *all.* This much is clear from his own words: "the words S and Pr... designate ... those concepts with their determinations... which are translated into the oral expression by means of the complementary signs "all", "some", etc. (syncategorematic terms). Thus in the propositions "some man is unjust", "this man is culpable", the S is not exactly the concept "Man" taken by itself, but is, to be precise, the term "some man" and the term "this man"" (126). This theory already formed the background of Albert of Saxony's question, whether "all" could be said to be a *part* of the logical subject or only a word bringing about a modification of the noun or adjective that followed it (cp. IV-8).

Albert held that the *significatio*, or function value, of a quantifier or other logical constant when it is applied to a noun or adjective is *a certain suppositio* of the latter. I have not been able to find any explicit mention of this connection in Maritain's logic. In the philosophically important case of essential or natural *suppositio* he informs us that "the proposition expresses an eternal truth, and affirms... the relation (of identification) of the object of thought signified by the Pr with the object of thought signified by the S" (270).

This may not be very enlightening from a theoretical point of view, but it at least shows clearly enough that no clear line of demarcation can be drawn between an inherence theory of the copula and the comprehensional identity theories of the copula, to be discussed below. There is not even a sharp difference between the former "theory" and (1b), the subsumption theory. Erdmann does not even mention an inherence theory in his survey of the proposed theories of the copula. Jacoby, who, like von Freytag, claims to support a comprehensional identity theory of the copula, speaks of "the participation of the carrier of inherence [the logophore!] in the identity between that which is inherent in it and its genus" (1962/13, 17). Ziehen is of the opinion that "the term "logic of inherence" which is often heard ought to be reserved for that theory of the judgment which, generalizing *in a clearly untenable manner*, regards every judgment as a statement about a substance" (1920/612 n. 31; my italics). As a result of our study of Maritain and Jacoby we conclude that the word "inherence" is no indication that the author in question rejects the conception of the copula involved in identity theories of the comprehensional kind, which we shall now attempt to characterize.

4. Comprehensional Identity Theories of the Copula: Logic as the Study of Indefinite Propositions

(3) Of the comprehensional identity theory, too, there are two variants, (a) and (b). Those authors who support the (b)-variant express themselves more or less systematically in terms of "*partial* identity", while those who speak in terms of "identity" *tout court* are supporters of the (a)-variant.

In the (a)-variant it is said that the comprehensions or senses of the (logical) subject and predicate in a true (or, as is often said, in a valid) judgment are identical (or: identically the same). Precisely what may be meant by this will be discussed in later sections.

As Honecker says, this theory is best made plausible by reference to definitions (Honecker 1927/122). It is likely that Honecker has in mind the traditional notion of real definitions, being descriptions of a *res universalis* or *forma* or *Wesen*. We shall in fact have to delimit the class of "judgments" to the class of *completed real definitions* if this doctrine of the copula is to have any plausibility at all. In this doctrine, and only there, any statement (*das*) *S ist P* is seen as an application of "the law of identity, *A is A*". In the philosophy of Leibniz, such completed real definitions constitute real knowledge (cp. Loemker 1969/23).

Finally there is (3b), *i.e.* the partial intensional identity theory, which Erdmann calls "die Einordnungstheorie" and also "the theory of logical *immanence*" (cp. "inherence") (Erdmann 1907/359; cp. Ziehen 1920/612). This is the theory that the content of "the" concept '*S*' includes the content of "the" concept '*P*', without necessarily covering it exactly, that the copula of the proposition is in general an expression of this inclusion, and that this should form the point of departure for the theoretical logician.

That was exactly Leibniz's outlook. In formal demonstration he prefers the method *secundum ideas* above the method *secundum individua*, and he does so *because* the former is *independent of the existence of individuals* (cp. Leibniz 1969/238). Kauppi regards this comprehensional basis in Leibniz's logic as an expression of the Cartesian mode of thought and opposes it to the tradition of the schoolmen (Kauppi 1967/14). She thereby seems to overlook the "metaphysical" trend, especially in late scholastic logic, and the support given there to the inherence theory of the copula.

In order to show that a comprehensional account is likely to give rise to a profusion of undefined articles I quote from the *Nouveaux Essais*: "Car, disant Tout homme est animal, je veux dire que tous les hommes sont compris dans tous les animaux; mais j'entends [!] en même temps que l'idée de l'animal est comprise dans l'idée de l'homme. L'animal compréhend plus d'individus que l'homme, mais l'homme compréhend plus d'idées ou plus de formalités; l'un a plus d'exemples, l'autre plus de degrés de realité; l'un a plus d'extension, l'autre plus d'intension" (Book IV, Ch. 17–18; quoted from Spencer 1971).

There are three really important points here. The first is that Leibniz introduces the technical term "intension", with an "s", where Arnauld and Nicole had used "compréhension". This passage may well be the source of the use of the word "intension" for the alternative of the extension of a word or other denoting expression (Spencer 1971). Leibniz even speaks in terms of *more or less* intension, thereby recalling the meaning of "intensio" as used by the schoolmen,[7] thereby suggesting an *intensive* conception of the comprehension of terms. Spencer (*op. cit.*) thinks that this is no more than a suggestive metaphor, although one which is deeply rooted in Western philosophy. I shall try to make it plausible in this book that this conception is much more than a metaphor, that it is of fundamental systematic importance in traditional logic and all of the philosophy that goes with it. A recent edition of the *Essais* has: "... l'autre plus d'intensité" (1882/436).

The second and systematically more important thing to notice in this passage is that Leibniz talks in terms of "the" idea of an animal ("l'animal") and "the" idea of man ("l'homme"). This is certainly not unconnected with the fact that the whole passage is clearly written with an interior monologue in mind: "Car disant... j'entends...". It may be possible for someone who only has to analyse his *own* statements to find out whether the meanings *he* associates with his terms are related as part and whole, but it is as a rule not possible for somebody else to "hear" such relationships among the "ideas" of the speaker. Leibniz, however, clearly assumes that a universal proposition derives its truth from a more fundamental fact, *viz.* that one idea, '*the P*', is contained in another idea, '*the S*'. Of course, if someone else tells him: *all A is B*, he can "hear" nothing of the sort unless he happens *to know in advance* (1) that this statement is true and (2) that this is so because according to the speaker's terminology, an A is B by that definition of the word *A* which the speaker is known to assume. In such a case the speaker might have saved himself the trouble of making the utterance at all (cp. X-9, *sub* (i)).

Honecker thinks that this partial identity theory of the copula may sound fairly plausible as long as one focusses one's attention upon generic ("generelle") positive judgments such as "Der Mensch ist ein Lebewesen ("Homo est animal")", but that it can be refuted by reference to negative judgments like "Kein Reptiel ist Säugetier". One may agree with the latter, which recalls Lotze's harangue about 'Not-Man' (cp. Schröder 1966/99, Kauppi 1967/16), but we cannot very well agree with the first part of Honecker's statement, since we are here, precisely, raising the problem about the logic – if there is any – of statements of this form. In order to make the partial intensional identity theory plausible, it will be necessary to assume all generic judgments to be *uncompleted or completed real definitions* and to be *analytic*, in every conceivable meaning of that word (cp. Section 13 below). Like the inherence theory, the intensional identity theories are closely tied up with the idea that indefinite propositions are more basic in logical theory than quantified propositions. The theory of partial intensional identity assumes "the" concept 'Man' to be a class of properties or "marks" and "the" concept 'Animal' to be another class of marks. Its supporters hold, for instance, that 'Animal' ⊂ 'Man', and that this is most adequately expressed by means of the indefinite proposition "Man is an animal", in German: "Der Mensch ist ein Lebewesen". Supporters of (1b), on the other hand, regard

$$\hat{x} \text{ Man } x \subset \hat{x} \text{ Animal } x$$

as the logically fundamental truth. Summing up, we can say:

(i) that those philosophers who adhere either to the inherence theory or to one of the variants of the identity theory are all primarily interested in complete or incomplete definitions (desciiptions) of a *res universalis*, *viz.* a *forma*, *species*, or *intentio*, e.g., of 'homo', or ɩHomo, that is to say: of "the" meaning of a logophoric term;

(ii) that these theories are not designed to give a clear analysis of particular propositions or of universal propositions to the extent that these are not assumed to be mere stylistic variants of indefinite propositions; and

(iii) that given two general terms (expressions) S_1 and S_2, these theories do not even suggest a criterion by means of which one can determine whether both terms do or do not signify one and the same concept, which is to say that the expression *the concept S* in fact has no meaning whatsoever unless pragmatic specifications (a certain person, group of persons, or written text(s)) are provided – which is never done.

In so far as they are made explicit by their supporters, all five theories of "the copula" are of a simplicity which in all probability cannot be upheld in theoretical logic. The extensional identity theories have no contemporary supporters, but the inherence theory and the comprehensional identity theories have had an unsuspected philosophical influence, although, as theoretical logic, it has, except in a few cases, "gone underground".

5. SCHEMATIC SURVEY

In:	the copula of a judgment is a relation between	
	of the (logical) subject	of the (logical) predicate
1. the extensional identity theories ("Umfangstheorie") a: identity of extension (William of Ockham etc.; Hamilton) b: partial identity (inclusion) of extensions (Aristotle?)	the extension	and the extension
2. the inherence (attribution) theory (Maritain etc.)	the extension (substance, *matière*)	and the comprehension (*forma*, universal nature)

Schematic survey (continuing)

In:	the copula of a judgment is a relation between	
	of the (logical) subject	of the (logical) predicate
3. the comprehensional (intensional) identity theories ("Inhaltstheorie") a: *Das Subjekt ist das Prädikat*[8] (Hegel etc.) b: *Praedicatum inest subjecto* (Leibniz etc.; Erdmann)	the comprehension (*Begriff, Vorstellung,* or *Gegenstand*)	the comprehension (etc.)

6. EXPECTED PROPERTIES OF THE NOTIONS OF IDENTITY AND PARTIAL IDENTITY IN TRADITIONAL LOGIC

The use of "identity" and "partial identity" in the very basis of the traditional logic makes it important, if we are going to understand that logic at all, to determine what the properties of these notions are assumed to be. To day, "identity" is used for a relation which is reflexive, symmetric, and transitive and which satisfies what is sometimes called "Leibniz's law". This principle has two components which we shall call "PIi" and "PiI". One of them is often formulated as follows: if everything that can be truly said of the entity A can also be said of the entity B and *vice versa*, then A is identical with B (A is the same entity as B). This is an unprecise formulation of Leibniz's *Principium identitatis indiscernibilium*, or PIi, of which a more precise formulation will be given in Section 16. The converse of this principle is of no less importance: if the entity A (the entity called *A*) is identical with B (with the entity called *B*), then all that can be truly said of A can also be truly said of B and *vice versa*. This latter principle is sometimes called "the Principle of the Indiscernibility of Identicals", which we shall abbreviate to "PiI" (the capital "I" being short for "Identity" or for "Identicals").

At this point we shall not go into the difficulties raised by this principle, which explain why it is often called "the principle of extensionality"; we will return to it in Chapter IX in connection with the logic of analogy and substitution. Suffice it here to say that it would be very strange indeed to say that the copula is, or expresses, a relation of identity unless it satisfies the three conditions of reflexivity, symmetry, and transitivity.

A relation with these three properties is an *equality*, or, alternatively, an *equivalence relation*. A relationship which is characterized as an "identity" may, then, be expected to be *at least* an equivalence relation. Should we have to deal with an equivalence relation which does not even satisfy PiI in a context free of modalities it would be preferable not to speak of an identity, but of an equality.

Probably no one has ever used the word "identity" or even "partial identity" without holding that it satisfied at least some of the above-mentioned properties. We shall define three relations, all of which we shall meet in future sections under the name of "(partial) identity". Let 'S' be a concept, in the sense of a class of characteristics or properties or marks ("Merkmale"), which somebody associates with a term S. The relation of identity or partial identity to be defined will for short be called "τ" (short for "Teilhabe" or "Teilidentität"). We assume, then, that

$$S \text{ is } P \text{ is true if and only if } 'S' \tau 'P'.$$

Various definitions of "(partial) comprehensional identity" can then be imagined:

(3a) $'S' \tau 'P' \underset{\text{Df}}{\leftrightarrow} 'S' \cap 'P' \neq \Lambda$
        $\leftrightarrow$ there is a concept 'X' such that 'S' contains all marks in 'X' and 'P' contains all marks in 'X'
        $\leftrightarrow (E'X') ['X' \subset 'S' \;\&\; 'X' \subset 'P']$.

This is to say that the class of marks 'S' bears the relation of partial identity to the class of marks 'P' if and only if 'S' and 'P' have some marks in common. The judgment-copula so defined will be reflexive and symmetric, but not transitive. A transitive relation would be

(3aa) $'S' \tau_M 'P' \underset{\text{Df}}{\leftrightarrow} 'S' \cap 'P' \supset 'M'$
        $\leftrightarrow 'M' \subset 'S' \;\&\; 'M' \subset 'P'$,

for if 'S' τ_M 'P' and 'P' τ_M 'Q', then 'S' τ_M 'Q'. We might have used the symbol "$\sim_M$" in place of "τ_M", for this is indeed a copula with the properties of an equivalence relation. However, if 'S' τ_M 'P', this cannot, or rather, ought not to be expressed simply as S *is* P, for here 'M' is entirely suppressed. An adequate means of expression would be: S *and* P *are both* M. But unfortunately, *propositions of this form are not well-*

formed in Aristotelian logic, since they contain three terms – an important fact for the interpretation of much traditional philosophy, as we shall see later.

Another candidate for the name "partial identity" is the relation

(3b) $'S' \tau 'P' \underset{\text{Df}}{\leftrightarrow} 'P' \subset 'S'$
            $\leftrightarrow 'S' \supset 'P'$.

This relation is reflexive and transitive, but not symmetric, just like the relation of class inclusion which we have used in the *definiens*. It has recently been discussed by Kauppi (1967), who offers the same simple exposition of the traditional notion of partial identity as we have done here in (3b). Lambert and van Fraassen take *'P' is intensionally included in* *'S'*, hence *S is P* in one important traditional sense, to mean the same as *Necessarily, all individuals which are S, are P*. This suggests two different definitions of the traditional τ, *viz.*

(3c) $'S' \tau 'P' \underset{\text{Df}}{\leftrightarrow} \Box(x) [Sx \supset Px],$

and

(3cc) $'S' \tau 'P' \underset{\text{Df}}{\leftrightarrow} (|x) [Sx \supset Px],$

to be read: *all possible individuals which are S are also P* (Lambert and van Fraassen 1970/1). A relation defined by (3c) or (3cc) is reflexive and transitive, but – of course – not symmetric. One of the aims of a calculus of modalities of possible objects would be to ascertain that τ satisfies certain interpretations of Leibniz's law.

7. FICHTE'S LOGIC: A SYMMETRIC COPULA

J. G. Fichte's *Grundlage der gesammten Wissenschaftslehre* ("Foundations of the Complete Theory of Science") leaves no doubt that he held an identity theory of the copula: "Everyone acknowledges the statement *A is A* (which is to say A=A, for that is the meaning of the logical copula), and that even without having to think about it" (1794/6). "By the statement A=A *judgments* are made" (9).

Hegel is no less positive. In the first part of his *Enzyklopädie der Wissenschaften* he states that "the subject is the predicate", so that "*every* judgement pronounces them to be identical" (my italics). It will be

remembered from I-2 that, according to Angelelli, the traditional doctrine
since Aristotle has been that the predicate is identical with the subject
when it gives the essence of the subject – and hence not always, as Hegel
says is the case. Hegel's position, then, is a simplification of the traditional doctrine. In Hegel's opinion, "the copula: is, derives from the
nature of the concept, to be identical with itself in its alienation
("in seiner Entäusserung")" (1843/327). And then: "In the copula... the
identity of subject and predicate... is posited" (*op. cit.* /332). Precisely
the same can be found in his *Wissenschaft der Logik* ("Science of Logic"):
"... the *copula* expresses that the subject is the predicate" (II/270).[9]
     What are the properties of the relation which is here called identity?
     In Leonard Nelson's opinion, the characteristic feature of the logic
of Fichte and Hegel is that every judgment is assumed to be based upon
a *comparison* ("Vergleichung", "Gleichsetzung") of a subject-concept and
a predicate-concept, in the sense that the two are found to be, or not to be,
similar. In this logic, says Nelson, "an expression like $a=b$ in the first
place means nothing else than that the two entities, called a and b,
resemble each other in some respect, which obviously does not exclude
that they differ from each other in other respects" (1962/513).
     Nelson here writes as if in the logic of Fichte and Hegel we have to do
with that symmetric relation between intensions (comprehensions) or
concepts which we defined in (3a) of the last section. One might think
that this is too fantastic to be true, and that Nelson therefore must have
made a mistake, but this is not so. In the work mentioned above, Fichte
says: "X=X. A=X. B=X. mithin A=B, insofern beides ist =X" (32).
From two premises '$A=X$' and '$B=X$', to be read: *A is X* and *B is X*,
Fichte infers the judgment '$A=B$', to be read *A is B*. The symmetry of his
judgment relation cannot, therefore, be doubted; and the same holds for
its reflexivity.
     In order to reach his conclusion, '$A=B$', Fichte does not use the property
of transitivity; he seems to arrive at this result by *substituting* 'A' and 'B'
for 'X' in '$X=X$'.[10] He can do this, or so he thinks, because he has already
conceived of the judgments '$A=X$' and '$B=X$' as "identities".[11]
     This logical howler and its consequences are expounded by Nelson, who
offers a number of examples. It becomes clear that the properties of the
Fichte-Hegel logic are not mere accidental features of their general
philosophy which we can subtract from it leaving the core of that

philosophy intact. It is not hard to draw the most spectacular consequences from this logic:

1. Take two true judgments, *e.g.*, 'Gold is (a) Metal' and 'Silver is (a)
Metal'. Fichte interprets them as identities:

'Gold is Metal' becomes: 'Gold = Metal'
'Silver is Metal' becomes: 'Silver = Metal'

(the single quotation marks being added by me).
In the opinion of Fichte, this entails:

'Gold = Silver',

that is to say,

(i) Gold is Silver ["in a certain sense", but this is not added].

At the same time it is true that

'Gold is Yellow', *i.e.*, 'Gold = Yellow'
'Silver is not Yellow', or 'Silver ≠ Yellow',
or, in Fichte's own notation:

'Silver = – Yellow',
from which it follows that:
'Gold = – Silver',
i.e.,

(ii) Gold is not Silver.

The two "identities" (i) and (ii) contradict each other, as the reader
will hardly deny. There *is*, as many a philosopher can still be heard to
say, a "dialectical strife" between these two judgments....[12]

2. "For instance: Gold is Metal, Metal is not Gold, thus: Gold is not
Gold. Consequently, we shall have to drop the Law of Identity as well as
the law of Non-Contradiction" (Nelson *op. cit.*/515).[13]

3. "Coriscus is Socrates... Indeed: Coriscus is a man, and Socrates is a man. Hence Coriscus and Socrates are of the same kind, from which, for the mystic, the assertion follows of itself. The proposition: "Coriscus and Socrates are of the same kind" is a correct formulation of a comparison. But for this correct formulation of "Coriscus and Socrates are of the same kind", the logical mystic surreptitiously substitutes a false judgment, replacing the being-of-the-same-kind of the mere comparison by the identity of the objects that are being compared, stating that both men are in reality one man" (515f.).[14]

Summing up: for *every* judgment '*S is P*' which is held to be true one can, in Fichte's logic, demonstrate the simultaneous truth of '*S is not P*'; if '*S*' is '*P*', then '*S*' is also – *andererseits* – not '*P*'.[15] This may also, if one so prefers, be expressed by saying that neither judgment, and hence that no judgment is "really" true.

Fichte's logic may be regarded as a logic exclusively for Wolffian *propositiones propriae*. As we saw in V-11, Wolff considered these judgments as convertible.

8. REFLEXIVITY OF THE COPULA

Hegel states that "the law of identity", when expressed as: $A = A$, is nothing more than an empty tautology (1967, II/28). He rightly maintains that this law, in this form, is *without content* and cannot take us any further; it is of no help when we want to arrive at important results.

Hegel is here attacking the Leibniz-Wolff tradition, to which he himself belonged. Leibniz held that the necessary truths were precisely those of which it can be shown that their negations are self-contradictory, the necessary truths themselves being identities. He seems to have held that in order to decide whether a *given* proposition is a necessary truth or not, it is sufficient to investigate whether its negation involves a contradiction. Now in Hegel's general philosophy, though not in his logic, the problem is (as was Kant's; cp. IX-9 below) how we arrive at this proposition in the first place. One cannot very well investigate *all* meaningful sentences in order to find out, by means of Leibniz's criterion, which of them are necessary truths; there must be some principle or method by which the likely candidates can be singled out in advance. There is, therefore, no inconsistency in Hegel's position when he criticizes "the law of identity", *(the)* $A = A$, for not by itself *producing* the necessary truths, while at the same time accepting Leibniz's criterion for necessity or analyticity, as I think he does (cp. Russell 1947/758).

It was necessary to point this out first, in order to avoid misunderstandings and accusations of irrelevance when we now proceed to an investigation of certain technical features of Hegel's logic. The salient point is that Hegel is very far from rejecting the validity of $A = A$. On the contrary: as we saw in the last section he takes this formula to describe the structure of every judgment. This already allows us to state that, *for Hegel, the copula is reflexive.* He maintains this conception of the copula even though he goes on to say that a judgment *A is A* contains *more* than the empty tautology $A = A$: "In the form of the statement, in which the identity is expressed, more is contained than the simple abstract identity; the *pure movement of reflection* is contained in it, in which the other occurs only as appearance ["Schein"], as immediate disappearance ["Verschwinden"]" *(loc. cit.*; my translation and italics). Here "reflection" means "mirroring" (of the Absolute), which is to say that this passage neither directly supports nor contradicts our conclusion that he takes the copula to have the property of reflexivity. The meaning of the second occurrence of *A*, the predicate in *A is A*, mirrors the Absolute or essence in the meaning of the first occurrence, or subject, and since mirroring presupposes similarity we again have to do with a reflexive relation, but with another one, be it μέθεξις or μίμησις. The reflexivity of what Hegel calls "abstract identity" and of the relation which today is called "strict (strong) identity" does not suffice for his purposes.

It should be noted that Hegel, like so many others, assimilates the notions of identity, predication, and existence and combines them into one notion. He says about the formula *A is A*: "... *A* IS, is a beginning, before which something different flashes past...". This implies that he takes "is" to have a function such that the sequence *A is* is already well-formed and meaningful, which is possible only if the copula carries the connotation of existence. For Hegel (and others), then, *A ist A* implies a judgment of existence; this is more clearly expressed if one says *das A ist A*, or *the A is A* (cp. X-5 below).

Of no less importance is his next statement: "... *A* IS – A; the difference is only a disappearance; the movement returns into itself" (II/31).

Here "is" must be understood as expressing predication if this statement is assumed to contain anything else than the previous one.

This yields the conclusion that the reflexivity of the relation of predication, which is expressed in clause (iv) of our H-thesis, is of the greatest importance for the structure of Hegel's thought and for the conception of a Law of Identity which is something more than just tautological. As Hegel himself puts it: "The reflexion as such ["an sich"] is the identity..." (35). By means of A is $-A$ Hegel wants to express what we formulated in II-5: *the A has the property of being (an) A.*

It is worth while to note that from the combination of the id-logical comprehension axiom, the H-thesis, with the Law of Identity when formulated in the following manner: *for every A, A is identical with itself*, Russell's paradox follows. As will be remembered, clause (iv) of the H-thesis is this: *the/an M is M* is a tautology. Let M be the predicate "not being identical with itself/oneself". *The M* will then be *the not-being-identical-with-itself/oneself*. This is traditionally called *(the) Nothingness*, or in German: *das Nichts*. According to the H-thesis we then have:

(1) (The) not being identical with itself/oneself is not being identical with itself/oneself,

while the Law of Identity as a universal proposition yields

(2) (The) not being identical with itself/oneself is (being) identical with itself/oneself.

In short:

(2′) (the) Nothingness is, and also is not, identical with itself.

We have assumed that the second occurrence of "not" in (1) negates the copula. On this assumption we must conclude that the combination of the Law of Identity with the H-thesis yields *a contradictory theory* for the copula (cp. Faris 1968/14). Some traditional philosophers are led by (2′) into mysticism; others may prefer to deny that the second "not" in (2) operates upon the copula, by appealing to Kant's theory of negation with its distinction between "privative" and "limitative" negations. But that theory is too unclear to be taken seriously as a Way Out.

Instead of "is" we could have written "τ" in the above argument (cp. Section 6).

In modern logic one can define the empty set, as *the set of all elements which* are not identical with themselves:

(3) $\qquad \Lambda \underset{\overline{\text{Df}}}{=} \hat{x}[\sim x = x],$

without being led into a contradiction, precisely because of the distinction in modern logic between a reflexive identity and a non-reflexive relation Being an Element Of (which should be regarded as the converse of the relation of inherence). From (3), neither $\Lambda \, \varepsilon \, \Lambda$ nor $\sim \Lambda = \Lambda$ follows, in spite of the fact that (3) has precisely the same form as

(4) $\qquad$ (the) Nothingness $\underset{\overline{\text{Df}}}{=}$ the not being (identical with itself)

$\qquad\qquad\qquad \underset{\overline{\text{Df}}}{=}$ the $x[\sim x$ is (identical with) $x]$.

Of course it makes no difference philosophically whether we use the sign "$\hat{\ }$", as in (3), or the sign "the", as in (4). What matters logically is not the *shape* of the symbol but only the properties which are ascribed to it and to the relation called "is (identical with)". In modern logic, neither the relation of predication nor the set-theoretical relation of being an element of something is reflexive. For Aristotle's view of the matter, see II-5 above.

9. TRANSITIVITY OF THE COPULA

We have said already (V-12) that Kant assumed the transitivity of the copula when stating the adage *nota notae est nota rei*. This adage is often held to derive from Aristotle. Kant even regarded it as one of the most important of logical principles. It is important to realize that this principle prevents us from constructing any theory of *suppositio*, type theory, or other kind of stratification; in a logic based on this principle, all terms (predicates) must be regarded as being of the same kind.

Erdmann formulates the property of transitivity for the copula in various ways.

"The predicate of a predicate is a mediate predicate of the subject", and in a species-and-genus terminology: "The species of a species is a mediate species of the genus" (1907/708). Consequently we find no theory of *suppositio* in Erdmann's logic, nor in that of von Freytag, whose copula has the same properties as class inclusion and, hence, also the property of being transitive.

Wundt does not accept *nota notae est nota rei* as a principle of general validity (1893/314). Husserl, who for some time was his student, follows him here,[16] but does not construct an alternative.

Neither Maritain nor Pfänder can be said to employ a transitive copula, though it is most unclear what properties they consider it to have.

It is therefore somewhat surprising to find that Veatch, who, like Maritain, draws upon the logic of John of St. Thomas, explicitly states the transitivity property of the copula. He even discusses it at great length, thereby recalling De Morgan's work on syllogistic (Veatch 1950/358). It seems pretty clear that Veatch does not distinguish between (i) the relation *from ... follows ...*, (ii) the relation of class inclusion, and (iii) the relation of a name of an entity to the name of a property which can truly be predicated of this entity. Certainly the transitivity of the two first relations is of fundamental importance for logic, but the third relation, albeit a logical one, cannot be regarded as transitive since this would lead to mass-production of fallacies – as was seen clearly by the schoolmen (de Rijk 1962/1967).

10. TWO KINDS OF LOGICAL IDENTITY

At this point some remarks about the uses of the expression "logical identity" will be in order. Some modern logicians, for instance Beth, have used this expression as a variation of "logical truth", or "universally valid proposition". This can be historically misleading, since the definition of the latter expressions is such that a logical identity then becomes a trivial, tautological proposition.[17] Within the philosophical tradition, however, it became a common opinion that there are two kinds of logical identity, *real* and (merely) *formal* identities.[18] The former are the so-called real definitions while the latter, as we have just seen in our discussion of Hegel, were sometimes called "tautologies". In passing it may be remarked that the theory of tautologies was not clearly worked out, and that we have had to pay dearly for this seemingly innocent omission.

We have, then, the following division:

logical identities (logical
laws, valid judgments –
"gültige Urteile")

$$\left\{\begin{array}{l}\end{array}\right.$$

1. real identities ("real definitions", *the/an M is* $(P \& \ldots)$; "synthetic judgments *a priori*").
   These are indefinite propositions, held to express logophoric judgments.
2. formal identities, *e.g.*, $A = A$ ("tautologies", sometimes: "analytic judgments").

According to this classification, to identify the class of logical laws with the class of tautologies is to hold that real identities, whatever they may be, are in any case not *logical* laws. There has been, and still is, considerable philosophical opposition to the idea, expounded in Wittgenstein's *Tractatus*, that the class of logical truths is simply the set of tautologies. In the terminology of German idealism, to identify these classes means to delimit the domain of logical laws to that of the formal identities, thereby excluding the real identities, or real definitions. It may be debated whether the laws of higher-order logics can be said to be tautological, but in any case they do not resemble the kind of statements that earlier were thought of as valid real definitions.

The phenomenologist Pfänder takes a completely opposite standpoint. In his opinion, the formula $A = A$ is not really a *logical* proposition at all, since "it says nothing about any logical object ['Gegenstand']" (1963/182).

Logic is, however, no longer regarded as a system of truths about logical objects, but as the study of forms of inference. This is regretted by all those who would like to be able to issue irrefutable statements of far-reaching theological, ethical, or social import. In his essay on "the logic of logistic", Albrecht bases his hope that this will somehow be possible entirely upon the Aristotelean notion of exposition or ekthesis. This will be studied in a later chapter.

11. IDENTITY-LOGIC AND CONTRADICTION-LOGIC

(i) *Two senses of "logic of identity"*. We shall have to say something about the expressions "identity-logic" and "doctrine of identity". It has already become clear that the logic of Fichte and Hegel is no less a doctrine of identity than is the logic of von Freytag or of Veatch. Nevertheless the logic of the two former philosophers is sometimes characterized as a *logic of contradiction* and contrasted with *the logic of identity*. This is not, however, a paradox, for "logic of identity" is used in two senses. Fichte, Hegel, von Freytag, Veatch, and many others argue for a logic of identity in the sense that, for them, the *copula* is always a relation of identity or of partial identity. Contemporary logic is very far from being a logic of identity in this sense. Although a relation of identity is of systematic importance in modern logic, too, the copula in "this rose is red" has none of the properties of the identity-relation.

We saw that Fichte's logic leads to a mass-production of contradictions. When, for instance, Naville characterizes Hegel's logic as a logic of contradiction and opposes it to the logic of identity of which he himself is supporter, he understands something else by "logique de l'identité" than the theory that the copula is an identity relation. In Naville's sense of "logic of identity", modern logic, as well as the logic of Aristotle, is a logic of identity (Naville 1909/6, 8).

By "logic of identity" Naville understands any logic in which every entity which is denoted by a term is held to be, in first place, identical with itself[19] and, in the second place, never non-identical with itself, and in which a predicate cannot at the same time be (truly) predicated of and negated of the same thing.

In modern symbols this may be expressed as follows:

(1) $(x) [x = x]$,
(2) $\sim [Pa \ \& \sim Pa]$,
(3) $\sim [a \ \varepsilon \ P \ \& \sim a \ \varepsilon \ P]$,
(4) $\sim [a = b \ \& \ a \neq b]$

are *logical truths* (and hence tautologies). In modern logic such contradictions as

$$Pa \ \& \sim Pa,$$
$$a = b \ \& \ a \neq b,$$

can be rejected (as well-formed but certainly false) – because there is not the slightest need for them.[20] If symbols are used in the logic based upon the intensional (comprehensional) identity theory of the copula (the logic of contradiction), which is often done, then "is" is rendered by

"=". It then becomes impossible to distinguish systematically between the use of small letters, like "a", for individual terms and capital letters, like "P", for general terms (predicates), since "$a=P$" would immediately suggest that something is wrong, one way or the other.

The choice of principles in this logic has the consequence that whether one would like to do so or not, *no distinction can be made* corresponding to the distinctions drawn in contemporary logic between the logical relations expressed in $a=b$ or $I(a, b)$, $a\varepsilon B$, Ba, and assorted relations of equality (equivalences), $aR_{eq}b$.

Summing up: if the copula is understood as an (expression of an) identity or equivalence between concepts, then a logic is obtained which *nolens volens* will produce an infinite number of contradictions, and in a rather trivial manner at that. One may or may not attempt to turn this vice into a virtue.[21]

When the copula is not understood as identity or as similarity, one is in a position to adopt (1)–(4) as tautological truths of logic and as requirements to be satisfied by any future development, both of theoretical logic itself and of every other theoretical pursuit in which use is made of the science of logic.

(ii) *Plato's paradox.* The interpretation of the copula as identity or as similarity is not a necessary condition in older forms of logic for the generation of contradictions. The traditional systems of logic could not treat propositions with relational predicates without loss of information. As Russell stressed again and again, this is especially clear for asymmetric predicates. The consequences of this monadic basis are also discussed by Ubbink (1962) and Weinberger (1965). Ubbink writes about "contraction" and Weinberger about "reduction" of relational statements to monadic or "absolute" statements of the subject-predicate form. Beth has given a particularly clear formulation to the underlying assumption as far as non-symmetric relations are concerned, which he calls "Aristotle's Principle of the Absolute", and which will be discussed in Section 14 below.[22] For present purposes suffice it to say that traditional logicians tried, in one way or another, to reduce relational predicates to property-predicates. That this must lead, and has lead, to especially strange results when the predicates in question stand for asymmetric relations is a fact which is still too little heeded by continental philosophers; a lonely

swallow like Samuel (1957/126) does not make a summer. Some explanation is probably still not quite superfluous.

Anyone who wants to employ Aristotle's syllogistic as accurately as
possible will ascribe to a statement *S has the relation R to S'*, the *R* of
which we assume to be asymmetric, the logical structure: *S is/has (RS')*;
that is to say: the form of a subject-predicate proposition, with *(RS')* as
the predicate. Now in some cases of this kind it is possible to define a
quantitative (cardinal) scale of measurement so as to obtain statements
of the form *S is (n P-units)*. Take, for example, a statement *S is longer than
S'*, assuming *S* and *S'* to stand for two rods. It is then possible to abolish
the use of the predicate "longer than" altogether, since we can express
ourselves as follows: *S is (n length-units) and S' is (m length-units)*, for
instance, n and m centimeters, respectively. The relational statement is
here "analysed" as a conjunction of two subject-predicate statements
(and into the statement n > m, which is a mathematical statement and
hence, for many idealists, unimportant). This is not always possible,
although Leibniz clearly thought that a decomposition of this kind was
proper and feasible under all circumstances, provided one was allowed
to use *non-truth-functional connectives*, like "quatenus" or other "intensional" operators, the meaning of which, however, he failed to define.
Where such a decomposition turns out to be impossible, which happens
very often since not everything in this world which is susceptible of
degrees can be described with the aid of cardinal arithmetic (as philosophers of a more metaphysical bent will certainly agree), information
will be lost if we reduce the syllogistic premiss *S is/has the relation R to S'*
to a subject-predicate proposition *S is (RS')*. For we shall not get much
profit from such a premiss unless exactly the same combination of *R*
with *S'*, the term *(RS')*, occurs either in another premiss or in the conclusion we want to reach.

There is, however, another procedure which has played havoc in
European philosophy. It consists in simply dropping the relatum *S'*
from the "logical form" of the judgment. It is this procedure that Ubbink
calls "contraction", and Weinberger "reduction".

This procedure goes back to Plato's *Phaedo* (102 B7–C4) where he
discusses statements containing a predicate which is asymmetric and
transitive, *viz.* the predicate "greater than".[22] In his attempt to analyse
statements containing such predicates in a manner that will make *mea-*

surement understandable, Plato invoked an entity called "the Great-and-Small", which on occasion could take on the shape of the Long-and-Short, the Broad-and-Narrow, etc., and which he sometimes assumed to be justified by a logical principle, the Indefinite Dyad. In the *Phaedo*, comparative statements like

(5)   *S is greater than S'; S is warmer than S'*

or, in general, if *P* is the positive degree of an adjective:

(6)   *S is P-er than S', or S is more P than S',*

are said to have the following logical form:

(7)   *S has P-ness,*

or

(7')   *S participates* [read: *to a certain degree*] *in the P(-ness).*

Or, as one usually says,

(7")   *S is P,*

in our examples:

(6)   *S is great; S is warm*

(that is, as compared to *S'*).

A proposition

(8)   *S is less P than S"* (or: *S" is more P than S*),

for instance

(9)   *S is less great than S"; S is less warm than S",*

will then be understood in one of the two following ways: either (i) as a proposition of the form

(10)   *S is not P,*

in our examples:

(11)   *S is not great; S is not warm*

(namely, as compared to *S"*);
or else (ii) as a proposition

(12)   *S is Q*

in which *Q* is a predicate which gives the impression of being *contrary* to (inconsistent with) the predicate *P*. In our examples, *Q* would be the predicates "small" and "cold":

(13)   *S is small; S is cold.*

Ubbink calls the ensuing "contradiction" or seeming inconsistency "Plato's paradox".

When (9) is held to entail (10), this paradox takes on the form of an explicit contradiction:

(14)   *S is P and S is not P.*

This is clearly a negation of (2) above. *P* is here assumed to be a name for a property.

In order to derive a negation of (4), for instance: "Coriscus is Socrates and Coriscus is not Socrates", we shall need, in addition, an identity theory of the copula. In Plato's logic this requirement is fulfilled. Dürr derives a thesis from the *Sophist* which he regards as axiomatic in Plato's logic (Dürr 1945/185; cp. Coreth 1961/284f.):

(15)   *If (the) Being = the Identical, and if S and S' participate in (the) Being, then S = S'.*

The first part of the antecedent is a substantival formulation of the identity theory of the copula (assuming, as Plato certainly did, that to be *something* was to be in the sense of having some kind of being). When this clause is accepted as true, it may be dropped (by *modus ponens*) from (15).

Dürr discusses the case in which *S* is "Movement" and *S'* "Rest". His use of the word "axiom" is such that he clearly takes (15) to be a principle of universal validity in Plato's logic. This is to say that, in that logic, all things that *are*, are also "identical".

Inserting "Coriscus" for "*S*" and "Socrates" for "*S'*" it follows that Coriscus = Socrates (see further IX-10).

Readers who are interested in the use of these principles in what is today called "dialectical philosophy", and especially in its Marxist variant, should consult the critical study by Weinberger (1965).

12. NEO-PLATONIC FICHTEAN LOGIC (NPF-LOGIC).
THE INTENSIVE IDENTITY THEORY OF THE COPULA (ID-LOGIC)

In Platonic and neo-Platonic logic, an entity can *participate more or less* in any Platonic idea or universal: the μέθεξις-relation is susceptible of degrees. As Nelson points out, this form of thought in terms of gradable marks ("abstufbare Merkmale", 1962/522) is not only characteristic of Plato but also of the Chinese philosophers Lao-Tse and Confucius.[23] Beth (1965/9) points out that a variant of this position, in which its application is restricted to non-symmetric relations, is basic to Aristotle's logic as well; whence it is understandable that traditional logicians with less systematic and historical acumen than Beth do not always know where their own ideas derive from. The idea that *every* property can be possessed in degrees can, however, not be imputed to Aristotle, nor did he hold that the copula in a proposition could signify varying degrees of inherence, nor that it was symmetric.

The theoretical as well as the practical logic of Fichte is closely related to the logic of Plato, and in general to thinking in terms of *S more or less resembles P* (or *the P*, or *P-ness*), where *S* may indifferently be an individual term or a general one, and (*the*) *P* a general term.[24] In Platonic and neo-Platonic logic, the relation of an individual S to the S-idea is one of greater or smaller *similarity*, the S-idea being *completely identical* only with itself. Every proposition, whether the subject is an individual or a general one, expresses in this logic a greater or smaller similarity, as we saw in Section 7 above.

This may sound more like heuristic than like logic and it should, indeed, be remembered that the distinction between logic and heuristic was much less obvious at the beginning of the last century than it is today. That explains how the notion of a symmetric copula could form the basis of systems of thought purporting to offer alternative *logics*, like the philosophies of Fichte, Schelling, and Hegel.

Among these three philosophers, Nelson regards Fichte as the one who is primarily responsible for the re-introduction of neo-Platonic or mystical[25] logic into modern philosophy. For this reason, Nelson writes,

"I speak precisely of the opposition of the neo-Platonic-Fichtean logic to the Aristotelian-Kantian logic" (1962/518).

We cannot solve here the intriguing historical problem of Fichte's originality in this matter. At a guess, there has been an uninterrupted historical development, possibly via Wittenberg, in which, among others, Peter Ramus and other Renaissance logicians form important links. The logic of Fichte would then be a continuation of the "metaphysical" or "philosophical" logic which appeared, or re-appeared, in the thirteenth century. This "metaphysical logic or logicized metaphysics", as Moody calls it (1953/5, 36f.), presented itself as an alternative to the logic of the schoolmen, which had its roots in the logic of Aristotle, of the Stoa, of Abélard and of Boethius.[26]

As will be remembered from IV-14, this metaphysical or philosophical logic is characterized by the assumption of a "philosophical", "virtual", or "natural" *suppositio* of a *terminus per se sumptus*. In modern philosophy this philosophical *suppositio* of terms or of judgments was not mentioned before Maritain and Pfänder.

In the opinion of Nelson, Fichte has given *the* mystical logic its purest formulation. He characterizes Fichte's theory of science as the now classical attempt at systematizing mystical logic. Fichte found a follower in Schelling; and Goethe's profound sympathy for Schelling's "natural philosophy" was a consequence precisely of the mystical logic he found in all Fichte's and Schelling's works (Nelson, *op. cit.*/520).

How is this logic to be characterized? It is clearly a logic based on (i) an intensional (comprehensional) view of the copula, which is assumed (ii) to have the property of symmetry and (iii) to express *greater or less* similarity.

Because of (iii), we have to do not only with an *intensional* identity theory of the copula, but with an *intensive* one. These philosophers think and write in terms of lower or "external" degrees, *versus* higher or "internal" degrees of "real", logical identity. In addition to the "pure" identity ("reine Identität") mentioned above, Hegel, for instance, discusses an external, less essential kind of identity which he calls "similarity" ("Gleichheit"): "Now this external identity is the s i m i l a r i t y, and the external difference the d i s s i m i l a r i t y. Similarity is certainly identity, but only as posited, an identity which is not by and for itself" (II/35).

We shall call any traditional logic based upon an *intensive* intensional (comprehensional) copula-theory an "id-logic". Clearly the NPF-logic is one of the id-logics.

13. LATER GERMAN TRADITIONAL LOGIC AND ITS RELATION TO THE NPF-LOGIC

Erdmann dissociates himself in quite unambiguous terms from the mystical theory of negation which is a part of neo-Platonic thought, and in which Nothingness ("das Nichts") is conceived as the – logical – ground of everything, or, in Erdmann's own words, as "the absolute(ly) Infinite as an absolute indefinite (indeterminate) comprehension of all reality" (1907/509). It is clearly no part of his intention to connect the class of indefinite propositions, which he does not at all reject, with the mystical indefiniteness of Infinite Nothingness. The question remains whether the nature of the copula as seen by Erdmann and his followers differs really and clearly from the copula in the logic of Fichte. We shall have to return once more to the seemingly innocent difference between a symmetric and a non-symmteric copula, formulated and discussed *in puro* in Section 6 above, albeit as a consequence of certain remarks made by Fichte and by Erdmann.

It should be realized that although it is very easy for us today to formulate this distinction, since we possess generally accepted definitions of reflexivity, symmetry and transitivity, it was not so easy in the days of Erdmann or earlier logicians. I see no reason to assume that Fichte and Hegel, for instance, or even Erdmann, were clear about the difference between what today is called reflexivity and symmetry.

Erdmann usually expresses himself as a true follower of Leibniz[27] ought to do, but disregarding the mathematical vernacular of the latter. He says: "'To be predicated of' therefore logically means 'equal in content to a part of the content of the subject'" (1907/358).[28] He sees the "relation of the judgment" as "the relation of logical immanence", and this is, for him, "a relation of fitting the predicate into the content of the subject" ("Einordnung", *loc. cit.*). So this would be the reflexive and transitive, but not symmetric copula which we described in (3b) of Section 6. If this is taken to hold for *all true* judgments, then all true judgments must be analytic and ought to be called *logical* judgments. Erdmann does in fact

draw this far-reaching conclusion: "So there are no synthetic judgments of experience in Kant's sense" (296).

Erdmann's theory of the nature of true judgments is of course extremely
welcome to many philosophers who hope to be able somehow to bypass
both empirical investigation and a dialogical foundation for the meanings
of the terms used in a philosophical discourse. Note that Archbishop
Whately held a very different view (cp. IV-25).

But there are statements in Erdmann's *Logik* which have the effect of
obscuring the difference between him and the NPF-logicians in an even
more technical manner. These are remarks pertaining to similarity or
equality ("Gleichheit"). In spite of his explanation of the copula as an
expression of *partial* identity, he writes: "We have already seen that
elementary affirmations, logically speaking, express relations of equality"
(508). He uses the same word, "Gleichheit", which in the NPF-logic
clearly signifies a symmetric relation. As we saw, Erdmann invokes
Leibniz's "praedicatum inesse subjecto verae propositionis",[29] which
implies a non-symmetric copula and hence certainly not an *equality*
between subject and predicate. What is more, Erdmann takes what
Leibniz has to say about the matter to indicate that Leibniz's theory
already had representatives among the schoolmen (360). The remarkable
thing is that he goes on to say: "... if not for the purpose of logical
investigations, then at least for the purpose of metaphysical ones, and
especially for the needs of the ontological proof of the existence of God".
Erdmann himself thereby implies that his own view of the copula is none
too different from the inherence theory in the "philosophical" logic of the
thirteenth century (cp. Section 3).

So true elementary affirmative judgments are, from a logical point of
view, equalities. Let us relate this view to Erdmann's Principle of
Equality-with-a-Third ("Grundsatz der Drittengleichheit"): "If two objects are equal/similar to a third, then they are also mutually equal/
similar" (371). This is certainly Fichte's symmetric copula above. Only
one conclusion is possible: Erdmann is not at all clear about his own
definition of the copula or about what he understands by "Gleichheit".
His Leibnizian definition (3b) turns out to be, for him, indistinguishable
from the Fichtean copula (3a) (Section 6). His remark that "the geometrical
similarity in Euclid's sense is only a special case of the logical [similarity,
"Ähnlichkeit"]", points in the same direction. He lacked the intellectual

apparatus offered by the modern logic of relations and second-order quantifiers, which is needed in order to distinguish clearly between the notions defined in (3a), (3aa), and (3b).

Of course this has practical logical consequences. Erdmann describes the Principle of Equality-with-a-Third as a "variation on the Principle of Substitution". John of St. Thomas already wrote: "The first principle is: "Whatever things are identical with one third thing are identical with each other." On this principle the force of discursive proof rests" (John of St. Thomas 1955/122). The idea is clearly that when two objects are "equal", *then the name of the one may be substituted for the name of the other*. Now a very strong identity is required if this is not going to lead to fallacies, as Leibniz clearly saw. But Fichte did not understand this, and it seems that Erdmann did not either. We shall return to this topic in IX-10 below. [30]

Erdmann's theory of the copula underwent a revival in von Freytag's logic. Von Freytag develops a symbolism which he rightly holds to be "neutral in the controversy between a logic of comprehension/content and a logic of extension" (1961/43). This is to be expected if the copula of this subject-predicate logic is reflexive and transitive, but not symmetric.

In his book on "pure logic", which dates from 1955, von Freytag uses the word "diversity" ("Diversität") for the non-identity, in his sense of "identity", of concepts. But he has not always used the same terminology. During the congress for philosophy in Bremen in 1950, he proposed a number of theses about the relationship between what he called "pure, philosophical logic" and contemporary or modern logic, which he called "logistic" (cp. Plessner (ed.)/161f.). In these theses, the not being identical of two *concepts* is called "contradiction" ("Widerspruch"; *op. cit.*/162, 168). In 1955 he writes that this terminology led to misunderstandings at the Bremen congress (1961/195). The word "contradiction" is misleading because it is commonly used for a relation between judgments or propositions. "Difference" ("Verschiedenheit") is no good either, for the complementary notion would, he thinks, be equality ("Gleichheit") rather than identity. In 1955 he therefore chose still another expression, "diversity", for non-identity. But the word "Widerspruch", so beloved by Hegelian or "dialectic" philosophers, still seems to him the best word, and "on a higher level" he still uses this word for "the intertwining of identity with non-identity" (1961/15, 17). The reader will have no difficulty in recognizing the "dialectical" notion of "identity in difference".[31]

14. THE TRADITIONAL RESTRICTION TO SYMMETRIC NON-LOGICAL RELATIONS

(i) *Fichte.* As a justification of Fichte's logic and the way of speaking described in Section 7, *sub* 3, the following might be advanced. In a proposition *S resembles S' in as far as both are M*, there are *three* terms, *S, S'*, and *M*. Now if *M* is given, this propositional form is a two-place predicate expressing a dyadic relation with the properties: reflexivity, symmetry, and transitivity. We can shorten such a proposition to

$$S \underset{M}{\approx} S'$$

(cp. Def. (3aa) in Section 6; I have replaced "*P*" by "*S'*"). Since we assume *S* and *S'* to be different terms in the sense of different *words* (or series of words), we may also write

$$S_1 \underset{M}{\approx} S_2$$

without begging the question. As soon as S_1 and S_2 are specified, we shall have a three-term proposition, *M* being the name of the equality relation in question, with S_1 and S_2 as names for the arguments of the relation.

But in the traditional logical syntax only one-place categorical (simple) propositions were recognized: one subject (argument) and one predicate. It is, therefore, hardly surprising that difficulties arise for philosophers who want to accommodate the structure of all philosophically important statements within the confines of this logical syntax. The problem of expressing *relations of similarity* within the syntactical framework of subject-and-predicate logic is "solved" by Fichte and other neo-Platonists in the following manner: every relation of equality or similarity between two entities called S_1 and S_2, or *S* and *S'*, is *expressed by means of the copula*, which does not count as a term at all. So one of the two terms of the equality is regarded as a predicate – and vice versa.

Non-symmetric relations, and among them asymmetric relations, were considered only as non-equality and non-similarity. This is a blunder, although one that was also made by Leibniz. For if S_1 *is non-equal to* S_2 is true, then S_2 *is non-equal to* S_1 is also true: non-equality, or difference, is also symmetric. Asymmetry and non-symmetry in general, therefore, could not be treated without contradiction.

(ii) *Maritain* distinguished, as we have already seen, between modern and traditional logic in terms of "logic of relations" and "logic of inherence". Inherence is the relation of the logical predicate to the logical subject in a true judgment. It is quite clear from his examples that Maritain, too, assumes that the theory of the copula, supplemented perhaps by the Thomist doctrine of *analogy*, makes any other logic of relations, for instance that developed during the last hundred and thirty years, super-fluous, together with the new theory of "all" and "some" which was constructed on the basis of n-place predicates (propositional functions).

His theory of the copula turns out, however, to be exceedingly meagre. We know already that he splits up *suppositio personalis* into *suppositio accidentalis* and *suppositio naturalis* or essential *suppositio*. This is a distinction, he says, "with respect to the verb or with respect to the copula" (1933/87). But this is all he has to say about it, so we are certainly entitled to the judgment that this distinction, like most of the distinctions introduced by Pfänder, is merely a verbal one which cannot lay claim to any theoretical logical significance.

There is a certain systematic difference between Fichte's and Maritain's treatments of symmetries. Let S_1 and S_2 be individual terms and assume there to be a symmetric relation, M, between the entities they refer to. Fichte and other neo-Platonists and many Marxists as well will then promptly say that a relation of identity obtains between 'S_1' and 'S_2', or between S_1 and S_2. Fichte, as we know, expresses this as follows:

(1) $S_1 = S_2$, *in as far as both* $= M$.

A *logical* name, "identity", is thereby given to all non-logical symmetric relations. The appendix "in as far as both $= M$" is assumed not to be a part of the "judgment", or of the proposition, but rather its *logical ground*.

Maritain does not leave out the M in such cases. His analysis of propositions like "Peter and Paul are cousins" comes much closer to that of modern logicians than Fichte's. He states that S_1 and S_2 do not refer to *two*, but to one and the same logical subject, which he denotes as follows: (S_1 *and* S_2), e.g.: (Peter and Paul); the parentheses are added by me. The form which he ascribes to such judgments is clearly seen to be (cp. 1933/122):

(2) S_1 and S_2 – the logical subject – is M,

that is to say:

(3)   (S_1 and S_2) is M.

In Maritain's logic it is the theory of the *logical* relation of this subject 'S_1 and S_2' to the predicate 'M' which is supposed to replace the logic of relations and many-place predicates we know today. And it is this relation which is characterized as an identity,[32] or, in Veatch's publications, as an intentional identity. Subject and predicate both differ from the subject and predicate in Fichte's analysis of the same kind of statement. But the resemblance to a modern set-theoretic analysis is conspicuous. If a modern logician were to use capital letters as individual syntactic variables, he would write:

(4) $\langle S_1, S_2 \rangle \ \varepsilon \ M$.

The main difference from Maritain is that $\langle S_1, S_2 \rangle$ denotes an *ordered* couple, while in (3) (S_1 *and* S_2) denotes an *unordered* couple, due to the symmetry of the term-co-ordinating "and". Maritain does not realize that this reduction to complex terms by means of term-co-ordinating operators taken from natural language is possible only for relational terms M by which *symmetric* relations are expressed. This has been pointed out by Russell (1963a/44, 1964/335).

Maritain's discussion of certain kinds of *suppositio* is also conducted with the aid of complex terms of the form (S_1 *and* S_2) or of the form (S_1 *or* S_2),[33] that is to say in terms of *un*ordered couples or other n-tuples. Most natural languages do not possess an asymmetric or anti-symmetric word, corresponding to the figure $\langle -, - \rangle$ of set theory, by which an "enantiomorph" subject-term – an ordered couple – can be phrased;[34] and Maritain, who holds that the natural languages are ideal instruments for logical thought, does not know of such enantiomorph couples either. In natural language one resorts to spatial order (the order in which S_1 and S_2 are written down) or to temporal order (the order in which they are pronounced), putting the name of the non-symmetric relation in between: S_1 *is M (to)* S_2, e.g., S_1 *is Richer Than* S_2. A reduction of such propositions to subject-predicate form, with "Richer Than" as the predicate, is not possible.

Maritain's assumption that the logic of relations can be reduced to a theory of "is" remains, therefore, unproven. Further, if such simple and immediate reductions within natural language are meant as his examples imply, his thesis is plainly false.

Maritain in fact himself reveals that he has certain qualms about the problem of non-symmetric relations in syllogistic logic. In his *Petite Logique* ("minor logic"), after some none too successful attempts to accommodate them under the head of "oblique" syllogisms, he promises (1933/297) to return to the matter in his *Grande logique* ("major logic", or methodology). As it turned out, that book never saw the light of day.

Maritain's logic of (symmetric) relations and unordered complex terms has had considerable intellectual influence. His arguments against modern logic and his unsuccessful attempt to incorporate relations into a logic of inherence reappear in a textbook in traditional logic by the Dutchman van den Berg (1946/33), who refers to Maritain in his book, and more recently in the dissertation of the Dutch linguist Dik (1968), who does not.

(iii) *Modalities and relations: "intensional" logic.* In modern logic, modal operators are studied under the heading "intensional logic", because an identity does not always entitle us to substitute one term for the other in a proposition containing such operators. In traditional philosophy, the logic of relations was intensional (or: opaque) in this technical sense. This becomes especially noticeable where the copula is assumed to express a similarity or equality which is not complete sameness. That will become clear in Chapter VIII and especially in IX-10 below. I mention it here in order to point out that, for traditional logicians, relations in general had much in common with modalities.

It should therefore not surprise us to learn that Maritain, Pfänder, and others endow "the copula" with the double task of "expressing" both relations and modalities (Section 1). In traditional logic the logic of relations, the logic of modalities, and even the logic of higher-order quantification, are not distinguished from each other, but are all relegated, with a casual remark, to "the copula". Of course the late-scholastic Renaissance author John of St. Thomas from whom Maritain draws much of his inspiration could not, for chronological reasons, know these chapters of logical theory and Maritain himself wrote his logic before the first modern investigations into the logic of the modalities were made around 1920.

One and the same propositional copula may and also may not express a modality depending on whether the speaker intends (but does not declare) his proposition to have a subject term in natural (essential, logical) or only in accidental *suppositio*. For this reason we are entitled to rank the logics of Maritain and of Pfänder as id-logics.

(iv) *Aristotle's Principle of the Absolute.* In his *Metaphysics* and elsewhere (*Physics, De anima*), Aristotle expresses certain fundamental views that are highly relevant to the problem of the logic of relational predicates. But he expresses them in none too clear a manner, so that confusion with Platonic views and methods cannot be excluded. Beth, inspired by works by F. Enriques, L. Rougier, and B. Schulzer, gives a very precise formulation of what he calls "Aristotle's Principle of the Absolute" (1965/9). This principle, Beth says, lies at the root of a great number of arguments both in traditional philosophy and in relatively modern science as well, and "has been applied with remarkable virtuosity by Aristotle". In plain words, it can be stated as follows:

if R is any non-empty non-symmetric dyadic relation, then there is one entity A_R (*the* absolute entity corresponding to R) such that A_R bears the relation to every other entity (belonging to the field of R), while no entity bears this relation to A_R.

If we want to formulate this in the language of first order predicate logic, we shall have to use a small letter, "a", for the absolute entity which belongs to R:

$$R \neq \Lambda \rightarrow [(Ea)\,(x)\,\{x \neq a \rightarrow (aRx \,\&\, \sim xRa)$$
$$\&\,(a)\,(b)\,(x)\,\{(x \neq a \,\&\, x \neq b) \rightarrow [(aRx \,\&\, \sim xRa$$
$$\&\, bRx \,\&\, \sim xRb) \rightarrow a = b]\}].$$

This implies, among other things, that to every comparative relation, being a relation which may be used to define an ordinal scale, there belongs of necessity a final or maximal point on the scale. From the relation Warmer Than, for instance, one can easily construe a notion 'the absolute Warm(th)', being the notion of an entity with the property of being warmer than anything else; similarly, from the relation Better Than, the notion 'the absolute Good' can be defined, being the notion (or concept) of an entity which is better than everything else. Aristotle's Principle of the Absolute does not only say that these notions can be conceived and defined, but that they are logically fertile and even logically indispensable ("logically essential"), whatever relation R we choose to consider.

A particularly clear statement of this principle, followed by a number of applications, is formulated by St. Bonaventure: "Also, if there is *relative* being, there is *absolute* being, for what is relative can terminate only in what is absolute. But an absolute being dependent on nothing can only be something which has not received anything from another. But this is the First Being. All other beings, however, have something of dependence... Also, if there is being in an *imperfect* and *qualified* sense of the word, there is being in an *unqualified* sense of the word. For what is being in a qualified sense can neither exist nor be understood unless it be understood by that which is simply being; nor that which is imperfect being save by that which is perfect being, just as privation cannot be understood save in reference to that positive entity which is had... Also, if there is a being which exists *for the sake of another*, there is a being which exists *for its own sake*, otherwise nothing could be good... Also, if there is being *by participation*, there is a being that is by reason of *its essence*. For participation is spoken of only in reference to something which is possessed essentially by something, for whatever is *per accidens* can be traced back to something *per se*... Also, if a being exists *potentially*, a being that is *actual* exists; for no potency is reducible to act save by being in act nor would there be a potency if it were not reducible to act... Also, if there is a *composite* being, there is a *simple* being, for what is composite does not have existence of itself. It is necessary that it takes its origin from what is simple... Also, if *changeable* being exists, an *unchangeable* being exists; because according to what the Philosopher [*i.e.*, Aristotle] proves, motion is from a being at rest and for the sake of being at rest" (quoted from Wolter 1946/134f.).

Aristotle explicitly argued against the H-thesis (the Third Man argument). It should, however, be noted that the Principle of the Absolute supports the H-thesis and vice versa; for there can be no doubt that according to the adherents of the H-thesis an objective "concept" 'the absolute Good', if it exists, *is* good, where "is" is the "is" of predication.

Similarly, 'the absolute Warm(th)' must certainly *be* warm. This principle clearly produces an infinity of Whats, as Veatch calls them. Philosophers who assume this principle to be generally valid regard any relationship between an entity B and an entity C as having its logical foundation in an absolute entity A, in the sense that B and C *both participate in A, but not to the same degree*. "The" participation relation is said to be "a relation of identity". To Fichte, this means *similarity* between B and A and between C and A; and to all what-logicians it means that *identically the same* "logophore", essence, or nucleus is *in* all of the three. In fact, we may assume the logophore to be no other than A itself. "Identically the same as" of course designates a symmetric notion.

Beth applies this principle to a number of philosophically and scientifically important non-symmetric relations, and shows how the assumption of its general validity generates some very familiar Absolutes in scientific and philosophical literature, from the Stoic ἀρχή to Newton's Absolute Space and the *Arbeitsgallerte* or "value-jelly" of Marx.[35]

One important example should be added to Beth's list of famous Absolutes, and that is the notion of '(the Absolute) Logic'. It may be understood as the entity whose existence follows from Aristotle's Principle of the Absolute when this principle is applied to the comparative statement form *this logic is better than that* (cp. Barth 1972).

15. The Theory of the Copula and Reduplication in Individual Propositions

I shall now raise a question which is not dealt with in Maritain's logic. Can the "intensional" (non truth-functional) connectives "in as far as" ("en tant que", "in quantum") and "qua" (cp. IV-30) be preceded by an individual term? If so, the problem arises of the correct interpretation of individual reduplicative propositions.

Individual propositions containing the word "qua" are in fact quite frequently found. If we assume that they are well-formed in traditional logic, then we are entitled to apply Maritain's decomposition of reduplication to the following proposition

(r-sing)   Saul, qua Bantu, is primitive.

According to Maritain's analysis this means:

(1)   the Bantu is primitive,
(2)   all Bantus are primitive,
(3)   the Bantu is the Reason why Saul is primitive.

In connection with William of Sherwood's theory of *suppositio* W. Kneale remarks that from the proposition

(a)   Homo est dignissima creaturarum

it follows that

(b)   Iste homo, inquantum homo, est dignissima creaturarum.

For this reason Kneale inserts the expression "mobilis" in the name of the *suppositio* of the term "Homo" in the general proposition (a).

When the analysis (1)–(3) is applied to (b), the latter proposition is seen to entail

(b2)   All human being(s) is (are) the most dignified being.

This, however, implies the unqualified statement

(c)   iste homo est dignissima creaturarum.

But that is certainly not the intention. How did this mistaken result arise? It seems to me that the fault is Maritain's; it is not right to say, as he does, that

the *S qua M is P* ↔ ... *& all S's which are M, are P &*....

A faithful rendering of the traditional metaphysical logic – Maritain's own logic – would seem to be:

(r2)   *all M's are essentially P*

or

all M's are potentially P (and not: *all M's are P in actu*).

In fact Maritain himself talks about *puissance*.[36] If we understand this not only as a grammatical operator which yields a disjunction (*the*) *an M is P or Q* when applied to a term-conjunction *the (an) M is potentially P and Q*,[37] but also as a *logical force*,[38] then an expression like "iste homo, in quantum homo" must refer to a presupposed *logical ability* of the individual under consideration, to a *logical gene* in his logical Being.

From the statement

the Bantu as such *is* primitive

we then can – if our analysis is correct – draw the conclusion

Saul – who is a Bantu – in any case carries in him the gene or capacity (the logophore) of primitivity,[39]

but we may *not* conclude that he is primitive *hic et nunc*. If "in quantum" is used instead of "qua", the statement (r-sing) suggests a certain measure or degree ("quantum") of participation in the reference of a logophoric term: *in that degree, in which* Saul *is* a Bantu, he is also primitive, provided that (1) is true.[40]

The conclusion to which we have come is, then, that the use of reduplicative expressions in individual propositions presupposes an *intensive* conception of the copula. If it is allowed to use expressions of reduplication in individual propositions in Maritain's logic, then that logic certainly is an *id-logic*, according to the definition at the end of Section 12.

16. A NECESSARY CONDITION FOR ID-LOGIC: WEAK IDENTITY-
CONCEPTS AND LEIBNIZ'S PRINCIPLE OF THE IDENTITY OF
INDISCERNIBLES

(i) *Preliminary remarks.* We have seen that in the school of Fichte there is no systematic distinction between propositions of comparison, *A resembles B*, and categorical monadic subject-predicate propositions. Nelson (1962/518) imputes the absence of this distinction to the *Principle of the Identity of indiscernibles* (PIi). This principle, or, to put it more non-commitally, this expression, is usually connected with the name of Leibniz. Nelson, too, turns out to have a principle in the logic of Leibniz in mind (Nelson *op. cit.* 232f., 530f.). Nelson's analysis of the manner in which this principle has influenced philosophers so as to result in the logic of Fichte is suggestive, but does not penetrate deeply enough to be really convincing, although he certainly does make it plausible that a further study of this principle may throw more light on the intensional identity-theories of the copula. For this purpose Weinberger's monograph (1964) turns out to be very useful. Below, and especially in the second part of this section, I shall draw heavily upon Weinberger's study.

In order to understand Leibniz's principle we shall have to start out from a *definition* of "identity" as meaning "universal equality", "equality in all respects". One well-known statement in Leibniz's logic is the following: "Eadem sunt, quorum unum in alterius locum substitui potest, salva veritate" (Leibniz 1961, VII/219). Or in English: "Those are the same, of which the one *can be substituted in the place of* the other while preserving truth." This is often called "Leibniz's Law". It is reasonable to take it as an attempt to formulate a criterion of some kind of identity.

"Being the same" is here defined by means of a notion of substitutability. According to present-day logicians, a substitution is a replacement of an expression by another expression. The two expressions, then, are *two*, and hence cannot be the same. What *can* be the same, is the meaning (in some sense of "meaning") of the two differently shaped, or at least spatially or temporally separate, expressions. This is not clearly brought out in Leibniz's own formulation of this definition. A less confusing formulation of "Leibniz's Law" would be:

(1)   A is identical with B if and only if everything that can truly be predicated of B, can also be truly predicated of A, and *vice versa*.

The following discussion will be restricted to entities (a, b, ...) which do have spatial co-ordinates; numbers and other abstract entities will not be taken into consideration. That is quite sufficient for our purposes, for this class comprises not only human beings (Coriscus and Socrates) and planets (the earth and the moon; see Chapter VIII), but triangles drawn on a piece of paper as well. The discussion should therefore be relevant to our study of the Locke-Berkeley problem even granting this restriction.

Instead of (1), the following formulation of Leibniz's Law is sometimes found:

(2)   A is identical with B if and only if A has every property that B has, and *vice versa*.

If we want to express this in the language of (second-order) predicate logic, we shall have to use some kind of small letters instead of the traditional capitals:

(2a)   a is identical with b $\underset{\mathrm{Df}}{\leftrightarrow}$ (P) [Pa $\leftrightarrow$ Pb], where "P" is a property-variable. These formulations are still capable of several interpretations. We shall get a better survey if we use predicate-variables and variables for proper names or other uniquely referring expressions (syntactic variables):

(2b)   a is identical with b $\underset{\text{Df}}{\leftrightarrow}$ (P) [Pa is true ↔ Pb is true].

But what kind of entities are a and b? Our use of variables in (2b) suggests that *two* elements of a domain of discourse can be identical, which of course is nonsense. With support from a recent article by Feldman (1970), I hold that the formulation

(3)   'a' is identical with 'b' $\underset{\text{Df}}{\leftrightarrow}$ (P) [Pa is true ↔ Pb is true]

is as close as we can come to a historically faithful formulation of "Leibniz's Law": *eadem sunt* etc. and to a verbatim translation of (1) into the language of second-order predicate logic, as far as the *definiendum* is concerned. Even so, the truth-condition for an identity-*proposition* may be expressed as follows:

(3')   *a is identical with b* is true $\underset{\text{Df}}{\leftrightarrow}$ (P) [Pa is true ↔ Pb is true],

or perhaps

(3")   *a is identical with b* is true $\underset{\text{Df}}{\leftrightarrow}$ (P) [Pa ↔ Pb is true].[41]

For a proposition may be regarded as a description of a fact and as an expression of a judgment at the same time.

When we choose one of the formulations (2b), (3), (3'), or (3") it becomes clear that the truth-value of an assertion of identity will depend on the range of the syntactic variable P. The interpretations (3), (3'), and (3") of "Leibniz's Law" are in fact as unprecise as (2a) as long as neither the domain of the variable "P" for predicates or propositional functions nor the domain of the variable "P" for properties is specified. If this is not done, it is not clear what is meant by "for every predicate", nor by "for every property". It is clear that the domain of the predicate-variable "P" in (3), (3'), and (3") contains only monadic (unary) predicates and propositional functions.[42] But it is not clear

(1°) from which kinds of atomic predicates these propositional functions may be built up,
nor is it clear

(2°) by means of which connectives, quantifiers, or other operators, such as modal operators, these propositional functions may be constructed out of the atomic predicates,
nor it is clear

(3°) precisely what kinds of terms *a* and *b* may be; may we, for instance, choose any kind of definite description *the-so-and-so* as values for "*a*" and for "*b*"? Or must we restrict the application of this definition (3″) to statements containing only proper names and no definite articles, assuming our language to contain precisely one proper name for every entity (or concept) we want to be able to discuss? That would be an awkward restriction indeed, for most entities have no proper names at all in natural languages – which certainly makes the question a pressing one.

Modern logicians concentrate on the problems (2°) and (3°). It is well known that if *a* or *b* is a definite description, this principle ((3)-(3″)) breaks down if the property-expression (propositional function) *P*, or *Px*, is allowed to contain for instance modal operators. One of Quine's examples (1963/143f.) is that *b* is the expression "9", *a* the definite description "*the* number of planets", and *P* the property-expression "necessarily greater than 7". Then *Pb* is the proposition: "9 is necessarily greater than 7", and *a is identical with b* is the proposition "*the* number of planets is identical with 9"; if both these propositions are held to be true, (3″) entails that "*the* number of planets is necessarily greater than 7" is also true, but many logicians and most other philosophers, too, regard this proposition as false. Quine holds (as against Smullyan and others) that to invoke Russell's Theory of Descriptions and the notion of the *scope* of an operator is not sufficient; that theory alone is no solution to problems of this kind (cp. Linsky (ed.) 1971/4).

In all likelihood difficulties of this kind, which today are studied as problems of modal logic (cp. Hughes and Cresswell 1968, Ch. 11), have played their part in shaping traditional philosophical theory, though we do not as yet seem to be in a position to assess the precise manner in which philosophical and scientific thought has been influenced by them – if at all. For it is also conceivable that the philosophers of earlier periods had not yet advanced so far as to pay sufficient attention, consciously or unconsciously, to the intricacies of substitution in overtly modal contexts so as to be decisively influenced by them.

But earlier logicians had problems of their own; some of these were raised by the intensional (comprehensional) identity theory of the copula. Leibniz held that theory, in the partial-identities version. For him and for all others who adhered not only to this theory of the copula, with its implication that both terms in a judgment are *concepts*, but also to Aristotle's Principle of the Absolute (as we may safely assume the majority to have done), it would seem very hard indeed to escape from a general view of things very similar to that of the *intensive* intensional identity theory. From such an outlook, non-symmetric temporal relations, like Earlier and Later, as well as non-symmetric spatial relations like Longer and Shorter, To the Left Of, To the Right Of (with respect to a given orientation), are assumed to be grounded in "absolute" monadic concepts. Monadic concepts will then be given a higher general logical status than relational concepts.

(ii) *Leibniz's weak notion of identity: complete equality.* This becomes of the utmost importance when one wants to *use* definition (2a) or definition (3), or one of its variants, in order to arrive at a judgment of identity, *i.e.*, in order to *introduce* (be it in a monologue or in a dialogue) a statement of identity *a is identical with b*, or *a is identically the same as b*. One then uses one half of the definition called "Leibniz's Law", *viz.* ('f we take (3') as our point of departure):

(4) (K) [Ka is true $\leftrightarrow$ Kb is true] $\rightarrow$ *a is identical with b* is true.

Following Weinberger (1967/5), I have here replaced the variable "P" by a "K" for "Kriterium" ("criterion"). The following analysis is due to him; but I deviate from his notation in using syntactical variables throughout together with the "meta-level" predicate "true". A complete and definitive discussion of these problems will probably ultimately require a formalization of our syntax and a semantic apparatus more or less similar to that in R. Martin's work of 1963 (*q.v.*). But it would be misplaced and, at the present state of the history of philosophy, even impossible, to employ these techniques in full in a historical-problematical investigation like the present one.

Many people want to use (2a) or one of the variants of (3) as an epistemologically based *introduction rule* for expressions of identity. (The reader should keep firmly in mind that many traditional logicians con-

sidered all propositions of the form *the M is P* as expressions of identity.)
They will then have to make it plausible that

(5) (K) [Ka is true $\leftrightarrow$ Kb is true].

This raises the problem: which kinds of formulation, K, of criteria
should we accept as falling within the range of the quantifier in (5)?

Historically it is of the greatest interest that Leibniz's own Principle
of the Identity of indiscernibles – which we have not yet formulated –
assumes familiarity with an underlying "weak" notion of identity. Leibniz
constructs this notion by restricting the domain of predicates to be taken
into consideration in "Leibniz's Law" in a very special way. *He did not*
count predicates which describe the position in physical space of the entity
or entities in question, among the criteria at all. Such predicates did not
belong to the range of the variable "K" (or "P").

In the light of what was said in the first part of this section, this
becomes understandable as a consequence of the "irreality" of asymmetric
predicates. For a position in physical space is relative to some co-ordinate
system; and a co-ordinate system – whether it belongs to "absolute" space
or not – presupposes orientation and asymmetry ("farther from... than",
"closer to... than"). Leibniz's reason for omitting predicates of spatial
position, then, would seem to be of a theoretical logical nature, dictated
by Aristotle's Principle of the Absolute in conjunction with his own
account of judgments. It may have been strengthened by the notion of
"identically the same" entity moving from one position in physical space
to another; but this should not make us overlook the internal problems
of the theoretical logic of his time, especially of the theory of inference,
and in general the ways in which certain theoretical moves, once made,
compel one to make new moves, and even in certain directions.

The antecedent of (4), as understood by Leibniz, may be expressed
as follows:

(6) $(K)_{-L}$ [Ka is true $\leftrightarrow$ Kb is true].

The index "$-L$" (for "Lage", or "locus") is here, as in Weinberger's
study, used as an indication that predicates which determine positional
properties ("Lageprädikaten") are to be excluded from the domain of
"P" and "K".[42a]

Leibniz, then, assumed a kind of weak identity, to be called here "complete equality", which does not by itself exclude two things from being separated from each other in space, and for this reason be experienced as *two*, yet nevertheless *completely equal:*

(7)   a is completely equal to b if and only if a has every non-positional property that b has and conversely.

For this weak notion of identity which is defined in (7), *complete equality* or, as Leibniz called it, *indiscernibility*, we shall use the sign "$\underset{\text{CE}}{=}$". That an entity called *a* completely equals (is indiscernible from) an entity called *b* will therefore from now on be written: $a \underset{\text{CE}}{=} b$. That a and b coincide, or more precisely: that *a* and *b* refer to one and the same (concept of a) thing (with spatial coordinates!), will be expressed as $I(a, b)$. This latter identity may be called "strong total identity (STI)". It then becomes possible to make the unclear definition (3′) of identity as universal equality, *i.e.* "Leibniz's Law", more precise in two different ways, a "modern" interpretation (3a) and a "historical" interpretation (3b):

(3a) $I(a, b)$ is true $\underset{\text{Df}}{\leftrightarrow} (P)$ [*Pa* is true $\leftrightarrow$ *Pb* is true],

(3b) $a \underset{\text{CE}}{=} b$ is true $\underset{\text{Df}}{\leftrightarrow} (P)_{-L}$ [*Pa* is true $\leftrightarrow$ *Pb* is true].

Since Leibniz regards the truth of statements to be founded upon concepts and their logical relationships, a proposition $a \underset{\text{CE}}{=} b$ is for him true if and only if a certain logical relation holds between "the" concepts '*a*' and '*b*'. Let us call that logical relation "Leibnizian conceptual identity", or "LCI-identity" for short, and let us symbolize it as "$\underset{\text{LCI}}{=}$". Then

(3c)   '*a*' $\underset{\text{LCI}}{=}$ '*b*' $\underset{\text{Df}}{\leftrightarrow} (P)_{-L}$ [*Pa* is true $\leftrightarrow$ *Pb* is true].

(cp. (3) above). Clearly, LCI-identity is the highest degree of identity between *concepts of* things that have spatial co-ordinates.

(iii) *PIi: the Principle of the Identity of indiscernibles.* Now Leibniz was of the opinion that *two* things *never* have all other properties in common. In his opinion, two spatially separate objects will always show certain individual differences, if only we study them carefully enough:

(8)   if a and b are numerically two because they do not coincide in space, then a and b are not completely equal.

This is one formulation of his *principium identitatis indiscernibilium*. We
are not concerned here with the truth or falsity of (8), but with its relation
to Leibniz's *definition* of indiscernibility, or complete equality. As
Weinberger says, for a clear understanding of Leibniz's views on identity
it is of the greatest importance to see that in the works of Leibniz, (8)
occurs as a supplementation of (7), of which it is logically completely
independent. We can formulate (8), too, or rather its contraposition,
in a more precise manner, *viz.* as follows (cp. Weinberger 1964/9):

(9) $(K)_{-L}[Ka$ is true $\leftrightarrow Kb$ is true] $\rightarrow I(a, b)$ is true,

or, more briefly,

(9') $a \underset{CE}{=} b$ is true $\rightarrow I(a, b)$ is true,

that is to say

(9'')   'a' $\underset{LCI}{=}$ 'b' $\rightarrow I(a, b)$ is true.

(9) and (9') may be read: if the entity called a is indiscernible from the
entity called b, then a and b stand for one and the same object. This,
then, is Leibniz's *Principle of the Identity (Sameness) of indiscernibles
(complete equals)*, or PIi.

Those reasons for being interested in a notion CE of complete equality
which were earlier intrinsic to theoretical logic do not hold with respect to
modern logic. That the universal equality of something called a and
something called b entail that there is really only one object is today
therefore not an additional principle, but forms one half of the definition
(3a). This is the thesis

(4a) $(K) [Ka$ is true $\leftrightarrow Kb$ is true] $\rightarrow I(a, b)$ is true.

It expresses that if an entity called a is universally equal to an entity
called b, in a sense of "universally equal" that includes predicates
describing position in space (or time), there are not *two* things but only
one entity with two (or more) names. At present we therefore have the
situation that one half of the equivalence (3a) which is taken as defining
the truth condition of the proposition $I(a, b)$, *i.e.* (4a), is itself characterized as a Principle of the Identity of indiscernibles – and rightly
so. But Leibniz's original *principium identitatis indiscernibilium* is principle (9), which is stronger than the principle which differs from (4a)

only in the exclusion of locative predicates from the range of the quantifiers, *viz.* the principle

(4b) $(K)_{-L}$ [Ka is true $\leftrightarrow$ Kb is true] $\rightarrow$ $a = b$ is true.

Of course all the predicates P or K being considered in (3a) and (4a) will be monadic propositional functions, but as far as the modern theory of judgment and inference is concerned they may be constructed from (atomic) polyadic predicates, which even may contain quantifiers. And this was inconceivable in the days of Leibniz.

The present lack of interest in Leibniz's notion of complete equality is no doubt due to the fact that the world has come to realize the immense importance of non-symmetric relations for science and its philosophical foundations (see, *e.g.*, Dampier 1968/460). The fact that the theory of inference in the theoretical logic which preceded the work of De Morgan, Peirce, and Frege was unable to cope with predicates for asymmetric relations must be kept firmly in mind if we are to understand why earlier logicians cherished such a notion as CE. According to Russell (1964/227; cp. Barth 1970a, Bennett 1970), it was Immanuel Kant who, in his *Prolegomena*, for the first time explicitly pointed out the logico-philosophical importance of certain asymmetric relations, *viz.* the relations Right and Left. That may be so; but implicitly their importance was probably felt by Leibniz and all other philosophers who allotted a central role in their thinking to the concept of complete equality.

The frequent use of *vectors* and *tensors* in thought and in scientific language demonstrates that nowadays such asymmetric predicates play a very fundamental role indeed. But as long as the official theory of inference was unable to deal with non-symmetric predicates and with the need for multiple quantification that goes with the use of such predicates, it is quite understandable that most logicians came to hold the opinion that CE is a purely logical notion but that STI is not. This seems to have been the position of the majority of thinkers in the more recent philosophical tradition, at least on the European continent.

Leibniz's principle PIi seems to be a way of expressing belief in the possibility of basing non-symmetric spatial and other relational predicates upon monadic property-predicates; for (9″) says that if $I(a, b)$ does not hold, there must be some (logical) differences between the individual concepts 'a' and 'b'.

17. FROM OPAQUE SIMILARITY (ANALOGY) TO STRONG IDENTITY

(i) *Analogy of two terms defined as opaque similarity.* On occasion I have used the expression "analogy" as synonymous with "similarity". It will be necessary to say something more about this. Although two entities which are said in the philosophical tradition to be analogous certainly are similar in some sense or other, the converse does not always hold. To say of two red things that they are analogous since both are red is not to talk in accordance with traditional common philosophical usage, however vague that usage was. I shall here formulate a hypothesis about the traditional use of the terms "analogous" and "analogy".

In Section 14 it was noted that, for a traditional logician, relations were intensional in the technical sense of "non-extensional", "non-referential", or "opaque" (I shall not distinguish between the meanings of these three expressions here, but shall employ them all in the widest of their usual senses). Since the techniques for drawing valid inferences from statements with non-symmetric predicates were still unknown, a context containing such predicates may therefore be said to have been referentially opaque, or simply: opaque.

I first define a notion of opaque similarity, in the following manner:

two entities will be called "opaquely similar" when both of them satisfy the same opaque propositional function.

My hypothesis can now be given the following formulation:

(a) to a traditional logician, two entities satisfying one and the same propositional function or description with non-symmetric predicates and/ or modal operators are opaquely similar entities;

(b) two opaquely similar entities were called "analogous";

(c) this use of the word "analogous" was not conceptually distinguished from the traditional notion of an analogous use of terms interpreted by Bocheński as an attempt to treat typical ambiguity (IV-6).

Leibniz's notion of complete equality can then be understood as universal equality defined with respect to a domain of non-opaque criteria. Feldman comes to a similar result (1970/514), but he does not take up the question of which kinds of criteria Leibniz counted as opaque (or as Feldman says, as not referential). However, Angelelli has shown (1967b) that for Leibniz precisely those criteria are opaque which are expressed by *reduplicative* propositions.

Our conjecture at the end of the last section may then be formulated as follows: in Leibniz's opinion, opaque contexts can ultimately be analysed in terms of (second-)intentional (with a "t"!), or otherwise "higher", monadic predicates, if they can be logically analysed at all. Another formulation of what he had in mind for which there seems to be considerable historical justification is this: two entities are intentionally similar if and only if both of them satisfy the same opaque propositional function.

(ii) *A Principle of the Non-Distinction of the Distinct.* Assume that the entities called a and b satisfy the right-hand term (*definiens*) of (3b), but not that of (3a). Such a situation is compatible with the following mode of expression – assuming both entities to have the property S:

$$S \; nr. \; 1 \underset{\mathrm{CE}}{=} S \; nr. \; 2,$$

and also with

(α)   *the first* $S \underset{\mathrm{CE}}{=}$ *the second S.*

But suppose, for given expressions a and b, that the right-hand term of (3a) is satisfied; in that case we have to do with no more than one object (one human being, or one triangle), and the notation:

(β) I *(the first S, the second S)*

is therefore absurd, or at least highly misleading, unless this form was very explicitly introduced to the participants of the discussion in advance as a logically contradictory form. Unfortunately Leibniz's PIi, being very difficult to handle with the slippery instrument called natural language, seems to invite us to use the language form (β) whenever we judge the language form (α) to be appropriate; but this is of course a mistaken interpretation of what Leibniz wanted to say.

The reader will allow that a continual epistemological, in the sense of methodological, transition from the equality of *two* objects in one or more "re-spects" ("analogy") to strong total identity is literally a nonsensical idea. Because, *ex hypothesi*, we start out discerning *two* objects, we may rank them in some arbitrary manner so as to characterize them by means of expressions *the first S* and *the second S*, or *S no. 1* and *S no. 2* (*e.g.*, as "human being no. 1" and "human being no. 2"). That is to say, an argument

> $$\frac{\textit{The two entities S no. 1 and S no. 2 resemble each other}}{\therefore \textit{ possibly/perhaps the first S is strongly identical with the}}$$
> $$\textit{second S}$$

is completely illogical even with the qualifying "perhaps" or "possibly" prefixed to it, since *in an understandable argument the range of a bound variable must be the same in the premiss as in the conclusion.* Or, briefly, the inference form

(PND) $$\frac{\textit{S no. 1 resembles or is analogous to S no. 2}}{\therefore \textit{ possibly: I (the first S, the second S)}}$$

is logically invalid. Borrowing an expression of Frege's, which he and later Carnap used for a different purpose (Frege 1967/251, Carnap (ed. Schilpp) 1963/5, 995), I shall call this "the principle of the Non-Distinction of the Distinct (PND)".

Leibniz's principle (9), or (9′), if it is assumed to be a logical or a metaphysical truth, justifies an inference form which is treacherously similar to PND, namely

(PIi − L) $$\frac{a \underset{\text{CE}}{=} b}{\therefore I(a, b)}$$

It seems that while Leibniz's principle cannot be blamed, as Leonard Nelson thought that it could, for the extravagances of the logic of Fichte and Hegel, the similarities between it and PND may have misled those who came after him (and also, if the principle is older than Leibniz, before him) into thinking that the premiss of the former argument is "almost" the premiss of the latter argument. One could come to think so only by totally neglecting the *dis*similarities between *a* and *b*, or the "negative analogy" as one often says.

18. FROM OPAQUE SIMILARITY TO WEAK IDENTITY

If every predicate naming or describing a property N that was used to distinguish S_1 from S_2 (or: a from b) in the first place, has been excluded *a priori* from the range of "K", then we have an entirely different situation. Suppose that someone postulates: criteria of the categories $k_1, k_2, \ldots$

..., k_l are not required in order to determine some weaker kind of identity. One might, for instance, decide to exclude the categories pertaining to (certain kinds of) temporal co-ordinates, or to (certain kinds of) spatial co-ordinates, or both; but it will also be possible in principle to exclude quite other categories of predicates. It is then very feasible for us to have *two* entities or phenomena, S_1 and S_2, which we have been able to distinguish as two by means of criteria belonging to one (or more) of the rejected categories $k_i (1 \leq i \leq l)$, and which nevertheless may be called identical with respect to the restricted domain of criteria. An inference of the following form:

(An-WI) $\dfrac{\textit{the first S opaquely resembles the second S}}{\textit{possibly the first S and the second S are weakly identical}}$

is in that case not a nonsensical or illogical argument. But it is of course required that every one of the observed dissimilarities between the two S-things, *i.e.*, all those properties N_1, N_2,..., which S_1 does possess and S_2 not, falls within one of the excluded categories k_i. In order to verify that this is so, it will be necessary to determine the domain of the variable "K" in advance, at least in part; otherwise the conclusion of this argument is uninterpreted and cannot be tested or otherwise understood.

This form, then, which may very well be valid, is the form of *an argument from an analogy to some kind of weak identity* (An-WI). It is of the utmost importance to observe that *the conclusion of an An-WI argument may not be formulated with a copula of predication*: *S no. 1 is an S no. 2.*

I shall now, very briefly, sketch three applications of this principle of concluding from an analogy to the possible *weak* identity of the members of the analogy.

1. Astronomers have discovered that the phenomenon: the evening star (S_1) and the phenomenon: the morning star (S_2) resemble each other in respect of their position relatively to certain other celestial bodies, in respect of intensity, etc. Visibility in the evening (N) suffices to distinguish these phenomena, but this property does not belong to any of the categories which an astronomer considers as being of ultimate importance for the identification of celestial objects. This property, N, is too egocentric to be decisive in astronomy. The astronomical identity

of evening star and morning star can only be settled subsequently to the exclusion of this egocentric category. When predicates denoting distances are assumed to belong to the range of "K", but predicates denoting observability by an observer at a fixed point on the surface of the earth are not, then it is possible to identify the evening star with (or better: as) the morning star, but not the moon with the earth. In other words: in astronomy the moment at which a celestial object becomes visible to an observer may – under circumstances which I shall not here venture to describe – belong to one of the unimportant categories k_i of properties and the predicates used to express them to an astronomically unimportant category of predicates.[43]

2. By admitting the egocentric category of predicates mentioned above to the range of "K", we obtain a definition of a *Lebenswelt*-identity or a poetic identity, such that evening star and morning star will be non-identical, *i.e. two* objects or *Gegenstände* in a "phenomenal" egocentric world.

3. One might take the view that co-ordinates in physical time and/or space, even the egocentric co-ordinates, are without any deeper interest and that only those properties are ultimately to be reckoned with which are *completely independent of temporal and/or spatial observations*, with respect to any spatio-temporal reference system one might like to consider.[44] Suppose that we were consciously to neglect such characteristics of which it has already been settled that they depend, in some way or other, on space and time. The notion of weak identity at which we then arrive I shall call "essential logical identity", meaning, of course, the concept of identity of the "metaphysicized" trend of traditional logic.[45] Now it is hard to see how such a demarcation of the range of "K" is to be carried out. But it is not necessary for us to take up this problem; we start out from the assumption – in fact, from the truth – that there are philosophers who regard it as possible to discourse in terms of a range of values for "K" which is delimited in this manner. A predicate that stands for a property which, according to these philosophers, depends in some way or other upon the spatial position of its carrier, will be called an "L*-predicate". Properties which are designated by predicates that are *not* L*-predicates will be called "essential internal

properties". The definition of essential logical identity (EI) will then be:

(3c) $a \underset{\text{EI}}{=} b$ is true $\underset{\text{Df}}{\leftrightarrow} (P)_{-L*}[Pa$ is true $\leftrightarrow Pb$ is true].

Here we have a third, and philosophically important, interpretation of (3) in Section 16. This weak notion of identity is a truly *meta-physical* identity. It is applied in traditional epistemology in the following form:

(4c) $(K)_{-L*}[Ka$ is true $\leftrightarrow Kb$ is true] $\rightarrow a \underset{\text{EI}}{=} b$ is true.

The existence of an essential (ideal, or at least theoretical) world including no (measurable kind of) time and space must be postulated *in advance*. If one already has done that, then the subsequent use of the argument form

(An-EI) $$\frac{\textit{the first S and the second S resemble each other}}{\therefore \ \textit{possibly (perhaps) the first and the second S are essentially}}$$
$$\textit{identical}$$

is not an additional illogical or irrational step. What one ought to think about the postulate of an essential world is a question that may be left aside here. The *logical* problem is of course to determine the range of "K" at least so far that we can decide whether the names or descriptions of the observed dissimilarities N belong to it or not. If this can be determined, then an inference on the ground of an analogy, in the sense of a similarity, with a qualified proposition about weak *essential* identity as a conclusion will be *logically* acceptable. It would be illogical, however, to leave out the qualifying "perhaps" or "possibly" before all properties in the demarcated domain have been investigated; but since it may be assumed both that these essential properties are infinite in number and that not all of them are open to inspection, it will not be possible to investigate them all, so that the qualification cannot really ever be dropped.

We are now in a position to deepen our understanding of the possible influence of Leibniz's Law (3b) and his principle (9), or PIi, upon the later German idealists. Because of the difficulty of demarcating that domain of properties which underlies the notion of essential identity, we cannot exclude, and may even expect, that some philosophers who are well-disposed to the notion of an EI are unable to draw a systematic distinction between the *definiens* of (3b) in 16 and the *definiens* of (3c).[46] Anyone who is unable, consciously or unconsciously, to draw this distinction will conclude, since *ex hypothesi* the conditions for asserting EI and CE cannot be distinguished, that EI-judgments are indistinguishable from CE-judgments; that is to say: that

(10) $a \underset{\text{CE}}{=} b$ is true if and only if $a \underset{\text{EI}}{=} b$ is true.

If this is supplemented by Leibniz's PIi, then one has accepted the following principle:

(11) $a \underset{\text{EI}}{=} b$ is true $\rightarrow I(a, b)$ is true,

which is to say that essential identity excludes spatial two-ness. One then arrives at the following philosophical thesis: to a given essence, only one individual (in the metaphysical sense) can belong.[47] In this manner it is possible to explain the structure of Fichte's thought, which leads to statements about the identity of Coriscus and Socrates, as we have seen in Section 7.

In other words: (11) has the same effect as the Principle of the Non-Distinction of the Distinct and may be considered a variant thereof.

19. SIMILARITY, EQUALITY, AND IDENTITY IN LATER GERMAN TRADITIONAL LOGIC

It is not difficult to find the various notions of identity and of equality we have just discussed in the works of the logicians of two generations ago. For example, Eduard von Hartmann writes in his *Kategorienlehre*: "The highest degree of equality is *identity or sameness*, which excludes two-ness. There admittedly have to be two impressions for there to be anything to compare; but *the thing in itself* ["*das Ding an sich*"], to which these impressions are being referred transcendentally, may nevertheless be *one and the same thing*."[48] If the two impressions he refers to here may also be simultaneous impressions, and von Hartmann does not exclude this, then the passage contains a clear reference to a weak, possibly an essential, concept of identity.

Von Hartmann goes on to discuss the weak identity which Leibniz took as his point of departure when stating his PIi and which we have called "complete equality". He writes: "Equality ["*Gleichheit*"] is here understood *in the strict logical sense* as being of the same kind in all respects, or of identity in that wider sense of the word *whereby no numer-*

ical identity is required.... Similarity ["Ähnlichkeit"] is, according to the common [*i.e.*, the non-mathematical] use of language already a mixture of equality and difference; the same images which are *the same in some respects* [relations, "Beziehungen"] *and different in others.* They are called equal or unequal according as reflective thought picks out the one or the other kind of respects in explanation. Everything depends on the point of view" – compare Fichte's "*X*", and also Leibniz's thesis that relations are (merely) intensions – "taken by our reflection when it compares the objects or things and, accordingly, an opposite result may ensue... From this it follows that equality cannot be explained at all unless the concept of difference also plays a role" (italics mine). After that follows a – faulty, I think – passage on Leibniz's principle PIi; then he goes on as follows: "Real things which are identical in the wider sense, *i.e.* completely equal, but which are nevertheless numerically different, can only be so because ["dadurch... dass"] they are constituted of numerically different, albeit equal atoms.... The primitive atoms of the same kinds... become numerically different by being at different places, *i.e.* by having a different eccentricity *of position* in the space of the real world.... numerical identity and mere equality with numerical difference are often confounded..." (1923/30–34; italics his). Here von Hartmann clearly has STI and CE in mind. "Numerical difference in case of an otherwise complete equality has to be either a difference of place or of time... The numerical difference, then, depends on the *Principium individuationis*, that is to say upon the difference in spatial and temporal relations, in as much as it is not also guaranteed by other kinds of inequality" (34).

This shows, first, that von Hartmann regards the notion of the complete equality, or CE, of two things as philosophically interesting. Second, that he also regards as philosophically important a notion of identity or sameness ("Dieselbigkeit"), which he takes to refer "transcendentally" to a *Ding an sich.* Third, that he explains numerical *difference* as difference of spatial or temporal co-ordinates, which according to common philosophical usage are not "transcendental" characteristics, so that this notion of difference cannot be meant as the complement of the notion of sameness, the latter being not the notion of numerical identity but the notion of essential identity, EI. And, fourth, that he regards sameness, or EI, as the highest degree of CE. This last point has also been noticed by Windelband: "Thus E. v. Hartmann, for instance, treats identity or "sameness" as the "highest degree of equality", and so posits a mere relationship of degree between the two: on the other hand, Cohen likes to stress the fundamental difference between the two expressions..." (1910/4).

Erdmann agrees with von Hartmann in holding *Identität* to be the highest degree of equality: "It [the characteristic of identity] should therefore not be interpreted as a *kind* of equality, *e.g.*, as "absolute" equality.... It is not a kind, but *the limit* of equality, which the latter approaches more and more the slighter all possible differences between two objects become, and *which is reached when* all possible differences cease to exist ["fortfallen"] and *the two objects* ["Gegenstände"] *merge* into one and the same. As a limit it thereby remains just as peculiar and independent as every [other] limit is with respect to its variables, which may come infinitely near it without ever reaching it, except by giving up its being ["Wesen"] (1907/241; my italics).

Erdmann cannot very well have meant that numerically different things, two persons, for instance, could in principle merge into each other and form one physical thing. In all probability this is not meant as a conceptual transition from equality to STI; here, too, we have to interpret "identity" as EI if Erdmann's words are to have any philosophical significance at all. Here he does not seem to have the Leibnizian CE in mind, though later on he says: "But the content of two objects ["Gegenstände"] is the same ["der gleiche"], when the determinations that are presented in the one are also posited in the other. The equality is c o m p l e t e in so far as this concerns *all* [my italics] properties of the objects under comparison. In the case of simple objects the equality of their content in the narrower and proper sense... can only be a complete one; *the predicative equality of content, however, that is to say the equality of any kind of relation, may be incomplete* even in this case [my italics]. The incomplete equality may be called l o g i c a l s i m i l a r i t y ["Ähnlichkeit"]. Complete equality is the upper limit of the incomplete [equality]" (364).

Von Freytag's discussion of kinds of equality and identity is the last to be quoted here: "Identity is not equality. For the latter is a borderline case of similarity ["Ähnlichkeit"]. Similarity is a relation between two *relata*, but one thing can only be identical with itself. Therefore similar things may be almost equal, but nothing can be almost identical [what about partial identity?]. Two real objects can highly resemble ["ähneln"] each other, two objects of thought may be completely equal, but each of them is identical only with itself" (1961/16).

There are, then, degrees of equality (E), with CE as the upper limit. This is a fairly unproblematic notion. Neither of these authors offer a description of the systematic connection between CE and EI in traditional logic, and the difference between EI and STI is not sharply formulated. This may be explained in the manner of the last section.

What I have tried to show in the present section is, first, the presence of a notion of the complete equality, CE, of *two* things in the thought of some authors on traditional logic, and, second, to point out the necessity, whenever identity is discussed, in some sense of the word "identity", of a more precise use of language than is common in traditional logical circles.

20. "Is" AND "THE": THE NEED FOR AN ID-LOGICAL CONTINUUM OF ARTICLES OR COPULAS IN ORDER TO DISTINGUISH DEGREES OF ESSENTIAL IDENTITY

With respect to the traditional logical systems which we have called "id-logic" we are faced with a number of problems. One of these, problem (iii) below, has already been formulated earlier in this chapter, and we now bring it into connection with our investigation in Sections 14–17 above.

(i) Is the chaotic situation which we have observed in traditional literature with respect to the validity of the argument form *the/an M is P, S is an M, hence S is P*, and which we have described in previous chapters (especially in Chapter VI), systematically related to the fact that the connection between the notions of complete equality (CE) and essential identity (EI) is so unclear?

(ii) What is the nature of the conceptual relationship between CE and EI?

(iii) Which theory of 'the copula of the judgment' is hidden behind the chaotic state of affairs with respect to the argument form mentioned under (i)? We are primarily interested in systems in which indefinite propositions play a fundamental role and in which the H-thesis is assumed to be valid.

The survey of traditional theories of the copula which was offered in Sections 3–5 was built up from such explanations as can be found in the traditional literature. This survey is, however, quite insufficient when we want to explain the state of affairs in the traditional theories of inference. It will therefore be necessary to formulate supplementary hypotheses about traditional conceptions of "is". I have already hinted at these hypotheses in Sections 11 and 13.

It seems to me beyond doubt that question (i) should be answered in the affirmative. It will be remembered that Angelelli characterizes the traditional theory of predication as one in which a notion of something called "essence" plays a fundamental theoretical role. When in that logic the copula "is" is held to be an expression of something called "identity", then no doubt this has to be an essential identity in the sense described above. We saw, furthermore, that whichever of the traditional "theories" of the copula is chosen, the traditional interpretation of individual propositions *the S is P* invites one to rest content with an analogy – in the sense of opaque similarity – in cases where a strong total identity (*hic et nunc*) is required in order to avoid fallacies. This fits in with the hypothesis we formed on the basis of the material in the last paragraphs, *viz.* that a notion of essential "weak" identity is more fundamental to traditional logic than the notion of strong total identity.

All of this taken together suggests the hypothesis that, in id-logic, a proposition expressing a judgment of analogy is regarded as a proposition whose truth is *grounded in* an essential identity, and therefore as a proposition which expresses this essential identity. Concerning problem (ii) above, we assume that, in id-logic, a proposition of similarity *the first S resembles the second S* is regarded not only as an expression of a degree of "external" equality, with CE as the maximum, but at the same time, in another logico-theoretical dimension, as an expression of a degree of "internal" essential identity, with EI as maximum.

This degree of essential identity may then play the role of a *necessary condition* for that similarity or analogy which is perceived in nature or in the mind. Neo-Platonic logicians often speak and think in terms of *degrees of necessity*.[49] A judgment that possesses the highest degree of necessity can only be a judgment of, *i.e.*, about, maximal essential identity. In neo-Platonic logic every judgment is regarded as a judgment about an analogy or an identity. In that logic, therefore, there corresponds to each degree of essential identity another meaning of the word "is", the copula of the proposition, and hence another degree and/or kind of judgment-copula. The terms that are related by "the" copula of a judgment are the so-called logical subject and the so-called logical predicate of the judgment.

If our characterization of traditional notions of identity is correct, then clearly any one proposition (*the, an*) *S is P*, or (*the, an*) *M is P*, in which *S*, *M*, and *P* are terms of a natural language, *represents as many judgments as there are degrees of essential identity*. The most likely hypothesis is that the adherents of such a logic have presupposed a continuum of degrees of essential identity.

We must, then, assume that a logophoric proposition *ıM is P* is a proposition whose subject term *M*, or *ıM*, has a "philosophical", "virtual", "natural", "essential" *suppositio*, and at the same time *that there are infinitely many degrees of this kind of suppositio*. However in the various expositions of the traditional theory or theories this is brought out badly or not at all.

Our hypothesis immediately explains the problem stated under (iii) above. If there are infinitely many degrees of "is", then of course the argument form in question is not universally valid. Suppose that in a certain case it were established that the major premiss is an expression of the maximal degree of essential identity. One might then say that the inference is valid (although perhaps only after a certain modification of the conclusion), but we would in that case have made use of an extra assumption. Unless this assumption is added to the two premisses as a third premiss of the inference, it will be necessary to inform the listener, by means of a special sign ın one of the only two premisses which are allowed by the rules of syllogistic, whether the "is" of the major premiss is meant as an expression of maximal EI or not. Otherwise he cannot be expected to understand what is said to him or what he reads. But in the logical tradition this is never done. One can think of various means of making this explicit. We could, for instance, install *an ordered sequence of indexed propositional copulas*. Another possibility would be to make use of *an ordered sequence of articles*, and a third possibility is a combination of both devices at the same time. If we want to be able to speak of an essentialistic *logic* and not only of an amorphous piece of imagery, we should have to construct logical rules for each one of these copulas or articles.

Hartmann (1942/203–208) offers a good survey of the model behind the traditional logophoric logic, a model he personally rejects.

McTaggart's work is of great interest in this connection, although he did not present it as a study in logic but in metaphysics. McTaggart was, I believe, the first person to give a fairly accurate description of what I take to be the kind of order characteristic of the old metaphysical model. In the second volume of his work *The Nature of Existence* (1927), McTaggart discusses the kind of order that obtains among the terms in his C-series, as he calls it. He means by this a very fundamental series of metaphysical entities of some kind, about which he says explicitly that they are ordered by means of a connected, transitive, and asymmetric relation. This series clearly constitutes a special metaphysico-logical dimension. McTaggart owes his modern terminology to the philosophical works of Bertrand Russell, to whom he refers (*op. cit.*/51 n. 1; part I, 1921/83).

In McTaggart's view, we humans experience the direction of this non-temporal dimension as the difference between the temporal relations Earlier and Later. McTaggart is none too clear about the nature of the elements of the C-series. Nevertheless he relates the existence of this series to *error*, *i.e.*, to the fact that human beings may be and often are *mistaken* about many things. This entitles us to regard his description of the C-series as relevant to the history of the metaphysical trend of traditional logic.

21. ID-LOGIC AS TOPOLOGICAL LOGIC

A logic in which the truth or falsity of a proposition or judgment is relative to a parameter or "position" ("locus") in some space or dimension ("range of positions") is often called a "topological" logic (cp. Rescher 1968/229). Systems of chronological logic, systems of modal logics of "possible worlds", and many-valued systems as well may be subsumed under and treated as topological logics.

The hypothesis which we formulated in the last section as an explanation of the missing logic of the ἀδιόριστοι may now be expressed in the following manner: those logics in which ἀδιόριστοι play a central role may be understood as attempts at formulating a topological logic in a space with a dimension of a special "logical" or philosophical nature, which is to say that it was intended to be a dimension independent of physical space and measurable time. *Ex hypothesi*, an indefinite judgment has a "position" or co-ordinate in this logical dimension, and the same must be assumed to hold of each concept separately. I shall from now on use the letter "ι" as a variable for such co-ordinates, and the italicized letter "*ι*" as a variable for names of such co-ordinates. We shall have to think of this co-ordinate as a function of certain arguments, but leave open the question of what the nature of these arguments are. In an intensive identity- or inherence-theory of the copula, the range of positions, that is to say the set of values that this function can take, must clearly have infinitely many members. It will be safest to assume this range to have the power of the continuum. Furthermore, we obviously have to assume that it is ordered by a relation Higher Than, so that it will be possible to speak of higher and lower values of the function ι, or perhaps of more or less internal and more or less external values. We may call "ι" a level-variable. Finally we assume that ι has one extreme value, λ, being the "logical" co-ordinate of the innermost or highest level, the Logos-level, in a hierarchically structured field of judgments and concepts and, in addition, another extreme value, being the logical co-ordinate of the most external, lowest, or, in some meta-

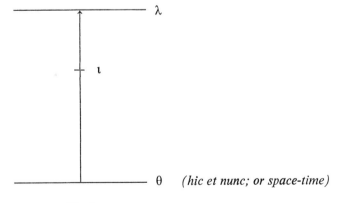

Fig. 3

physical interpretations, the actual level. Following Rescher we shall call this lowest extreme "θ".

The hypothesis that for every property M, designated by an expression *M*, there exists an extreme λM, or '*λM*', is in agreement with the classical philosophical principle which Beth calls "Aristotle's Principle of the Absolute", and from which Plato's theory of Ideas can be derived (Beth 1965/12f.).

We shall identify both the propositional form

(1a)   *the*$_{abs}$ *M is P*

and the propositional form

(1b)   *the M as such is P*

with

(1c)   *λM is P*.

But on the strength of the literature on traditional logic it is not possible to make out precisely what it is that is modified by the operator "ι". Using parentheses one can distinguish four possibilities characterized by four different forms of our symbolic language:

(2a)   *(ιM) is P*,
(2b)   *ι(M is P)*,
(2c)   *M is$_\iota$ P*,
(2d)   *M is ι (λP)*.

Should these linguistic forms be regarded as synonymous? The first suggests that a certain intensity, or degree of Being, or whatever one might like to call it, is attributed to the "logophoric" subject term. The second formulation expresses that the judgment '*M is P*', or perhaps we should say: the truth of this judgment, is realized at the level ι, or: to the degree ι. The third corresponds to the notion of a graded copula, and in the fourth, ι occurs as the name of a degree of participation ("μέθεξις", "Teilhabe") of M, or '*M*', in λP. Where the distinction between a concept and a judgment is considered to be of little or no importance, as is the case in the logic of Hegel and in that of Pfänder, the distinctions between (2a)–(2d) certainly collapse. But other traditional logicians may have been of the opinion that these distinctions are of some systematic importance. I shall write

(3) $\iota\, M \ is \ P$

without parentheses so as to leave open the question which of the four forms (2a)–(2d) the various logicians and other philosophers had in mind.

In id-logic, an individual proposition *this M is P* should be understood as an expression of a judgment at the level θ, that is to say as synonymous with

(4) $\theta M \ is \ P.$

The similarity between (1c) and (4) explains how the interpretation of individual propositions as propositions with a *vis universalis* could have been maintained for so long.

22. Concept-pyramids and internal relations: topological logic of internal genera and species

We are now in a position to explain the meaning of the expressions "logic of inherence or of predication" and "logic of relations" (Maritain), or "what-logic" and "relating-logic" (Veatch), as well as the original meaning of the expressions "internal relations" and "external relations", which were frequently used by Moore and Russell. To this end we first have to complete our sketch of a specifically logical space by adding a tree-like configuration or "pyramid" of concepts '*S*', '*M*', '*P*',..., or a system of such trees, each tree or pyramid being rooted in a *summum genus* at the level λ. The relation of a concept '*P*' to a concept '*S*' lower down on the branch of the tree is a reflexive and anti-symmetric relation which may be called "partial identity". In the majority of works on traditional logic it is assumed, mostly tacitly, that such (systems of) "concept-pyramids" form the basis of logical thought in all respects. Leisegang writes extensively about such pyramids (1928/200f.). In his textbook on the logic of concepts van den Berg, following Porphyry, connects this taxonomic form of thought with "degrees of assertability" ("graden van zegbaarheid"; 1946/139f.). But van den Berg does not attempt to develop a logical theory of such degrees either. Von Freytag's logic is founded upon the same model (1961/34f.). Jacoby puts it as follows: "... the ordering system of these identities is the concept-pyramids, in which the deductive inferences clamber" (1962/15). This should be compared with the reflexive and transitive relation (3b) which we defined in Section 6: 'S' ⊃ 'P'.

The logic of Fichte should also be understood against this background. The relation which Fichte takes to be the basis of all predication is the reflexive and symmetric relation of similarity ("Ähnlichkeit") between two concepts 'A' and 'B' situated on two different branches of a tree, both branching off from one and the same concept 'X'. The definitions (3a) and (3aa) in Section 6 belong to this conception of predication.

We now combine our hypothesis of a logical dimension in traditional logic with the traditional assumption that logic should be based entirely upon the idea of concept-pyramids. The over-all view which then results is that of *a topological logic of genera and species "internal" to a given concept (given by means of a term of some natural language)*. This is in accordance *e.g.* with the logical vernacular of Jacoby: "the identities and non-identities that are described in deductive inferences are internal ones" (16).

It is clear that the "logic of inherence" is a logic of *internal relations*, being the *logical* relations of inclusion, exclusion, etc. among "internal" genera and species, which are understood as classes of marks (characteristics, or properties).

We shall refrain from any metaphysical interpretation of this logic and try to propound the systematic differences between inherence logic and modern logic in metaphysically neutral terms.

Contemporary logic, the logic Maritain and Veatch call "relating-logic", is based upon the principle of polyadic predication: a predicate in a scientifically or philosophically fruitful language may have one, two, three, four, five, or still more "subjects", according to the needs of the users of the language. Predicates with more than one subject (or more correctly: with more than one argument) are employed in oider to express the existence of relations between the elements of the domain of entities under consideration (which may very well be of a high degree of abstraction).

Traditional logic rests upon the principle of monadic predication; this is to say that every proposition is assumed to contain only two terms, a subject term and a predicate term. Now such propositions certainly cannot be used to express and describe *all* kinds of relations between two or more entities. Let us call relations between the elements of the domain of discourse "external relations". We can then say that by means of subject-predicate propositions one cannot express or describe *all* external relations. If one still wants to do justice to the fact that relations, and among them non-symmetric relations, obtain between the entities in the domain under consideration, then obviously the postulate of a logical dimension, "internal" to the concept (to the meaning of a term), suggests a kind of solution. This postulate enables us to assume that every external relation "really" is, or "reflects", a relation between two entities (meanings, concepts) which are found at a logical level ι, higher than or internal to the logical level of the terms of the external relation. (From a strictly logical point of view, the latter level is always the extreme level θ, whatever the nature of the elements of the domain.) *Ex hypothesi* this logical relation between two higher concepts '*S*' and '*P*' can always be described by means of two-term propositions *S is P*. In this way the belief was established that relations between two things, say A and B, can be completely described by means of two-term propositions; but of course the terms of these propositions cannot then be simply the proper names of A and B, but must be some kind of description of the logical relationships in which '*A*' and '*B*' stand to other concepts. How these descriptions should be related to the names of A and B was never clearly worked out.

It was in this logical setting that earlier philosophers pondered about the Locke-Berkeley problem about the role of an example in the demonstration of a statement with general import.

We can sum up the above by saying that because of the dogma of monadic predication, traditional logicians were forced to find other solutions to the logic of relations and that this led them to a doctrine in which "*internal*", *logical relations* between "internal" genera and species played – albeit with moderate success – a role which in contemporary logic is taken over by polyadic propositional functions. Of such propositional functions it is sometimes said that they express "external relations" (Russell 1964/224).

23. THE THESIS OF ID-EXISTENCE

What was said about similarity ("Ähnlichkeit"), equality ("Gleichheit"), and identity in Section 17 can now be systematized in the following manner. At the θ-level there may be *similarity* (E) between two entities. Maximal θ-similarity is external *complete equality* (CE). We now assume:

to every degree of similarity (expressed as: $a_1 \underset{E}{=} a_2$) there corresponds a degree, ι, of essential identity, in this sense that some judgment 'ι...' is a necessary condition (Beth 1942/19) for (the truth of the) judgment '$a_1 \underset{E}{=} a_2$'. This is to say: *there is* an ι with this property.

This thesis, which we shall call "the thesis of id-existence", is presented here as a hypothesis about id-logic.

If we want to bring some order into the remarks about similarity, equality, and identity which we found in the literature, then the best possible account seems to be the following:

at *each* level, ι, one can have *similarity* between the concepts (or meanings, or other kinds of entities). For each level ι the highest possible degree of similarity is ι-*equality*, which again is a *degree* (ι) *of essential identity*. The series of equality-relations: θ-equality (CE), ..., ι-equality, ..., has λ-*equality* as its final or maximum term. λ-equality is nothing else than (*maximum*) *essential identity* (EI). Or, as Hegel puts it, "identity ... which is to and for itself" (1967, II/35). Equality at any other position than λ is more or less "external identity" (*loc. cit.*).

However, as long as it is not possible somehow to determine the ι-values of judgments or of concepts or both, or at least to *order* them, and as long as there is, consequently, no manner in which to communicate these values and their relative ranks to other persons, this topological conception of internal genera and species is not a theory but only a fairly primitive piece of imagery. If the same definite (or indefinite) article "the" ("a", "an") is used as a common name for every value of "ι", whether high or low, then the notion of a logical dimension simply has the effect of an escape clause. We shall return to that in VIII-16.

It is well known that many traditional logicians of a neo-Platonist bent were strongly influenced by the doctrine of the logical τοποι or *loci* in the classical theories of arguments and rhetoric. We cannot go into the history of that doctrine here, but it seems quite obvious that this doctrine must have been one of the causes of the doctrine of a logical space with one specifically logical dimension.[50] The doctrine of the logical τοποι became fused with the doctrine of a philosophical, virtual, natural, or essential *suppositio* of a *terminus per se sumptus*. Instead of the small number of discrete *suppositiones* assumed by the schoolmen, this led to what Ong aptly calls "a dialectical continuum" (Ong 1958/205).

One important question about this dialectical continuum, to which an answer is nowhere to be found, is the following. Take some general term M, say the term "man". We may safely identify the highest level λ at which "the" concept 'man' can be "taken", with the philosophical, essential *suppositio* of the term "man". But is this λ also the level of the *summum genus* in the (system of) concept-pyramid(s)? Or is it only one of the "local" λ's in a larger logical field, which is similarly ordered in pyramidal fashion and which takes its origin in the one Logos-level λ_{max}? For any logic which is built upon the topological concept-pyramidal conception it is clearly of paramount importance that this question be answered in considerable detail; but the answer is missing from every exposition of traditional logic.

24. PARTIAL IDENTITIES: WUNDT

I shall now try and give an impression of how the notion of logic as a theory of identity, or of "partial" identity, has fared in later German idealism, represented by Wundt, his sometime pupil Husserl, and Husserl's disciple Pfänder.

Wundt's systematic basis contains both the term "identity" and the term "partial identity". In his theory of judgments he speaks about "reduction to judgments of identity", and holds that "when we arrange all the relational forms of judgments [in the sense of the Kantian table of judgments, E.M.B.] according to their relation to the judgment of identity, then the alternative and the complete disjunctive judgment must be classified as special forms of the judgment of identity" (1893/234). About alternative and complete disjunctive judgments he further says: "They arise from the identity-formula $S=P$ by decomposition of the predicate-concept into two or more parts ["Glieder"]: $S=P_1+P_2$ is alternative, $S=P_1+P_2+\cdots P_n$ is disjunctive" (*loc. cit.*). Here "disjunctive" should clearly be read "*aut*-junctive", just as in the work of Boole and all other traditional logicians before De Morgan. A decomposition of *the* meaning of a predicate P should therefore yield:

S is either P_1 or else P_2 or else ... or else P_n.

Wundt does not discuss the relationship between this *aut*-junctive kind of decomposition and the conjunctive decomposition of concepts that can be found, for instance, in Hegel and in so many other traditional philosophies (cp. XI-1, 2 below).

In the opinion of Wundt, "a judgment of *subsumption* is closely related to a judgment of identity; it may be understood as a *partial* judgment of identity, as in this judgment the subject is posited to be identical with only a *part* of the predicate: "*S* is equal to a part of *P*"" (234). In our notation: "the" concept '*S*' is identical with a part of that class of marks which is "the" concept '*P*'.

Wundt introduces a special symbolism which I shall explain here because it is of some importance for the evaluation of his treatment of substitution in connection with analogies (opaque similarities), to be discussed in IX-10. He writes: "When we assert that the concept in question should be taken only partially, by adding the letter v to a concept-symbol, then accordingly the identity-formula $S = vP$ corresponds to the complete judgment of subsumption" (*loc. cit.*). We have earlier written "τ" where Wundt has "$= v$". Some authors, like Vaihinger and von Freytag, express the same thing by means of a line between "*S*" and "*P*".

Since Wundt offers no definition of his notion of *the* identity of a concept, it is understandable that he is extremely vague about the process of "taking" a concept "partially". He does not even illustrate the connection between the use of "$= v$" and the *aut*-junctive decomposition of concepts. He refers to Boole, who used this sign, "$= v$", for class inclusion. Nevertheless, he considers *the* content of concepts to be of primary logical importance. Our main objection is of course that Wundt gives us no criterion for when two terms represent the same conceptual content. He thus completely fails to identify the basic entities of his logic.

The specific mark of a *definite entity* ("eine bestimmte Grösse") is in Wundt's opinion its measurability. In addition to definite entities he recognizes *indefinite entities*, which therefore must be unmeasurable.[51]

This is to say that the "factor" v can in no way be *measured* and that it consequently remains undetermined. What one can do, however, is to indicate the extreme cases by "0" and "1": "The values 0 and 1 correspond to the borderline case in which the concept disappears and that in which it is thought in its entirety; *in all other cases it has an indeter-*

THE IDENTITY THEORIES OF THE COPULA   273

minate value v lying between 0 and 1. If we should want to represent only the quantitative aspect of the concept, we should therefore always have to put only one of the three symbols 0, v, or 1 for it. However, if the quantity and the quality are to be indicated simultaneously, then it has to be represented as a product of two factors, where the values 0 and 1 and the indeterminate value *v* are considered quantitative qualifiers of the concept. Thus *0.A*, *v.A* and *1.A* are the three forms into which a concept whose quality is *A* may occur, when we imagine it to be decomposed into its quantitative and its qualitative factors" (261; English italics by me). I have translated the word "Determinator" by "qualifier" and I shall translate "Determination" as "qualification".

Wundt takes the formula $S = vP$ to be the logical form of a universal judgment. A particular judgment is in his opinion an equation of the form $vS = vP$, in which "v" stands for an indefinite quantum (236).

Wundt's form *vA* is strongly reminiscent of our *ιA*. Now he also introduces a notation which is meant to represent the construction of complex concepts. He takes, for instance, the logical form of the expression "white sheep" to be *b.A*, *i.e.* a logical product of one variable for a "qualitative" qualification, the letter "*b*", and a second variable, "*A*", for the concept that is to be qualified. For these letters we may consecutively substitute "white" and 'sheep", or "good" and "man", which yields, Wundt says, "*a* good man". Similarly if we take "in the moon" as the value of *b* and "the man" as the value of *A*, we shall obtain the concept expressed by "the man in the moon" (253).

These examples illustrate the haphazard manner in which many traditional logicians used the definite and indefinite articles. There is no system behind their usage, and no marked difference between logophoric and other usage.

Although no definition of the expression "the concept '*A*'" is given, Wundt means that this general analysis allows him to split off a *part* of *the* concept '*A*': "The concept *b.A* separates, from the whole concept *A*, that which is at the same time *b*: every qualification thus presupposes that the qualified and the qualifying concept are divisible" (259f.). This, too, shows that Wundt certainly has more in mind than the divisibility of the *extension* of the concept which was discussed by Boole.

Wundt believes, then, that any *concept* can be decomposed into a *quality* and a "factor", v, which is of a *quantitative* nature, albeit undetermined and undeterminable; we might as well say: *indefinite*. The qualitative factor is itself often the logical product of a main concept '*A*' and a qualitative, or qualifying, "factor" b. Putting the quantitative factor, or rather its name, in front, he thus arrives at the expression "*vbA*", which he takes to illustrate the general construction of a concept, with all its logically relevant qualifications. It is noteworthy that "*b*" is a small letter, just like "*v*". Wundt himself says: "Thus *v* itself is connected with *A* in the same manner as a qualifying concept, *i.e.* *vA* may again be regarded as a product of the factors *v* and *A*. In fact it is clear that quantification may be regarded completely as a quantitative qualification. The difference between this quantitative qualification and the qualitative one, however, is that the latter requires a great number of different symbols, in accordance with the great variety of qualifiers, while for the former only one indefinite symbol *v* is required" (258). These words tend to confuse the meanings Wundt wants to give to the expressions "quantitative qualification" and the kinds of divisibility that go with them.

It seems that Wundt is drawing on two sources at the same time, *viz.* on the Boolean logic of classes and products of classes as well as on the Platonist logic of graded μέθεξις. But no connection between "*v*", "*b*", *and the definite article* can be made from the passages quoted here, nor from any other parts of his book. It seems quite clear, however, that the combination "*vb*" is intended to express what in our notation is rendered by "ι".

It should be observed that the notion of quantitative qualification of a concept which Wundt discusses here hinges, of course, upon the traditional interpretation of logical "quantification" as quantification of terms, which we have discussed in IV-8.

25. HUSSERL ON IDENTITY AND RELATIONAL PREDICATES

Husserl calls the relation from individual objects ("Gegenstände") to a general entity "methexis" and "participation" ("Teilhabe"). He characterizes it as "a quite peculiar relation of identity" (1948/392), but does not describe the properties of this relation. About equality he says that it is nothing but "a correlate of the identity of a generality" which may be "herausgeschaut" (*op. cit.*). That may be true, but illuminates nothing.

In his *Formale und transzendentale Logik* Husserl also seems to treat relational predicates as monadic, *i.e.* to analyse *aRb* as *a is Rb*, in accordance with traditional grammar. He first describes the logical forms of the judgments 'this paper is white' and 'this wall is whiter than this (same) paper', as *this S is P* and *this W stands in the relation p to this (same) S*. Strangely enough he says that these are the logical forms of the said sentences "in Aristotelian formalization". His subsequent comments concerning the logical grouping of the words in the second sentence betrays that he has indeed an Aristotelian analysis in mind. "Upon a closer inspection... formal partitions ["Formschichtungen"] also reveal themselves to us in [their] essential generality ["Wesenallgemeinheit"], in judgments of every form. ... *This paper* or the form *this S* occurs once as subject form, as the form of the substrate which is about to be made more specific... and a second time as an object form *inside the relative predicate*" (1929/267, last italics mine). This must mean that, after all, the term *this S* is not one out of *three* terms in a judgment of the form *aRb*, but that the logical form which he has in mind is the Aristotelian two-term form *a(Rb)*, more specifically: *this W (p this S)*, precisely as described in Section 11 above.[52]

26. PARTIAL IDENTITIES IN PFÄNDER'S LOGIC

About identity and equality as copulas Pfänder says the following: "Thus in the judgment "Gold is a metal", the predicate-concept "metal" is not identical with or equal to the subject-concept "gold", but is the higher species-concept belonging to it" (1963/193). This is not Fichte's copula (3a) or (3aa) (Section 6) but the "Leibniz-copula" (3b). With a certain reservation Pfänder supports the partial intensional identity theory of the copula; his view is that "the law of identity... may be extended yet further" (*loc. cit.*). He offers the following description of the copula in the judgment 'Gold is metal': "The relation of subject-concept to predicate-concept which we have before us here, may be called partial identity, in so far as a "part" of the subject-concept – in our example the concept "metal" – is identical with the full predicate-concept. However, here the relation in which the "part" stands to the whole in the subject-concept is a special one, namely that of the higher species-concept to its subordinate concepts. In other cases this relationship of "partial identity" has another sense, but even so *the extended Law of Identity* holds" (202f., my italics).

But Pfänder does not formulate any extended Law of Identity. It is impossible to guess from his text what principle he has in mind. This is all the more regrettable as many traditional logicians hold very strange opinions concerning substitution, opinions which in all likelihood are based upon some unclear extension of the notion of identity and upon equally unclear, more or less Leibniz-like principles and definitions.

Following Leibniz, Pfänder holds that when the partial identity is due to the predicate-concept being a part of the subject-concept, then the judgment at issue is necessarily analytic. In his opinion the Law of Identity says precisely that these logically analytic judgments are necessarily true.

But since to Pfänder "partial identity" does not always mean the same thing, he cannot, and does not, regard all true judgments as analytic in the logical sense. There are also judgments which are not *logically*, but *ontologically analytic*. Pfänder's example here is a mathematical one: '*The plane triangle has three inner angles.*' The (logophoric) subject-concept of this judgment does not contain the predicate-concept, "but on the other hand one only has to analyse more closely the very object [*the triangle* – E.M.B.] that is meant by the subject-concept in order to recognize that it necessarily has three inner angles" (160). – "These essence-analytic or ontologically analytic judgments, however, do *not* fall under the Law of Identity, since their predicate-concept is neither wholly nor partially identical with their subject-concept" (*loc. cit.*).

Finally Pfänder assumes the existence of judgments that are synthetic (in the sense of non-analytic) both from a logical and from an ontological point of view. About these judgments it is also wrong to say that the predicate-concept is partially identical with the subject-concept. An example of such a judgment is: 'The moon is now hidden by clouds' (197).

Assuming that Pfänder presupposes a logically relevant hierarchy of levels, this would clearly be a judgment at the θ-level, while ontologically analytic judgments may be ranked as judgments at the λ-level. In order to investigate what the properties of the copula in the latter two kinds of judgment are, i.e. in the ontologically analytic and the ontologically synthetic judgments, we turn to that section of Pfänder's work where he maintains that in addition to a subject and a predicate, a copula is required. This section is part of his second chapter, called "Nature and construction of the judgment in general". His exposition turns out to consist of a series of impressive but ultimately empty statements. No theory of the logically important copula or copulas is presented here either. The only information the author gives us is that *the* copula ("der Kopulabegriff") has a double function. It has "the function of relating the predicate-determination to the subject-entity ["Subjektsgegenstand"] in general", and "at the same time it carries out the assertive function" (43).

This is followed by some unclear remarks about "claims", "sealing", and the "imperative sealing" of judgments, and by the statement that "every dictatorial gesture, the slightest oppression of the object by the judgment, is a sin against the spirit of the judgment and pollutes the intellectual conscience" (43f.). That may be so, but no conclusions can be drawn from this about the properties of "is" and their implications for the theory of inference.

The same verbal hollowness is found in Pfänder's reflections on "the so-called modality of the judgment" (92f.). I have already referred to this section above, in connection with the problem of relative generic judgments (VI-7). Here he mentions (i) the logical modalities of the judgment, (ii) the ontological or objective modalities of the state of affairs ("Sachverhalt"), and finally (iii) the psychological modalities of the act of asserting. About "the objective possibility" Pfänder offers no information, not to mention a useful theory; we learn that this is still another question belonging to ontology and not to logic (94). In fact, Pfänder is here only echoing Kant. The rest of this chapter is formulated in terms of "the logical weight of the assertive beat", "the damping ["Dämpfung"] of the assertive beat" and "degrees of this damping", "size of the still remaining excess of logical weight of the assertive beat", and so on. More interesting is the information that "this logical damping function, then, joins the copula" (96) and that the logical "damping" has a certain *size*. His only description of the logical "damping" function is the following: "If one of the premisses of an argument is problematic, then, even if the other one be assertoric or apodeictic, the conclusion will be only problematic, and it will be problematic to the same degree as the problematic premiss" (346). Arguments (inferences) show a *modality fall*: the modality of the conclusion is clearly the lowest modality occurring in

the premisses and is therefore equal to the modality of the most problematic premiss. This pertains to problematic modality. When the conclusion follows from the premisses with necessity, an apodeictic modality
is, according to Pfänder, superimposed upon the problematic modality.

All this induces a mental picture quite similar to the one that I sketched
in Section 21. What Pfänder says about a logical "damping" function
joining the copula is, in any case from a dialogical point of view, equivalent to saying that the copula of predication has degrees, and that the
"is" of the conclusion is of the same degree as the "is" in the most
problematic premiss.

But we are not told how to determine the degree to which a premiss
is problematic. Pfänder avoids every controversial or difficult question
by referring to a discipline which he calls "ontology".[53]

The notion of a problematic judgment and the use of this expression
by traditional logicians will be discussed in Chapter X.

27. REVOLT AGAINST THE INTENSIVE IDENTITY-THEORY: LOTZE ON THE COPULA AND THE GENERAL DOG

In the first volume of his *System der Philosophie*, the *Logik*, Hermann
Lotze (1817–1881) makes some very critical remarks about the id-logical
account of the judgment-copula. Personally, he adopts the extensional
identity theory of the copula. This becomes clear from his analysis of
the proposition "some human beings are black". Lotze holds that
"one... *means* from the start only those who are black, in short: the
Negroes; only they are the true subject of the judgment" – and not, as
the id-logicians would say, some aspect or modification of a concrete
universal concept 'Man'. "It is also quite clear that what is meant is
not the predicate in its generality, but only that blackness which occurs
on human bodies..." (1912/80). In his opinion the sense of this judgment
is that "some human beings, by which only the black human beings
should be understood, are black human beings". It is a *complete* identity;
the judgment-relation accordingly has no degrees, in Lotze's opinion.
Contrary to the schoolmen and others, Lotze regards this as an identity
of *content*; but this seems to be a mere matter of words. Using scholastic
terminology one may say that the extension of the grammatical subject
term in personal *suppositio* and restricted in accordance with the predicate

term is identified with the extension of the predicate term, *viz.* the class of Negroes. Lotze reminds us that a Latin rendering of this judgment would have the predicate in the plural: "nonnulli homines sunt nigri". He concludes that it is "the lack of inflexion in the German expression" which prevents some people from seeing that this interpretation of "is" is the right one" (*loc. cit.*).

Lotze interprets even indefinite propositions *the/an M is P* extensionally: "Furthermore we say: the dog barks.[54] But the general dog does not bark: only a definite individual dog, or many or all individual dogs, are the subject of this proposition. But the predicate, too, is meant differently from what the form of expression might suggest:... it does not simply bark, always and eternally, but now and then... in short: the dog which we mean in this judgment is really only the barking dog, and the same barking dog is also the predicate" (80f.).

Most of the examples which are found in logic books are, Lotze thinks, categorical judgments: "Gold is heavy", "the tree is green", "the day is windy". It is sometimes said that the predicate of a categorical judgement is "simply" ("schlechthin") predicated of the subject. But it is not immediately clear what this is supposed to mean. Lotze's answer here amounts to a denial of the Fichte-Hegel view. He holds that "among all connections of S and P the complete identity of the two ("beider") would be the one which would most clearly deserve to be called simple. But it is just this connection which is generally *not* meant in a categorical judgment; the proposition: Gold is heavy, does not state that gold and heaviness are identical, nor do the propositions: the tree is green, the sky blue, state the equality of the tree and greenness and of the sky and blueness" (74). As Peirce puts it: "Lotze and Trendelenburg represent the first struggle of German thought to rise from Hegelianism" (II/234; para. 2.387). Scholtz's critique of Lotze's logic certainly does not do him justice (Scholz 1967/61). Since Lotze wrote his *Logik* before Erdmann wrote his, it is not surprising to find that he does not express himself about the partial intensional identity-theory of the copula.

Contrary to Husserl, Lotze sees Plato's doctrine of a "participation" of things in eternally immutable general concepts or ideas as an inadequate answer to a metaphysical question rather than as a piece of information about the logical relation between the subject and the predicate of a judgment. In Lotze's opinion, Aristotle did not solve this problem either. True, Aristotle laid the foundation for a workable future theory by stating "that words are above all *said* of their subjects".[55] In the works of Aristotle one does not find the mistake made by later logicians when they weakened the connection between S and P to a connection between the image which the speaker holds of S and his image of P. But Aristotle did not really formulate a theory of "is", and Lotze expects the meaning of the copula "to remain for a long time the motive force behind future developments of logical activity" (72).

Lotze, then, certainly does not consider "pure logic" a completed science. At the time of the second edition of Lotze's *Logik* in 1880 the evolution which he had predicted a few years earlier was already well under way. One of Lotze's pupils was Gottlob Frege.

28. REJECTION OF PARTIAL IDENTITIES: SIGWART

Christoph Sigwart (1830–1904) also finds it necessary to express himself about the difference between identity and predication in his *Logik*. In the first part of the third edition he writes: "This relation of the thing to its properties and activities has likewise already been subsumed under the concept of i d e n t i t y; but here, too, one has required of this concept a versatility which it does not possess... the thing... is not identical with its properties, nor with its activities; it is not these properties and activities themselves; the vermilion is not identical with its redness, and the sun is not identical with its shine; and the principle which is supposed to legitimize the judgments: vermilion is red, the sun shines, cannot be called [a] Principle of Identity" (1904, I/114).

This makes it clear that he is no follower of Fichte's and Hegel's logic. But he goes further than this. In a polemic footnote directed against the logic of his contemporary Wilhelm Wundt, Sigwart explicitly declares that, contrary to Wundt, he does not recognize any partial identity either (486 n). He is prepared to use the expression "identitas partium", for instance with respect to certain parts of Europe and certain parts of the Russian empire, but not "identitas partialis". He obviously fears, or perhaps he even observes, that the notion of identity between a part of one thing and another thing (*viz.* (3b) in Section 6) is confused with the notion of a degree of similarity between these two things (or concepts; *viz.* (3a) in Section 6). Identity has, Sigwart says, no *degrees*; the expression "partial identity" as well as the expression "relative identity", which other authors use to indicate degrees of identity, are in his opinion self-contradictory (110).

29. CONCLUSION

This clearly refutes von Freytag's and Jacoby's description of traditional logic, a description which implies that it was one homogeneous problem-free science, namely the science of identity and partial identity between intensional concepts.

In fact, some of the most serious traditional logicians of the nineteenth century were pronounced antagonists of the intensional, and *a fortiori* of the intensive identity-logic. "Pure logic" never was the completed science von Freytag, Jacoby and others are trying to re-instate.

30. APPENDIX: VEATCH ON IDENTITY AND SENILITY

In Chapter I we already met H. B. Veatch as a protagonist of some variant of traditional logic which he does not describe systematically. In one of his attacks on modern logic, however, he demonstrates a surprising lack of knowledge of that science. I am referring to the following argument, taken from his most recent work.

Veatch starts out from the assumption that someone were to say: "Bertrand Russell is senile", and immediately adds: "... what surely no one would ever say" (1969/32). He then asks what Russell's own analysis of a proposition of this simple kind would be. His answer is that according to Russell himself, if this proposition is true, then Bertrand Russell has the quality Senility. Veatch, however, rejects this interpretation of the given proposition as "patently ridiculous" (*loc. cit.*). As it turns out, Veatch believes that what he takes to be the Russellian understanding of this sentence can also be rendered by the proposition "Bertrand Russell is senility". But the latter proposition is ridiculous only if "is" is interpreted as identity (which, incidentally, is what we are accustomed to do in ordinary language when both terms are grammatically nouns, neither of them an adjective). Now to interpret "is" as identity is, as we have seen, a characteristic of great parts of *traditional* logic, and one which has been severely criticized many times by modern logicians. It is to be regretted that Veatch did not – at least not visibly

– take notice of Angelelli's work of 1967, where this is discussed in several places. And in the study by H. Lenk, published the year before Veatch's, von Freytag is criticized for holding the predicative relation to be "a relation of identity" (Lenk 1968/32). But predication is no longer seen as the act of expressing an identity, and the set-theoretical version of Veatch's example, *viz.* "Bertrand Russell *is an element of the* class of all senile people", is neither more nor less ridiculous than "Bertrand Russell is senile". Neither predication nor the ε-relation has the properties which may be expected of a relation called "(partial) identity".

In Chapter I we met Veatch himself as a supporter of the identity-interpretation of the predicative "is". It is possible that since the publication of his first book Veatch has changed his opinion about "is"; certain places in his last book may seem to point in this direction (1969/33). It seems more likely, however, that he holds an *intensive* identity-theory of the copula, and that he makes the mistake of imputing to modern logicians, who do not recognize degrees of identity, the view that if "Bertrand Russell is senile" is true, then Bertrand Russell $\underset{\text{EI}}{=}$ senile. In the notation we introduced above, this may be written: "Bertrand Russell is$_\lambda$ senile", or "λ Bertrand Russell is senile". Veatch himself, on the other hand, holds – if my analysis is right – that the example in question only warrants the interpretation that Bertrand Russell is senile at the θ-level, which in the same notation should be expressed as follows: Bertrand Russell is$_\theta$ senile".

NOTES

[1] I here use the words "intensional" and "extensional", and not "intensive" and "extensive", because I want to reserve "intensive" for those theories of the judgment-copula in which it is an intensive, *i.e.* a gradual relation.
[2] Cp. Geach 1968/18f.
[3] Cp. Section 28 below.
[4] Cp. Bocheński 1956/62. See Erdmann 1907/335f. about W. S. Jevons.
[5] Cp. De Rijk 1967 II, 1/183ff.
[6] Cp. Section 4 of the introductory chapter.
[7] Cp. VIII-n. 23 below.
[8] "The subject is the predicate": See Section 7 below.
[9] See the commentary by L. Nelson (1962/468).
[10] Cp. also Lenk 1968/202, 207, and X-6 below. I have not adapted Fichte's use of symbols in my quotations from his work to my own use of symbols, nor vice versa.
[11] This policy of carrying out substitutions on the basis of arbitrary judgments which

are not distinguished from proper identities can be found in so recent an author as W. Wundt: cp. IX-9. See also note 30 below.

12 This is also the opinion of R. Havemann, who is an admirer of Hegel: "The development of thought starts with processing the elements of formal logic.... Thinking starts with the insoluble dialectical contradiction between identity and difference. All natural phenomena contain this insoluble contradiction. They are identical with one another and still different from each other. As we have already seen in our exposition of the Platonic doctrine of ideas, trees are admittedly all different from one another, but at the same time they are also identical with one another. In a certain respect, in a certain abstract sense they are equal... For the logic of our thinking a real comprehension of the dialectical unity of equality and difference is an elementary condition, the fundamental first step..." (1964/48f). Like Fichte, Havemann, who is a professional physicist and not a logician, does not distinguish between "identical" and "equal in a certain respect". The fact that trees resemble each other in certain "respects" (or "aspects"), and differ in other "respects" can perfectly well be dealt with without resort to expressions like "a dialectical unity of identity and difference", at least if one is prepared to part with the Aristotelian framework of monadic terms. Cp. n. 31.

13 "The riddle of the dialectical method is herewith completely solved," Leonard Nelson writes. "That which at first sight seemed paradoxical, nay completely absurd, now turns out to be quite comprehensible, even self-evident. For indeed: "There is everywhere nothing at all, in which" according to this method "the contradiction, that is to say opposite determinations cannot and should not be pointed out" "(1962/469). Nelson is here quoting Hegel. What Nelson says about the logic of Hegel is the truth, but perhaps not the whole truth; cp. n. 25 below, and Chapter VIII, n. 16.

14 One might think that since Fichte presupposes degrees of identity, Nelson's interpretation of this example does not entirely do him justice, for given Fichte's "elastic" notion of identity it is not necessary to assume that he meant the "identity" of Coriscus and Socrates in the sense that they are one person and totally indiscernible. But we are here concerned about the consequences of Fichte's theoretical logic and not about his personal ability to count up to two. Fichte and others have produced systematic errors which indeed lead to the theoretical result that Coriscus and Socrates are one individual; cp. also Angelelli 1967/119. We will discuss this further under 3, in Section 18 of the present chapter. See also n. 47.

15 C. S. Peirce takes "therefore", not "is" as the logical relation *par excellence*. "Consequently, the copula of equality ought to be regarded as merely derivative" (1967, III/299). "The forms of the words *similarity* and *dissimilarity* suggest that one is the negative of the other, which is absurd, since everything is both similar and dissimilar to everyting else" (I/304). According to Peirce, dissimilarity is not the same as otherness.

16 Cp. Angelelli 1967/142. The fact that Husserl was a student of Wilhelm Wundt for some time should not be forgotten.

17 According to Beth, in alethic logic the logical truths of second and still higher orders are exceptions to this rule (Beth 1965/362, 1967/102f.; Barth 1968a).

18 Cp. Wundt 1893/193f.

19 Most present-day logicians are inclined to regard identity as a relation between elements of language rather than as an "ontological" relation or property; see, *e.g.*, Lenk 1968/332. Weinberger holds a related view: "I am of the opinion that the identity-relation cannot be defined in the object-language, but that it can nevertheless be described and defined by means of equivalent conditions in the metalanguage" (1964/7).

In order to make this clearer I shall use syntactic variables and also explicitly take up the predicate "true", in Sections 16–18 below.

[20] For P one may choose any propositional function with one free variable. Thus (2) may, for instance, have the following form:

$$\sim [R(a, b_1, ..., b_n, t) \ \& \ \sim R(a, b_1, ..., b_n, t)].$$

Cp. n. 45 below. Cp. also von Wright 1969, Prior 1969, Barth 1969.

[21] Hegel tried to do just that; cp. Barth 1970b. Cp. n. 13 above.

[22] After the publication of the original (Dutch) version of this book (1971) there appeared a most interesting study of Plato's theory of relations by H.-N. Castañeda (1972). Castañeda's article contains (see especially pp. 473f.) obvious points of similarity with the attempt to understand the neo-Platonic "romantic" theory of relations which is undertaken in Barth 1969 (especially Section 5). I shall not comment further upon Castañeda's article here, except to remark that he does not incorporate the notion of *gradual* participation into his reconstruction of Plato, which therefore seems to me to be incomplete. – An interesting discussion of important phases in the history of the logic of relations is also to be found in Weinberg 1965.

[23] It is worthwhile to compare this with the intellectual struggle to work out a theory of change in the fourteenth century; cp. *e. g.* Wilson 1960/18. Cp. VIII-n. 23 below.

[24] "In fact Hegel retains only a gradual quantitative difference between the general concept and the particular being" (L. Nelson 1962/460).

[25] About mystic "logic", see Stace 1961/213f. Underlying the tradition which Stace characterizes as the "mystic" tradition there is in my opinion a motive which deserves to be called rational, namely the importance of asymmetry, *i.e.* of the "direction" between different logical rôles; cp. Barth 1970b. But compare this also with (11) in Section 18 below.

[26] Ong holds that the logic of Peter of Spain already bears the marks of decline: "The key word in Peter's explanation of ratio is thus not truth at all, but confidence or trust (*fides* – understood as the rhetorical and dialectical term, not the theological virtue). Here, already, reason appears in almost its full eighteenth century panoply. It is a way of getting at truth which is sure, common sense, and by implication and association, thoroughly scientific; but when actually pinned down to explanation, somehow it is only instinctive. Here in the thirteenth century, when the goddess of reason makes her most definitive appearance in scholastic philosophy in the most distinctive and influential of all scholastic manuals, she is supported not on the pillars of science, but on the topics or arguments of a merely probable dialectic or rhetoric" (1958/65). Peter of Spain, then, mixes topical and rhetorical elements into his logic. In this Boethius preceded him; cp. the remarkable article by Bird (1960). Moody sees the origin of what he calls "metaphysical logic or logicized metaphysics" primarily in the neo-Platonists Augustine and Avicenna. Assuming all of this to be correct, we must conclude that metaphysical logic originated through influences both from neo-Platonist philosophy and from rhetoric. This process culminated in the sixteenth century: "In the 16th century Cicero's teachings are often combined with Platonist dialectic" (Risse 1964/72 and *passim*). Cp. IV-14.

[27] Erdmann refers, among other things, to Aristotle's *De Anima* III, 6 430a 27. Cp. n. 4 of our introductory chapter.

[28] An orthodox contemporary discussion of this can be found in Geach 1968/37.

[29] See Nelson's criticism, 1962/142–146.

[30] It seems to me that in contemporary textbooks not nearly enough attention is paid

to the problem of deductive substitutivity and the way in which this problem differs from the problem of describing heuristic methods.

31 "Phenomenology, including that of Husserl, differs from the *Logische Propädeutik* of Kamlah and Lorenzen precisely in this, that in the former, identity is identity in difference" (Eley 1969/373f.). Cp. n.12.

32 Cp. Section 3. Cp. also Maritain 1933/270.

33 Cp. Russell 1964/224f. For a discussion of disjunctive combinations of proper names, cp. Geach 1968/68.

34 The expression "enantiomorph" is taken from Bennett 1970. I have discussed the convertibility (symmetry) of logical and of non-logical relations in a number of traditional schools elsewhere (Barth 1970b) and shall not take up this topic again here. See also John of St. Thomas 1965/496, 619 n. 18. Cp. n. 16 of Chapter VIII below. See, above all, the remarkable result reached by Cobham (Cobham 1956), based on an article by Quine from 1954. Cobham proves that under certain set-theoretical conditions, every theory which is formulated in a language with the structure of first-order predicate logic (quantificational logic) may be translated into a theory containing as its only primitive predicate a *symmetric* dyadic predicate. According to Quine 1954 this had to be an asymmetric predicate. Professor M. Löb drew my attention to this article by Cobham. The philosophical import of Cobham's result has not yet been worked out. In any case D. Scott's recommendation (in Lambert 1970/143) that we should not always dismiss a certain language because a simpler language can be shown to fulfil the same needs (usually in a more complicated way) seems relevant also in this connection.

35 Beth's list of Absolutes is discussed by Zelený (1962/45f., n. 20), who accepts Beth's analysis of their genesis in every case but one: Marx's value-jelly concept. Unfortunately Zelený fails to offer a single argument against Beth's analysis here.

36 Cp. Maritain 1933/37, 41. Veatch neatly describes what I take to have been the philosophical interpretation of the theory of the logical *vis universalis* of singular terms, for he speaks of "causation as a suggested means of logical mediation in the syllogism" (1970/302f.). He holds that "the function of the middle term is to intend the cause (or effect)" (308).

37 Cp. XI-8.

38 Cp. also the expression "intellectual energy" in the recent commentary on the material logic of John of St. Thomas (see John of St. Thomas 1965/625).

39 This is precisely what Sartre opposes: "Behind the act there is neither power, nor *"exis"*, nor virtue" (1957/12). Sartre uses the same word for power, "puissance", as Maritain. "... it often seems that what Sartre is trying to do, while continuing to use for his own purposes the philosophical vocabulary of rationalism and idealism, is to work out ideas that at many points have clear affinities with tendencies in contemporary pragmatism and sometimes even in analytical philosophy" (Olafson 1967). Sartre rejects the ontology (the model) behind the traditional logophoric philosophical language, while continuing to use this language himself. Cp. also Naess 1967.

40 Cp. John of St. Thomas 1965/617, n. 11, 12, 13.

41 One might also think of replacing the sign for "material" equality in the right-hand term of (3) by a sign for "strict" equality (see for this notion *e.g.* Hughes and Cresswell 1968). If one does, Mates 1968 should be taken into account. I shall not do that here, but will follow in broad outline Weinberger's exposition, in which no modal notions are discussed. I do that in order to bring certain of the problems with which Leibniz and his contemporaries were struggling, like the strictures imposed by the syntax of Aristo-

telian syllogistic, more clearly to the front. As a result of these strictures many earlier logicians took refuge in vague notions of modalities with which an arbitrary use of the articles was attended.

[42] When o and a are given, the predicate "Dist (x, o, a)", to be read: the distance from x to o is a, is a monadic predicate, while "Dist (x, y, z)" is a triadic one.

[42a] It is certain that Leibniz tried to analyse propositions with non-symmetric predicates as *reduplicative* propositions; cp. his own definition of "reduplicative proposition" (see Angelelli 1967b) and, *e.g.*, the examples discussed by Mates (1968/521). As Angelelli points out (*op. cit.*), at certain places in his oeuvre Leibniz makes it quite clear that he wants to exclude precisely all reduplicative expressions from the range of "*K*" (or "*P*"). This makes Weinberger's analysis all the more interesting.

[43] The Marxist notion of the "dialectic" of the process of acquiring scientific knowledge should be analysed in terms of the domain of the predicate-variable "*P*". The revision of a scientific theory usually implies that the range of this variable (or: these variables) is changed, and therewith the identity of the objects under investigation. Cp. Geach 1972/238ff., on "relative identity".

[44] It would be a worthwhile task, but a difficult one, to investigate the relationship between theoretical logic and opinions about space and time (absolute, relational, and relativistic theories) in the works of important philosophers. "Locus et tempus non variant rem interne" (Baumgarten in 1739; 1963/99). Cp. Aristotle, *Topics* VI 6 144b32. Cp. also Bennett 1970/178.

[45] The equivalence-class

$$\{\text{Saul-at-the-point-a-at-the-moment-t}\}_{\text{STI}}$$

may, if one so prefers, be called "*the* being-Saul-at-a-at-t". This entity, if we may call it that, has nothing whatsoever to do with an eternal essence, a "timeless validity of logical truths" in the sense of an "identical, enduring" *the A* in the individual Saul. I here make use of the vernacular employed by van Peursen in his description of Husserl's ambitions (van Peursen 1967/35, 42). Bolzano was the first to point out that (in traditional logic) time-determinations ought to be taken up into the subject term and not, as was usual, into the predicate term (cp. III n. 13).

[46] Hegel betrays that he finds this distinction difficult, by saying that the principle: "there are no two things, which are equal to one another", *i.e.*, Leibniz's principle PIi, is "contrary (opposite) to the Law of Identity" (1967, II/38).

[47] Peter of Spain, *Tractatus Syncategorematum*, p. 91: "... all particular men, due to the fact that they are men, participate in one nature and are reduced to the unity of the species. Whence Porphyry says that "by participation in the species many men are one man". See V-7 for the non-distinction of the meanings of "individual" and "species" in the logic of Ramus. Fonseca, however, "affirms that there is no identity between singular substances and their essences". Angelelli holds that in this respect Fonseca is an exception among the commentators on Aristotle's *Metaphysics*, book Z, 6 (Angelelli 1967/118).

[48] I believe that such a statement should be related to the topic of X-5 and X-6 below.

[49] Zabarella characterizes the apodeictic statements of Aristotle's proof theory as different degrees of necessity. "Ramus takes over this view from his teacher Sturm, after having at first rejected it. But he carries these statements from the theory of judgment and the theory of proof over to the whole of logic and makes them its material principle, as distinct from method, its formal principle" (Risse 1964/157).

[50] Cp. n. 26. Cp. above all Ong 1958/104f.

[51] Cp. IV-19.
[52] Cp. Pivčević 1967, Føllesdal 1958/41.
[53] Cp. Angelelli 1967/11, 27 n. 13; Specht 1967; Wesly 1970.
[54] I have had to change Lotze's example: "Der Hund säuft", since in English dogs simply drink, just like human beings. The original text is admittedly wittier.
[55] Cp. Ong 1958/107f.

PART 3

DESCENT

CHAPTER VIII

ARGUMENT BY ANALOGY

1. INTRODUCTORY REMARKS

Several authors have taken an interest in the use of the articles in Hegel's works, among them Leonard Nelson, M. Heidegger, and the Dutch authors J. A. Oosterbaan and W. van Dooren. Heidegger and Oosterbaan are both convinced that there is some system in Hegel's use of articles, especially in his use of definite articles in titles of chapters, but van Dooren has no confidence in the possibility of discerning any systematic features here (van Dooren 1965/143).

In fact Hegel expressed himself, though not in great detail, about the kind of importance he ascribed to articles and other constants for logic: "It is an advantage of a language when it possesses a large fund of logical expressions, *viz.* property expressions and abstract expressions, for the determination of thought itself; many prepositions and articles, even, belong to such relationships which rest upon thought; the Chinese language is said not to have developed this feature so far or only poorly; however these particles occur in a completely subordinate manner and are only slightly less extricable than the prefixes, the conjugation signs and so on. It is of far greater importance that in a language the determinations of thought can be *exposed* ["herausgestellt"] as nouns and verbs and thus be stamped as substantive ["gegenständlich"] forms; in this the German language has many advantages ["viele Vorzüge"] above other modern languages" (1967, I/10; my italics).

Hegel's preference for substantive or "objective" forms implies, for him, that in propositions aimed at describing the essential natures of things, "alle" must be replaced by a definite article "der", "die", or "das".

Nelson reproaches Hegel for not making use of an *indefinite*[1] article where he says about a certain animal "es sei Tier". "When we are talking about an animal, then, in as far as we have already learnt to talk at all, we certainly do not say, as Hegel maintains that we do, "es sei Tier" ["it is animal"]. We rather say that it is an animal. ... This meaning, which according to the grammar of our languages belongs to the article, namely the relation of a concept to objects falling under it, but which are not identical with it – this meaning of articulated language is unknown to Hegel" (Nelson 1962/460). This criticism is just, but not yet very penetrating. Nelson explains what he has in mind as follows: Hegel assumes that there is only a gradual – Nelson says: quantitative[2] – difference between a general concept and an individual that falls under it. If so, Hegel is no better and no worse than all other philosophers who assume a topological id-logic with a specific logical dimension. But in Nelson's opinion "no concretization of concepts will lead us to the individual, but only the direct knowledge given in observation. It is this [knowledge] to which the [indefinite] article preceding the name of the concept refers. And so we understand how, due to his own complete grammatical confusion, Hegel is prevented from understanding the relation between concepts and observation and that without it no knowledge is possible" (*loc. cit.*).

Now in modern logic the role of the indefinite article in a sentence of the form *S is a P* is expressed by means of the use of small letters for individual terms: *s is P*, or *P(s)*, and secondly by means of a – sometimes ingenious – use of quantifiers in combination with copula and identity-sign. In individual propositions the indefinite article in "this is an animal" may therefore be omitted. It is true that if in some systems of logic and language, *e.g.* in that of Hegel, quantified propositions are considered to be of less fundamental importance than indefinite propositions *der (die, das) M ist P*, it may reasonably be expected that those nuances in the use of the articles which belong to an articulated use of the chosen language, will not be jettisoned. But Nelson's criticism is not really to the point. For although we do say, in English as in German, "this is *an* animal", we also say simply "this is water", "this is mud". Nelson's remark about the importance of the relation of concepts to observation is no less relevant in these cases, although no use of the indefinite article is grammatically appropriate here. So Hegel may be forgiven for saying "this is animal" as long as he does not go from there to "this is *the* animal". That step, as we know, would take him directly into the unconditional validity of the argument form *the/an M is P, S is (an) M, hence S is P* and primitive paradigmatic logic.

2. THE ARGUMENT BY ANALOGY

We saw in Chapter VI that several more recent traditional logicians hesitate to pronounce themselves clearly on the validity of this inference form, and understandably so.

In Hegel's *Science of Logic (Wissenschaft der Logik)*, however, an argument of this form is presented *expressis verbis* and its validity is discussed at great length. The second part of this work, which Hegel called *Subjective Logic, or Theory of the Concept* treats of logic in the narrower sense. Its third chapter is *The Argument (Der Schluss)*, and here Hegel discusses the various kinds of argument which he recognizes as philosophically important. Starting from the *Argument of Existence*, of which there are four variants, he proceeds to more and more fundamental argument forms, via the *Argument of Reflexion* with three variants, to end up with the three variants of the *Argument of Necessity*. The third variant of the Argument of Reflexion is the *Argument by Analogy*, of which he gives us the following astounding example:

> Die Erde hat Bewohner,
> Der Mond ist eine Erde,
> Also hat der Mond Bewohner.

In English:

> The Earth has inhabitants,
> the moon is an earth,
> hence the moon has inhabitants.

The spacings are Hegel's own, so that we can be quite confident that his Argument by Analogy is precisely the first figure form whose validity has been the main problem in the traditional logic of the articles throughout this book, namely the form

> *the M is/has P*
> *S is an M*
> _____
> ergo, *S is/has P*.

Under the Argument of Reflexion fall, first, the *Argument of Allness*, or *Universal Argument*, secondly, the *Argument by Induction*, and, finally, the *Argument by Analogy*. Hegel's example of a universal argument is this: "All human beings are mortal, now Cajus is a human being, hence Cajus is mortal" (II/336). He regards an argument of this kind as trivial: "if, *by accident*, Cajus were not mortal, then the major premiss would not be correct... before the major premiss can be held to be correct, there is the question whether the proposition in the conclusion might not itself be an instance counting against it which must be answered in advance" (*loc. cit.*; my italics). Only a person who hopes to acquire what Nelson calls "knowledge from mere concepts", a knowledge which does not need to be *tested* in any way whatsoever, can use this as an argument against universally valid deductive argument forms.

The second variant in this group is the Argument by Induction. Hegel gives a much narrower meaning to the expression "argument by induction", or "inductive argument", than we do today: it turns out that in his vocabulary this comprises only the inductive method called *inductio per enumerationem simplicem*. It is therefore not surprising to find that he considers the use of still other kinds of argument to be required. In this we can certainly agree with him. The problem is how to give an account of those argument forms which are either deductively valid or useful in some other manner, without falling into PPL or otherwise blurring the distinction between valid and invalid arguments and also the very meaning – his own – of the logical constants which are in question.

3. THE LOGOPHORIC INDEFINITE MAJOR PREMISS

From the example about the inhabited earth and its similarity to the moon it is obvious that the definite article does have a function in Hegel's logic which it does not have in modern logic, but it is not immediately clear what this function is. About this argument Hegel says: "However its Middle [concept] is no longer a single quality, but a generality ["Allgemeinheit"], which is the Reflexion-in-itself of a concrete [entity, concept], hence its nature; – and conversely, because it is thus the generality as of a Concrete [entity], it is at the same time this Concrete [entity]. – Here, then, a single thing is the Middle, but according to its general nature" (339).

It should be clear from this quotation that Hegel characterizes the middle concept '*(the) M*' as general and at the same time as concrete; it is what in English is called "a concrete universal". His major premiss does not mean: "the one and only earth in the universe has inhabitants", since he uses "earth" both for this earth, Tellus, and for its moon, as becomes clear in the second premiss. It seems clear also that in Hegel's logic, too, the possibility of understanding or "taking" a term in a "philosophical" *suppositio* is a necessary condition.

Hegel does not present this argument as a proof that the moon is inhabited. He does not hold that an Argument by Analogy is reliable. "The analogy is the more superficial," he writes, "as the general [entity], in which both singulars agree ["eins sind"], and according to which the one [singular thing] becomes the predicate of the other, is a mere q u a l i t y or, as the quality is taken subjectively, some c h a r a c t e r i s t i c or other, [in other words:] when the identity of the two is taken as a mere s i m i l a r i t y". And then: "That kind of superficiality, however, to which a form of common sense ["Verstand"] or of reason ["Vernunft"] is carried by being degraded into the sphere of mere i m a g i n a t i o n, ought not to be brought into logic." The premisses clearly ought to express something more than mere θ-similarity, namely (some degree of) essential identity, between the subject-concept and the predicate-concept in the premisses.

4. ANALYSIS OF SIMPLE ARGUMENTS BY ANALOGY

The following definition is taken from Rescher: "An analogy, in the most general sense of the word, is a comparison revealing the resemblance between two or more entities, and to *draw an analogy* is to make a comparison that calls attention to such a likeness" (1964/276). This definition is very wide indeed, and wider than my definition of analogy in VII-17 as intensional similarity. But according to both definitions, to draw an analogy is to make a *comparison*, to point out an agreement or similarity between two or more entities; anyone who formulates an analogy formulates a comparison drawing attention to this similarity. An analogy, then, is not in itself an inference, but simply a statement of a certain kind. "An *argument by analogy*, as the name suggests, is one which is founded upon the drawing of an analogy" (Rescher, *loc. cit.*). Haenssler too, is of the opinion that we can reason *from, i.e.* on the ground of, an analogy, but that an analogy generally is not itself the result of an inference.[3] An argument by analogy is an argument in which at least one premiss is (or: describes) an analogy.

In the simplest form of argument it is assumed of two things, a and b, that they have the properties X, Y, Z in common, and that a also possesses still another property P, and from these two premisses the conclusion is drawn that b perhaps (possibly,...) has this property too:

(A) (1) Xa & Ya & Za &...
     (2) Xb & Yb & Zb &...
     (3) Pa,
     _____
     (4) *perhaps*: Pb (conclusion)

I shall say that an argument of this form is *a simple argument by analogy*.

Of such an argument it may be said that the individual a is chosen as a *paradigm* and that the argument as a whole is a *paradigmatic inference*. Aristotle himself called such arguments by the name "παράδειγμα" (*Prior Analytics* II 24), and reserved the word "ἀναλογία" for proportions between *four* entities: "the relation of a to b is the same as the relation of d to e".

The Latin translation of "πάράδειγμα" being "exemplum", it is not unreasonable to assume that the theory of simple arguments by analogy and the scholastic and Renaissance theories of expository syllogism and *exemplum* were closely and systematically related to each other, if indeed they can be distinguished at all.

5. REDUCTION TO TWO PREMISSES

Observe that apart from being deductively invalid, this argument form is not well-formed according to Aristotelian syllogistic. True, only monadic propositional functions occur (although they may of course be complex and/or "intensional", *i.e.* relational or modal). But there are at least four of them: a, b, X, and P, and the number of the premisses is three, not two. However, the first two premisses can easily be contracted into one:

(B′) (1′) *a resembles b in so far as both are X, Y, Z,...*
      (2′) *a is P*
      _____
      (3′) *perhaps/possibly b is P.*

In modern notation:

(B)   (1) $aR_{eq}b$ or: $bR_{eq}a$

(2) Pa

_____

(3) *perhaps/possibly: Pb.*

This argument, too, has four terms: a, b, P, and R_{eq}. The first premiss in schema B expresses that a and b stand in an equivalence relation R_{eq} to each other ("eq" for "equivalence"). Fichte treated the copula as a reflexive and symmetric relation. As we already know, he attributed the following verbal form to (1'): *a ist b*, or *a=b*. In the school of Fichte, therefore, the idea of taking the copula as a suitable name for R_{eq} of course naturally occurs to one.

(i) If the premisses of B' are written down in converse order, with *a is P* as the major and $aR_{eq}b$ as the minor premiss, then B' is seen to be an expository quasi-syllogism in the third figure, with *a* as the exposed middle term. But instead of a particular conclusion *some ... is P*, like in Darapti and in the *dictum de exemplo*, or an indefinite proposition *(the) ... is P* which some logicians allowed as a conclusion in Darapti, the conclusion in B' is *perhaps/possibly b is P*.

(ii) If the minor premiss in this expository quasi-syllogism is converted into $bR_{eq}a$, which is allowed since R_{eq} is symmetric, we obtain a first figure arrangement of the three individual terms:

(1) *a is P*	Or: (1) *the (individual) M is P*
(2) $bR_{eq}a$	(2) SR_{eq} *the (individual) M*
(3) *perhaps/possibly b is P*	(3) *perhaps/possibly S is P*

6. HEGEL'S STRESS ON THE IMPORTANCE OF FORM

Hegel's Argument by Analogy has a most surprising verbal form. It is not Darapti, even if the copula is assumed to be symmetric so that the minor can be converted; for then the subject term in the minor premiss becomes the term "*an* earth" while that of the major premiss is "*the* earth". But if we define an expository syllogism as any two-premiss three-term argument with one or two individual premisses in one of the Aristotelian figures, allowing the premisses to contain articles, as in the logic of the Renaissance, then his Argument by Analogy certainly is some kind of expository syllogism. The minor premiss, "The moon is an earth", is in any case an individual proposition.

Hegel made a number of statements about the form of his Argument by Analogy which are of some interest for us. He says: "It is also inappropriate to represent the major premiss so that it would sound as follows: What [*e.g.*, the moon] resembles an object [the earth] in some characteristics [X, Y, Z,...], also resembles it in other [characteristics; *e.g.* P]." If we were to start the argument with this premiss, we would be close to the analysis of simple arguments by analogy which is shown in schema A. We are not here looking for the right meaning of this premiss, however, nor are we arguing for or against this way of formulating an argument by analogy. We are only concerned with Hegel's own reason for rejecting this formulation: "In this way *the form of the inference* is expressed as a content [*i.e.*, as the content of the major premiss], and the empirical content which is rightly so-called is buried in the minor premiss... *But in the inference itself the empirical content is irrelevant*, and to make its own form the content of a major premiss, is as insignificant as if any other empirical content were chosen in its place... *What is important, is always the form of the argument*, whether it has this [form] itself or something else as its empirical content. Thus *the Argument by Analogy is a characteristic form*..." (340; my italics).

There is no denying that Hegel thought in a most formalistic manner.

One might think, he then says, that in arguments by analogy one is not dealing with a special form at all: "What may lead to this idea in connection with the Argument by Analogy, and perhaps also in connection with the Argument by Induction, is the fact that in these [arguments] the Middle, and the Extremes as well, are further determined than in a *merely* formal inference and [that] therefore the form-character, because it is no longer simple and abstract, must also appear as a characterization of the content" (my italics). This shows how important it is to him that the term "(die) Erde" should be able to stand for the "internal", or "higher", nature of our Tellus, while at the same time having a formal function in the argument, in the sense of being a recognizable part of it, occurring at definite places in the propositions from which the argument is made up.

But his notion of a form as the *formation of* the terms of a proposition or of an argument is tied up with the Platonic notion of the *Forms behind* these terms. The assimilation of these two concepts of form is most certainly inspired by Kant's program of developing a material or, as Kant preferred to say, "transcendental" logic, taking his table of (forms of) judgment as the starting-point. This program has been admirably analysed by Swing (1969) who holds that only two of Kant's commentators seem to have taken seriously his claim that he carried out his programme successfully. These two are Hegel and Peirce (Swing, *op. cit.*/vii, 26). I think it may be said that, in a critical sense, this also holds of Aebi (1947).

Hegel's Kant-inspired assimilation of the two meanings of "form" is unmistakeable when he concludes: "But this [fact], that the Form thus determines itself as content, is a necessary development of the Formal in the first place, and concerns the essential nature of the argument..." (*loc. cit.*). This notwithstanding, he demonstrates an almost unbelievable lack of freedom vis-à-vis the formal framework of Aristotelian logic, which he tries to retain even in arguments that are not deductively valid (*i.e.* not in the modern sense of "deductive"; cp. IX-11). It is worth-while to take a closer look at his struggle.

7. THE LIMITATION TO THREE TERMS

It is not immediately clear what the form of Hegel's Argument by Analogy has to do with an inference from premisses one of which is an analogy. But the fact that Hegel makes no attempt to justify the name of this argument form makes it unlikely that he can have chosen it without support from traditional principles. He does not seem to regard this name as problematic at all.

In a simple argument by analogy, an argument of the form A, there are at least four terms: a, b, X (possibly a conjunction of X with Y, Z,...), and P. In the modern B-form there are exactly four terms: a, b, R_{eq}, and P. The difference between individual and general terms does not concern us at the moment.

Hegel's respect for formalism, the quasi-syllogistic form of his own Argument by Analogy, and the presence of four terms in a simple argument by analogy call for a detailed comparison of the two argument forms and their relation to Aristotelian syllogistic, the strait-jacket within which European philosophers have been trying to perform their intellectual gymnastics for more than two thousand years. That system rests upon the principle that a basic form of argument which is what we today

would call deductively valid must contain three terms. Every deviation
from this principle was considered as a fallacy, the fallacy of four terms
or *quaternio terminorum*.[4] There had to be exactly two premisses, and
exactly three terms. The validity of Aristotle's inference rules, the valid
syllogistic forms, can be demonstrated only upon the assumption that
there are at most three terms each of which occurs twice, and each time
in precisely the same sense. In other words, if one term occurs twice but
in different senses, this also counts as a *quaternio terminorum*, and the
inference is invalid. By a (deductively) valid inference form we understand
a *completely safe* inference rule, a rule with the property that it will under
no circumstances lead to a false conclusion provided the premisses are
all true. Such an inference form (rule) is therefore *absolutely dependable*.

However, if one drops the requirement that the rule in question be
absolutely dependable, one cannot necessarily speak of a *fallacy* of four
terms, even if the Aristotelian framework is retained. The use of four
terms is only fallacious *when* and *because* one wants the rule to be
absolutely dependable. If somebody continues to speak of the fallacy of
four terms although this requirement is dropped (as in arguments by
analogy and other inductive arguments, by definition of "inductive"; cp.
IX-12), then this can only be explained as caused by irrational cultural
ties with the syllogistic tradition, perhaps reinforced by a personal
addiction to the number 3.

In Hegel's own discussion of his Argument by Analogy we read:
"When the form of an argument by analogy is considered in the said
formulation of its major premiss, that when two objects agree in one
or more properties, then any ["eine"] further property that one
of them has also belongs to the other, then this argument may seem
to contain four determinations, the *quaternio terminorum*; – a
circumstance which would entail the difficulty of bringing the analogy
into the form of a formal argument" (341). Here we have it in black
and white that Hegel regards the use of four terms as a transgression
of the conditions of well-formedness to be satisfied by the premisses. He
correctly observes four terms in a simple argument by analogy: "There
are two singulars ["Einzelne"], thirdly a property which is immediately
assumed to be common to them, and fourthly that other property which
one of the singulars has immediately, but which the other only obtains
through the inference" (*loc. cit.*). Since four terms are unacceptable on

the formal (syntactical) principles he adheres to, he cannot assume schema B′ to be logically fundamental. He is looking for a solution which will allow him to express arguments by analogy as arguments with only two premisses, and yet no more than three terms.

8. THE TAUTOLOGICAL PREMISS

Hegel explains the presence of four terms in the following manner: "This is due to the circumstance that, *as it has turned out*, the Middle of an analogical argument is *posited* as a particular entity ["Einzelheit"], but immediately also as its true generality… In the example given above the *medius terminus*, the Earth, is taken as a concretum, which according to its truth is just as much *a general nature* or genus as an individual ["Einzelnes"]" (*loc. cit.*; my italics).

In the first place the phrase "as it has turned out" is completely misplaced, for it has no more than a rhetorical function in this connection. Hegel's purpose is to explain why other people have thought that in order to express an argument by analogy four terms are needed. He explains it by pointing to the middle term in his own formulation, which he intends as a term with double meaning. This is to say that he reduces the two terms *a* and *X* of the A-schema to one equivocal term "die Erde" *by using the definite article in two senses*.

Hegel, being a post-Renaissance philosopher, does not offer a theory of *suppositiones* at all. If he had done so, he would clearly have had to say that in the major premiss his middle term "(die) Erde" has at the same time both *suppositio personalis* and some kind of "philosophical" *suppositio essentialis* or *naturalis*. From the point of view of the history of ideas and their cultural impact it is of great importance to recall at this point that the neo-Thomist philosopher Maritain postulates a *suppléance essentielle* which he classifies *under* his *suppléance personnelle*, and not, like the schoolmen, as contrary to it. His account of *suppositiones, i.e.* of the functions of terms in use, is clearly also a necessary condition for accepting Hegel's Argument by Analogy on his own terms. According to scholastic terminology the special kind of equivocity to which Hegel resorts in order to solve his problem is a *fallacia secundum univocationem*. The traditional meaning of this expression seems to have been the following. Even when each term is univocal in the sense that it has only one *significatio*, the syllogism in question can still be fallacious, namely when the *suppositio* of one of the terms is not the same in both premisses, or if its *suppositio* in the conclusion differs from its *suppositio* in the premisses. Risse (1965) explains the meaning of this expression as a fallacy "in which one homonymous ["gleichlautend"] concept can mean either a species or an individual".

Our last quotation from Hegel supports the impression that the reference of Hegel's middle term is both a general nature and an individual thing as well. And this "is" means, as we know, "is identical with", in some sense of "identical". The "identity" of the individual and the general nature can be a high one or a low one, or in Hegel's terms: it may be an *internal* or an *external* identity. (Note that Hegel's terminology is dichotomic rather than comparative.) This entitles us to blame him for making the fallacy of four terms, in spite of the said "identity" between the two meanings, the individual earth and the Idea in which it participates. Nevertheless Hegel sticks to the quasi-syllogistic form of his Argument by Analogy.

Pondering over this problem, I came to the conclusion that there are a number of principles in traditional logic, (a) to (d) below, which he may have confused in order to give an irrefutable proof that, formally speaking, there is after all no illegitimate *quaternio terminorum*.

(a) In his attempt to find a solution to the (sham) problem of the four terms, Hegel was no doubt helped by the circumstance that in the logophoric traditional logic every logophoric proposition *der (die, das) M ist (ein, eine) M* is a logical, and even a tautological truth. For this is *Plato's Principle* which we know from II-5, as clause (iv) of the H-thesis. One may add a tautological truth to the class of premisses of any argument without thereby influencing the inference relation. If we add a premiss of this form to the two premisses Hegel explicitly mentions, then we obtain an argument (C below) which turns out to be a special case of a simple argument by analogy, as characterized by schema A:

C.	Compare:	A.
(1) *the M is M*		(1) *a is X*
(2) *S is (an) M*		(2) *b is X*
(3) *the M is P*		(3) *a is P*
(4) *S is P*		(4) *(perhaps, possibly) b is P*

Here (1) is supposed to be (an application of) Plato's Principle. We then have an argument of the same verbal form as a simple analogy. In this argument, however, *the moon is not compared with this earth, Tellus, but with a logophoric concept 'the Earth'.*

The first premiss in C is not explicitly stated in Hegel's argument, but as I said this may be explained by reference to the fact that the premiss in question is a tautology, a "formal" identity. Hegel himself characterizes $A = A$ as a tautology, an empty identity, a law of thought which is without content (II/28f). He offers the following examples of this tautological form: "A is A", "a tree is a tree", "a plant is – a plant", or in the original German: "eine Pflanze ist – eine Pflanze", and "God is ["sei"] – God". All of a sudden he moves on to "die Pflanze". I shall quote a few lines from his discussion of this tautological identity: "For when, to the question what is a plant? the following answer is given: *a* ["eine"] plant is – a plant, then the truth of such a statement will immediately be admitted by the whole company in which it is tested, and at the same time it will unanimously be said that Nothing has been said thereby. ... ... When we take a closer look at the effect of boredom [resulting] from such a truth, then the beginning: the ["die"] plant is, – makes ready for saying something, for producing a further determination" (30). This shows that Hegel makes no difference between "eine Pflanze ist..." and "die Pflanze ist...", and that for him "die Pflanze ist eine Pflanze" is certainly tautological.[5] But if the identity relation between 'die Erde' and 'eine Erde' is a tautological one, it need not be expressed in an additional premiss; what is more, this tautology allows Hegel to identify the subject of (3) with the predicate of (2), so that he is left with exactly what he wanted, a two-premiss argument with three terms.

But if we assume that in Hegel's opinion the information expressed in the third premiss of C, *i.e.* the major premiss of his Argument by Analogy, results from "empirical" acquaintance with *this* earth, Tellus, then there is still the problem that the logical subject of (3), which is Hegel's major premiss, is not the same entity or concept as the logical subject of (1). If we write *this M is P* instead of (3), then clearly we have a fourth term after all. The assumption may be false or in need of a qualification, but even if it is true there is in Fichte's and Hegel's logic a way of getting around this difficulty.

(b) Hegel may have thought, and rightly so, that *this M is (an) M* is a tautology, and in his logic this means that it is a "formal" (read: verbal) *identity*. That implies, to start with, that "diese Erde ist eine Erde" could also be taken as an additional premiss, instead of (1) in C. This, too, results in an argument of the form A; but *a* is now the term "diese Erde", for which we often use the expression "die Erde". The sentence "die Erde ist (eine) Erde" may be interpreted as "λ Erde ist (eine) Erde" and also as "θ Erde ist (eine) Erde", both of which are in Hegel's opinion tautological identities. I think we may safely assume that he considered any statement of the form ιM is M to be a tautological identity, for any ι. The number of "identities" which are available to us according to "pure" logic, and which in all probability were assumed by Hegel, is in any case considerable, and suggests conclusions which otherwise could not be drawn. As the Hegelian van der Meulen puts it: "The earth is, to be sure, its [own] general nature, it is [a] celestial body, hence *an* earth" (1958/81). All these "identities" together make it possible to manoeuvre in such a manner that from a "formal", or verbal, point of view one cannot be accused of a *quaternio terminorum*. By calling all symmetric relations "identities" one can delude oneself into believing that important theoretical or practical results have been obtained.

9. ANALOGICAL TERMS AND ANALOGICAL LANGUAGE

(c) In IV-6 and in VII-17 we mentioned the doctrine that "analogical" terms and an "analogical" use of language is of great positive value in philosophy. "Analogous" is here a monadic predicate, applicable to terms in the sense of parts of language, an analogous term being one which is neither really equivocal nor unequivocal. According to this doctrine there is a continuous transition from completely unequivocal to completely ambiguous terms, the field in between being the domain of analogical language. A term is used analogically when it is ambiguous in such a way that its two or more meanings somehow resemble one another. The nature of this relationship is not easily described. It seems that the following senses of "an analogical(ly used) term" are all part of the later traditional meaning:

(1) a term which is used metaphorically,
(2) a phrase or a context (propositional function) defining an inten-

sional, or opaque, similarity (VII-17),

(3) a term which is used with more than one *suppositio*,

(4) a systematic ambiguity (type ambiguity, in the sense of the *Principia Mathematica*; cp. IV-6).

More likely than not (1)–(4) were conflated in the minds of traditional logicians,[6] including Hegel. (2) can probably be subsumed under (1). In connection with Hegel's Argument by Analogy (2) and (4) are of special interest. The traditional understanding of what I have called "intensional similarity" is relevant, since one cannot expect to be able to discuss the similarity of the moon and the earth without using spatial and other relational predicates, including non-symmetric ones. A sentence *the M is P* indeed shows a kind of systematic ambiguity, for the definite article is interpreted as λ but also as θ; however, unlike the Russellian theory of types, later traditional – but not scholastic – logic assumes a continuous transition from θ to λ.

It is highly probable that Hegel, in search of the right analysis of arguments from analogies between two entities, considered "analogical" use of language to be a necessary requirement which is in no need of a special justification.

It ought to be possible to reason about relations and to investigate relations. The instrument called Aristotelian syllogistic is however entirely unsuited to these tasks, and Hegel and many other traditional philosophers seem to have felt this, rather than understood it. That explains why many Thomists and other philosophers could become so interested in "analogical" language, which undeniably is a kind of ambiguous language, and also why they regarded it as a metaphysical, not as a logical topic and technique. In fact it *is* possible to reason about and to investigate relations of many kinds. If one believes that this *can only* be done by speaking in an "analogous" or otherwise imprecise manner, then the idea may occur that the use of ambiguous language in argument is a rational and highly productive method.[7]

10. APPLICATION OF THE OCKHAM-WALLIS-KANT-WOLFF REDUCTION TO THE ARGUMENT BY ANALOGY

(d) In Chapter VI we studied the conditions for the validity of the argument form in question when it is assumed that singular propositions

can be reduced to universal propositions. We concluded that it is a
necessary and sufficient condition that *the M* and *an M*, or *der (die, das)*
M and *ein(e) M*, are interchangeable terms, *i.e.* that one can be substituted for the other whenever we might wish to do so. Applying this to the
example of the moon and the earth we arrive at the condition that "the
earth" and "an earth" must be interchangeable. In Section 1 and in
Section 8 above we have seen that Hegel indeed makes no systematic
distinction between these two kinds of expression.

Summing up our remarks under (a), (b), (c), and (d), we conclude that
in the eyes of Hegel historical, terminological and even "formal" factors
reduce the danger of his ambiguous use of the term "die Erde" to a minimum. That "analogical" language is required precisely in arguments by
analogy, if at all, is no more than one could expect; each of the two
interpretations of *the M is an M* is a formal identity, so that 'the earth',
'this earth', and 'an earth' must be identical concepts. Applying the
Ockham-Wallis-Kant-Wolff doctrine of the *vis universalis* of singular
propositions we can therefore reduce the Argument by Analogy to a valid
Aristotelian syllogistic form. Thus Hegel can say: "In the above example
the medius terminus: the earth, is taken as a concrete [entity] which
according to its truth is just as much a general nature or genus as an
individuality. [Seen] from this side the *quaternio terminorum* would
not make the analogy an imperfect inference" (341).

As far as I can see Hegel here is stating that there are no *formal*
objections to his argument form, when the fundamental principles of his
logic are accepted. *Formally* an Argument by Analogy is, for Hegel, a
valid argument; the fact that it is, after all, not a dependable argument
must be explained in another way than by reference to its "external"
form. We will return to this question in Section 12.

So the argument: "The Bantu is primitive, Saul is a Bantu, ergo Saul
is primitive" can, in Hegel's logic, be characterized as a *formally* valid
argument, although it is possible that it is a superficial one, too "external"
to be dependable.

In other words, *formally* speaking – both in his sense and in the present
sense of "formally" – Hegel's logic can be subsumed under PPL (cp)
VI-2), which is to say that from the standpoint of a (critical or uncritical.
listener, Hegel's logic cannot be distinguished from a primitive paradigmatic logic.

11. BURBURU

Hegel's Argument by Analogy is formulated in that Aristotelian figure which Kant held to be the only "perfect" one, hence the only figure suitable for the formulation of fundamental *vernünftige* philosophical inferences (cp. V-12). Since Hegel's philosophy is an attempt to carry out Kant's program of constructing a science of material ("transcendental") logic, this ought to be no surprise (cp. Swing 1969/vii).

Both Überweg and Trendelenburg maintain, however, that Hegel considered his Argument by Analogy as an argument-form in the second Aristotelian figure, and that this is called the third figure in Hegel's own numbering (Überweg 1882/442, Trendelenburg 1862, II/332). Trendelenburg further holds that in classical logic an argument by analogy cannot be formulated in the second but must be expressed in the third Aristotelian figure. This is, as will be recalled, the figure of Darapti and the primary figure of expository syllogisms. But Trendelenburg rightly goes on to say that Hegel quite clearly *formulated* his Argument by Analogy in the first Aristotelian figure, whatever his intentions may have been. It seems, then, that in Hegel's logic not only the first and the third, and the second and the third, figure are being confused, conflated, or given each other's numbers, but, what is more important, also the first and the second figure. It is therefore extremely fascinating to learn from Kneale and Kneale that already Petrus Ramus, "for no obvious reason, makes Aristotle's second figure his first and Aristotle's first his second" (1962/304).

Hegel characterizes scholastic philosophy as a whole in terms reminiscent of Ramus as "eine barbarische Philosophie des Verstandes", *i.e.* "a barbarious philosophy of the intellect" (1928/198f.; cp. Ong 1958/59), and he does not seem to exempt the logic of the schoolmen from this verdict. Ramus, on the other hand, receives favourable treatment: "He... contributed greatly to the simplification of the formalism of the dialectical rules", Hegel says (*op. cit.*/252). One could not agree more. Ramus's contribution to logic was, however, a "random simplification" (Ashworth 1970/20). In all probability Hegel was strongly influenced, consciously or unconsciously, by Ramus in his logic, both in its dichotomic basis and with respect to his use of articles.

The verbal form of the Argument by Analogy may correctly be called by the name which Piscator constructed for Ramus's *syllogismus proprius* in the first figure: "Burburu" (cp. IV-15, V-7). The following facts demonstrate that this name is appropriate. Ramus spoke of the dialectical art as *ars disserendi*. In the German translation of Ramus's *Dialectique* by Beurhaus in 1587, "disserere" is translated by "die Vernunft... zu gebrauchen" (cp. Ong, *op. cit.* /180). In the seventeenth and the eighteenth centuries the term "Vernunftlehre", or "Vernunftkunst", *i.e.* "the art of Reason", became the usual name in Germany for that complex of problems which we now call "logic". Incidentally it was Hegel who brought about the rehabilitation of the word "Logik" in Germany by choosing this word and not "Dialektik" or "Vernunftlehre" for the title of his *Wissenschaft der Logik*, notwithstanding his great liking for and his frequent use of the word "Vernunft" (cp. Scholz 1967/10f.). Hegel's logic was, one might say, *eine burburische Logik der Vernunft*.

12. THE IMPERFECTNESS OF THE ARGUMENT BY ANALOGY AND THE PROBLEM OF DEPENDABILITY

Hegel did not regard his Argument by Analogy as formally imperfect. However he adds: "But from another point of view it [the *quaternio terminorum*] makes it so..." (341), *i.e.* in a certain sense the presence of four terms makes this argument form imperfect after all. He goes on to explain to his readers the kind of imperfectness that ensues: "... for when the one subject [this earth] admittedly has the same general nature [to be an earth] as the other one [the moon], then it is undetermined ["unbestimmt", cp. "indefinite"] whether that determination [of being inhabited], which is also made available to the latter, belongs to the former in virtue of its nature, or in virtue of its special features ["Besonderheit", particularity], for instance whether the earth has inhabitants *qua* ["als"] heavenly body in general ["überhaupt"], or only *qua* this particular heavenly body" (341).

Hegel did not, however, include in his conclusion any phrase qualifying the truth of the conclusion. This could have been done by prefixing either the word "probably" or the word "possibly" to the conclusion that the moon is inhabited, or a synonym of either of these. But for Hegel, to do this would be to act contrary to his opinion that his Argument by Analogy is "formally" valid.

Since Hegel does not discuss the problem of a suitable qualification of the conclusion at all, let us consult the *Collegium Logicum* of Bolland, in order to see whether this Dutch Hegelian is able to throw some light upon the matter. Bolland informs his disciples that "the modality of the judgment does not permit of being expressed in the form of the judgment: it rises above the schema of the judgment" (1931, II/x). "The quality, the quantity and the relation of the judgment: these already lie, if one may put it that way, in the very connection between subject, copula, and predicate, but this is no longer the case with respect to modality: we are here concerned with such notions as 'in my opinion', 'probably', and 'unquestionably', hence with ways of thought in which we go beyond [lit.: above] the simple schema of the judgment" (782). If we may rely upon Bolland's interpretation of Hegel's logic, then it is clearly no accident that a word like "probably" or "possibly" is missing from the conclusion of Hegel's Argument by Analogy. According to Hegel and Bolland, what the latter calls "modality" *cannot* be expressed in the proposition which is to express the judgment in question.

As it turns out, this attitude towards modalities in later European philosophy derives from Kant, who in his *Critique of Pure Reason* says: "The modality of judgments is a quite peculiar function of these [*i.e.*, of judgments], which has the distinguishing characteristic of contributing nothing to the content of the judgment (for besides quantity ["Grösse", magnitude], quality, and relation there is nothing else which constitutes the content of the judgment), but only concerns the value of the copula in relation to thought in general" (1968/144).

Here, as in so many other cases, Hegel shows little logical originality; but for this very reason his many mistakes should not be blamed on him alone, for most of them are mistakes of his predecessors and reflect the spirit of his age.

13. A COMPARISON BETWEEN HEGEL'S AND MILL'S ANALYSES OF THE EXAMPLE OF THE INHABITED EARTH

"There is no word, however, which is used more loosely, or in a greater variety of senses, than Analogy", John Stuart Mill remarks (1965/364).[8]

As a justification for continuing my own analysis of Hegel's Argument by Analogy I quote Haenssler, author of the monograph *Zur Theorie der Analogie und des sogenannten Analogieschlusses (On the Theory of Analogy and of the So-called Argument by Analogy)*. Haenssler introduces his investigation by saying: "The indisputable fact that this logically quite ungrounded concept ['analogy'] is used in all fields of knowledge as a bridgehead for inferences of the greatest range, exempts us from the duty of giving a justification for the present study" (1927/7).

Mill's description of arguments by analogy are of special interest to us since he discusses exactly the same example as Hegel. His formulation of that example is as follows: "For example I might infer that there are probably inhabitants in the moon, because there are inhabitants on the earth, in the sea, and in the air; and this is the evidence of the analogy.... Now the moon resembles the earth in being a solid, opaque, nearly spherical substance..." (*op. cit.* 366). He then cites yet other similarities between the earth and the moon which we need not mention here. But it should be mentioned that his conclusion does contain the word "probably", which according to Kant, Hegel, and Bolland does not contribute to the form of the judgment.[9]

Mill's formulation of this argument is thus quite different from Hegel's; it has the form of schema B' in Section 5, and contains four terms.

In connection with the modifier missing from Hegel's conclusion it is of importance to observe that contrary to Hegel, Mill immediately proceeds to discuss the arguments that may be put forward *against* the conclusion: the *negative analogy*.[10] He mentions the *dissimilarities* between the earth and the moon, which make the conclusion less plausible or probable. If X, Y, and Z were the only known properties of the moon, then the conclusion that the moon is inhabited would indeed be plausible. But even then it would not be more than plausible; the evidence that the earth is inhabited together with this positive analogy does not warrant a certain conclusion (unless the range of the variable P were limited to just these property-predicates; but in that case the earth and the moon would be (essentially) one and the same entity). But the fact that the moon has a great number of properties which the earth has not, and *vice versa*, is an important counter-argument. In Mill's opinion this makes the argument in favour of the conclusion weaker than it would have been if we had not been in possession of the knowledge contained in the negative analogy.[11]

Hegel demonstrates no interest in the problem of how it may be avoided that the "identity" in question is so "superficial" that the argument is "relegated to the sphere of mere imagination" (340). All he can suggest in this connection is the following: "In the exposition of the nature of the inference and its various forms given here we have incidentally also taken into consideration what constitutes the primary interest in the usual study and treatment, namely how in each figure a correct inference can be made; only the main point, however, is mentioned, and the cases and intricacies are passed over which arise when the difference between positive and negative judgments together with their quantitative determination, *especially of particularity*,[12] is also taken into account" (328; my italics). He adds to this that logic has been developed in such detail that "its so-called subtleties have resulted in general vexation and disgust".

There can be no doubt that this is true. Hegel, however, draws the conclusion that as a rule "the natural intelligence", which "has asserted itself, in all directions of mental culture, *against the insubstantial*[12] *forms* of reflection" (my italics), is in no special need of any further reflection upon logical, as distinct from illogical, thought. The natural intelligence, he says, "feels that it can ignore such a science for this reason, that it already by itself and naturally carries out the various operations of thought, without any special training". Hegel seems thoroughly to agree with this. There is, however, little reason for his optimism.[13] On the other hand he, too, realizes that just as the sciences of anatomy and physiology are beneficial to our gait and our digestion, in the same way "the study of the forms of reason will undoubtedly influence the correctness of thought to a still greater degree".

But Hegel sees the development of man's critical logical capacity wholly as a task of *education*. This is another Ramistic feature of his thinking.[14] Hegel's own *Science of Logic* shows, contrary to that of Mill, few traces of critical reflection upon the forms of intellect and reason propounded by him.

14. A CONTEMPORARY INTERPRETER OF HEGEL ON THE ARGUMENT BY ANALOGY

J. N. Findlay's discussion of Hegel's treatment of arguments by analogy is rather surprising. He speaks with great sympathy for the form of Hegel's Argument by Analogy, but deals with this feature of Hegel's logic, which in the opinion of the present writer is of the greatest systematic importance, so summarily and briefly that we may quote him here in full: "The "truth" of the Inductive Syllogism is accordingly the Syllogism of Analogy, which reasons from particulars to particulars, from certain Universal features common to a set of Individuals to certain Specific features only known to be present in some of them. The pattern of this Reasoning is I-U-S ["I" for "individual", "Einzelnes"; "U" for "universal", "Allgemeines"; "S" for "special", "Besonderes"]. Hegel gives the now well-worn example of the moon's probable habitation on account of its somewhat superficial analogy to the inhabited earth. What is interesting in this part of Hegel's treatment is the accuracy and brevity with which he both anticipates and resolves the difficulties afterwards hit upon and puzzled over by Mill" (1964/242).

This quotation shows that Findlay's evaluation of Hegel's argument form must rest upon a misunderstanding. The phrase "on account of its somewhat superficial analogy to the inhabited earth" gives the impression that Hegel at least pays serious attention to those factual properties X, Y, Z,... which the two entities have in common, although he disregards the dissimilarities between them.

But as a matter of fact Hegel, contrary to Mill, does not mention a single one among the properties which are common to the two heavenly bodies, except that he calls both of them "earth", although the definite description "the earth" is usually employed in a unique sense, as a proper name. No (other) property of the two entities is mentioned, yet the exegesis of his Argument by Analogy occupies four and a half pages of his book, with the example of the inhabited earth as a recurrent theme. This is bizarre. In these pages we frequently find the word "nature", often "general nature" ["allgemeine Natur"], and still more often the word "genus" ["Gattung"]. The genus in the example of the inhabited earth is of course a concept 'Earth'. It seems that Findlay did not penetrate deeply enough into the characteristic features of Hegel's logic, nor Mill's, for it is certainly misleading to suggest, as he does, that the differences between Hegel's and Mill's formulations are of little or no importance. Hegel is so exclusively interested in Aristotelian essences and internal properties and relations (cp. VII-22) that he does not even stop to investigate, or to recommend that it be investigated, precisely which properties the two entities do have in common. Mill, on the other hand, holds that "individuals have no essences" (1965/73).[15] True, Hegel took the similarity between the moon and the earth as a source of inspiration, but in the argument which is assumed to reflect his mental endeavours this similarity is not mentioned. As he himself puts it, he is not concerned with a "mere similarity" between two observable entities; he is concerned with relations of "essential" identity, EI, which are expressed by the copulas of the two premisses. Where, in his argument, the properties X, Y, Z, common to the two heavenly bodies, would have to be mentioned in order that we might liken him with Mill, as Findlay does, we find the same predicate M which is used in his major premiss in a *definite description* of that entity which is taken as an *exemplum*. When read in its uniquely referring sense, this premiss expresses what Mill calls "the evidence" of the argument. In Hegel's example, not in Mill's, this is the predicate "earth". Hegel employs it to formulate a premiss which he intends to be uniquely referring and logophoric at the same time.

It should now be clear how mistaken Findlay is when he says that Hegel "anticipates... the difficulties afterwards hit upon and puzzled over by Mill". Hegel, on the contrary, works within "*another* theory of predication where quite peculiar systematic elements are at work (essence)" (Angelelli 1967/108f.; cp. I-5 above).

Findlay's assertion that Hegel, contrary to Mill, *solves* these difficulties is downright grotesque. Findlay writes as an apologist, rather than as a scientist. What Russell calls "the scandal of induction" was certainly not abrogated by Hegel's Argument by Analogy. It can even be said that Hegel, contrary to Mill, does not even see that the problem of inductive logic is the same as that of deductive logic, *i.e.* the problem of how to distinguish good from bad reasoning. He does not attempt a discussion of how one might reduce the chance of drawing a false conclusion from true premisses – a problem which is still open today. Mill did at least try to contribute to this distinction. This difference between him and Hegel is systematically connected with the latter's efforts to cast all his arguments (inferences) into forms with two premisses, three terms, and only monadic predicates. Since he mentions only one form in which an argument by analogy can be stated, the result is that his arguments by analogy are all treacherously similar to Aristotelian syllogisms.

15. Interpretation of Hegel's Argument by Analogy

As a solution to the problem of how one might arrive at concepts and
judgments which would serve us as starting-point of a philosophical
sorites, that is to say: how one might arrive at judgments expressing
maximal, or nearly maximal, essential identity, Hegel offers us the postulate of a phenomenon or process which he calls "Aufhebung". This is a
substantive term with the double connotation of "raising" and of
"abolition" or "suspension", much like the English "lifting".[16] The
history of his idea 'Aufhebung' and its function in intellectual life around
1800, fascinating as it is, cannot be gone into here; I hope to discuss it in
some detail in later publications.[16] Suffice it therefore to say that *Aufhebung* is an inductive process in the traditional, but *not* in the modern
sense of "induction" (cp. IX-12).

What is abolished in the *Aufhebung*?

Hegel answers this question clearly enough: "As, therefore, the Argument by Analogy is its Intermediary's requirement against that immediacy, to which its Intermediary is bound, so it is the factor of *individuality*
["Einzelheit"] whose lifting is required. Thus [!] there remains to the
Middle the objective universal, the *genus*, purged of the immediacy"
(II/342). And then: "... as its immediacy has raised/lifted *itself*, however,
and the Middle has determined itself as a universal being to and for itself,
so... the Argument of Reflection has passed into the *Argument by
Necessity*" (343; my italics).

The middle concept, originally the double meaning of the ambiguous
logophoric term "the Earth", or speaking generally: 'ιM', with an
unspecified ι must somehow be transformed into the purely logophoric
'λ Earth', or 'λM'. It is the individual component of the concept behind
the ambiguous "the Earth", the logical subject of the major premiss
and the middle concept in the Argument by Analogy, which is abolished
in that process of purgation or purification which Hegel calls "Aufhebung", and which must somehow be carried out. Hegel in fact suggests
that the concept carries out the *Aufhebung* itself (last quotation). However that may be, the outcome of the process is a pure genus-concept,
without specific individual components.[17] And, Hegel says, this purified
genus-concept resulting from the Aufhebung is nothing else but "the
Ideal" (I/39f.). A Marxist may here want to substitute "essential Matter",

but the possible kinds of metaphysical interpretation of the Hegelian theory of predication and of inference are of no importance to us here; we are only concerned with that theory as a network of interlocking semantic and logical assumptions and doctrines.

I shall now attempt to show how the Argument by Analogy fits into that setting.

To start with I shall recall some of the facts and conclusions which have been revealed in this chapter so far. In Section 3 we learnt that an argument by analogy may be *superficial*. Clearly it must also be possible for such an argument to pertain to more deeply-lying structures, in some sense of "deeply". Secondly it became clear that in Hegel's own opinion, nothing is wrong with the *form* of the argument, as formulated by him. Thirdly we learnt that in the Kantian logical tradition, to which Hegel certainly belongs, the kind of words by means of which we usually express the untrustworthiness of a "superficial" argument do not contribute to the form of the argument.

In VII-23 I formulated the following hypotheses about id-logic: according to id-logicians, from a proposition a_1 *resembles* a_2 it follows with necessity that *there is* a (higher) ι and a judgment – possibly, but not necessarily with the same grammatical subject and predicate as the premiss – which is true, or is realized at (or with respect to) this level ι. This latter judgment is therefore a *necessary condition* for the former judgment,[18] but not a sufficient condition. Many id-logically oriented philosophers, especially in the German-speaking countries, call it "a prejudice" ("prejudgment", or "presupposition"; "Vorurteil", here used as a laudatory term, meaning something that ought to be reached; cp. Gadamer 1960/250f.). Hegel makes use of this id-logical assumption in the following manner:

(1)   this Earth is inhabited/has inhabitants (premiss)

Hegel's interpretation of (1) is:

(2)   this Earth *resembles the* Inhabited.

This is a "mere similarity". In the terminology we have introduced in this book:

(3) θ Earth $\underset{\overline{G}}{=}$ the Inhabited.

Now if we apply the thesis of id-existence to (3), and if, in addition, we
assume that the nouns in the grammatical subject and predicate of (2)
may also be taken as names for the logical subject and predicate of the
"higher" judgement or *Vorurteil*, then we obtain:

(4) $(E\iota)_{\iota > \theta}$ [ι Earth *is the* Inhabited].

However, Hegel drops the existential operator "there is…" or "for
some…". In X-5 below we shall see that the omission of quantifiers
pertaining to "higher" entities is a feature of traditional logic which is
not restricted to Hegel's logic. *Hegel drops the quantifier and replaces the
instantiated ι by the definite article.*

If he assumes that on the grounds of a similarity ("Ähnlichkeit")
like (3) he can arrive at a judgment like (4), expressed as "the Earth *is the*
Inhabited", then he certainly assumes that, in general, that ι to which
"the" refers is lower than the extreme value λ; and it is precisely for this
reason that the Argument by Analogy is in his eyes untrustworthy or
imperfect ("unvollkommen"; Section 12). The subject-concept of the major
premiss is partly, but not completely *aufgehoben*. In the language of
"pure" logic: the higher the concepts and judgments are "localized" or
"placed" in the logical field the purer ("reiner") they are. In the process
called "Aufhebung" it is the so-called Quantity and the Individuality
of the subject which are "lifted", or in other words: the initial judgment
is purged of its quantificational determinator. Given the Ockham-
Wallis-Kant-doctrine, this includes its individual character, if it is an
individual proposition. Given, furthermore, the traditional understanding
of a quantifier ("all", "no", or "some") as an operator working upon
the first noun in the proposition, it follows that it is *the logical subject* of
the initial proposition which is so purged and brought to a higher level.
For Hegel this means that in propositions about essences "alle Menschen"
should be replaced by "der Mensch" (cp. Lenk 1968/317).[19] In the
metaphysical interpretation of this logic the categories Quantity and
Individuality belong only to the θ-level, which in that interpretation is
physical space. These categories are fully developed only at this lowest level.

There are many quite different philosophical interpretations of such
traditional logical expressions as "essential", "internal", etc.; more
often than not various meanings of these and related words are found
within the works of one and the same author, with little or no explanation

of their interrelatedness. This technical multiple interpretability is
certainly no less conspicuous in Hegel's work than in other traditional
logics. Having noted it, we can now give the following analysis of his
Argument by Analogy relative to the logical dimension assumed in his
logic:

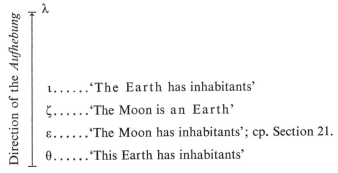

Fig. 4.

The definite article in the subject term of the major premiss refers to a
logical degree between θ and λ, which indicates the degree of essential
identity between the logical subject and the logical predicate in the
judgment. That ι in Figure 4 is higher than ζ is incidental.

Neither Hegel nor any of his followers ever described a function
$\varepsilon = \varepsilon(\iota, \zeta)$, *i.e.* the way in which the logical degree (the degree of essential
identity) of the conclusion (ε) depends upon the logical degrees of the
premisses (ι and ζ). But it is reasonable to assume that the conclusion
cannot be of a higher degree than the premisses. If this was Hegel's
assumption, we have $\varepsilon(\iota, \zeta) = \min(\iota, \zeta)$, just as in the case of the *logical
damping* in Pfänder's logic (VII-26). Hegel's Argument by Analogy is
then "deductive" in the traditional sense of showing a "descent", as
was often said: from the more to the less general, while not being deductive in the modern sense of completely dependable.

How this is to be reconciled with our assumption that for all traditional
logicians 'the earth (Tellus) is inhabited' is true on the lowest level, θ,
is hard indeed to understand. But that is Hegel's problem, not ours. If ε
is understood as a function of time, then that function cannot be described in a logic which is systematically based upon three terms and two
premisses.

16. WHY THE FOURTH TERM CANNOT BE EXPRESSED

Let us return to (4) in the last section. This judgment contains an existential operator for the level-variable "ι".

Now the problem of the four terms can be formulated as the question why Hegel does not incorporate this existential quantifier, or an expression with a similar meaning, in the form of his major premiss. How, after all, could he and his followers convince themselves that the inference form which he calls "Argument by Analogy" is formally correct and contains no *quaternio terminorum*? For the instantiated "ι" is of course a fourth term.

In the opinion of the present author there is, in addition to what was said in Sections 8, 9, and 10, still another explanation for this, which is important enough to deserve a section of its own. We have said already that id-logicians assumed a set, probably infinite, of ι-values, which they assumed to be ordered in some way by a relation Higher Than, or More Internal Than, or More Essential Than. In other words, essential identity was assumed to be a comparative concept. For a logic based upon this idea an ordinal scale and a ranking of the names (ι) of the elements (ι) of this set is absolutely necessary. Otherwise the verbalized judgments will be so indefinite that one cannot reasonably speak of a logic at all.[20] To this end one must, furthermore, be able to determine which of two essential identities which are not equally high or equally internal is *higher than* or *more internal than* the other.

The point I want to stress in the present section is the following. Hegel has to omit the existential quantifier in (4), and the instantiated ι as well, because he does not understand such rankings and wrongly associates them with metric scales and cardinal numbers. He connects the cardinals conceptually with multiplicity, further multiplicity with physical space, which is (or: which characterizes) the logically "external" θ-level, which also is the "lowest" level of his metaphysics.

Instead of (cardinal) numbers, Hegel greatly prefers an ordering by means of colours (colour-labels), using the order in which they occur in the spectrum. And in fact the function of ordinal numbers may well be carried out by colour-labels, at least if we do not want to distinguish very many of them. But Hegel seems to think that every ranking is an interval scale, just like the centigrade, the Fahrenheit, and the Réaumur temperature scales; for he does not discuss any scale which is not an interval scale. Interval scales, however, are not correctly described either: "the Warmth has a degree; the degree of warmness, be it the 10th, the 20th, etc. ..." (I/219), and: "The twentieth degree contains in itself the twenty; it is not only determined as different from the nineteenth, the twenty-first, and so on, but its determination is its own [cardinal] number ["Anzahl"]. But in as much as the [cardinal] number is its own, – and the determination [definiteness] is at the same time essential[ly?] qua [cardinal] number, – so it is extensive quantity" (I/216f.). As Russell said: "...quantity and magnitude, in Hegel, both mean "cardinal number"" (1934/346).

It is quite false, however, to say that the number twenty which by means of a convention is associated with a position on an ordinal scale such as the Beaufort scale, or Moh's hardness scale in mineralogy, is *the* number belonging to this position (cp. Russell 1964/243). That does not even hold of an interval scale. It is therefore quite groundless and even false for Hegel to state: "Extensive and intensive magnitude, therefore, is one and the same determination [definiteness] of the Quantity" (I/217). Consequently Hegel cannot and does not accept Kant's opinion that the mind, *i.e.* the Spirit, and hence also logic possess a "degree of reality" in the sense of an "intensive magnitude" (220). His own opinion is that "to be sure, the Spirit has Being, but of a quite different intensity from that of the intensive magnitude".[21] Of the kind of intensity pertaining to Spirit and logic he only says that it is one from which not only "the category of the extensive Quantity, but that of Quantity in general" has been eliminated (221; my italics).

I repeat that a ranking satisfies this demand. Only two things are required: firstly, that one is able to operate logically with an *asymmetric and transitive relation*. Secondly, either (i) that there is immediate intersubjective agreement – as in the observation of colours – with respect to the question which one of two essential identities which are not equally high is higher than the other (*quod non*), or else (ii) that we can find an asymmetric relation in intersubjective "external" space which may serve us as a *reflection* or *mirror* of the asymmetric relation Higher Than (or More Essential Than, or More Internal Than). To find such a relation may turn out to be difficult,[22] but such an asymmetric relation need not have anything to do with quantities.

Hegel cannot find such a relation; one wonders whether he ever looked for one. In his opinion, ordering of continua and of other kinds of sets clearly need not be regarded as being of fundamental logical importance, for in the neo-Platonic "pure logic" they cannot be treated without contradiction (cp. VII-11). When this is applied to essential identity, it clearly means that one *has* to content oneself with a dichotomy or opposition between the Internal ("Inneres") and the External ("Äusseres"),[23] in the fashion of Ramus. A "quantity-free" ranking, which might suitably be dubbed "Das Transitiv-Asymmetrische überhaupt",[24] is for Hegel – but not for McTaggart (1927/258) – a completely strange category.[25] He consistently reduces all asymmetries, including the transitive ones, to polarities,[26] which he conceives in the manner of Ramus as dichotomies, and which he discusses by means of the *classificatory* connective "either-or".[27]

The degrees of identity ζ, ι belonging to the premisses in the Argument by Analogy, are not themselves degrees of probability; they belong to a specifically logical theory, the theory of internal genera and species, which has now been abandoned. The degree of untrustworthiness of the major premiss is a degree of superficiality, of externality. However the logical dimension in which the internal genera and species were "placed", as τοποι for a philosophical discussion, eludes everything Hegel ever understood by verbalized order.

A fortiori Hegel cannot express the dependability, acceptability, or probability of the conclusion as a function of the degrees of identity of the premisses.

17. ARGUMENTS BY NECESSITY

Hegel's doctrine of ascent in a logical field, the procedure called "Aufhebung", will not be discussed further in this work.[28] If the *Aufhebung* is completed, an Argument by (or: from) Necessity becomes a possibility. Here again Hegel distinguishes precisely three forms. The first one is the Categorical Argument, so called because one or both premisses is "the" categorical judgment (344). Here, too, the number of premisses is two, in agreement with the demands of Aristotelian syllogistic. By "a categorical judgment" traditional logicians usually understood a judgment which was neither hypothetical nor disjunctive; Hegel seems to demand in addition that the judgment be *indefinite*. The predicate of such

a judgment expresses the "immanent nature" of the subject and is therefore a general entity (294). I shall use λM *is P* as the logical form of a
judgment which is categorical in Hegel's sense. The subject term of such
a judgment, λM, is a name for "the genus purged of its immediacy"
(cp. Section 15).

This time Hegel offers no example. But we understand from his discussion that a categorical argument may have one of the following two
verbal forms:

the M is P	*the M is P*
the S is M	*S is M*
the S is P	*S is P*

Of these two forms the one to the right, at least, coincides with the form
he calls "the Argument by Analogy", in so far as that form is made
explicit in his work. In order to explain what the difference is between the
Argument by Analogy and the Categorical Argument one can, to start
with, use the signs introduced in this book:

Argument by Analogy	*Categorical Argument*
ιM *is P*	λM *is P*
S is (an) M	*S is (an) M*
S is P	*S is P*

Still more fundamental in Hegel's logic than the Categorical Argument is
however the Hypothetical Argument. As usual in traditional logic the
argument form so-called is *modus ponens*: "When A is [the case? exists?],
then B is. Now A is, hence B is" (II/346).

18. EITHER-OR AND *definitio rei*

Yet the hypothetical argument form is not the ultimate basis of rational
thought in Hegel's logic. That honour is bestowed upon the Disjunctive
Argument. Here Hegel recognizes two variants:

A is either B or C or D	*A is either B or C or D*
but A is B;	*but A is not C, nor D*
hence A is not C, nor D	*hence it is B.*

Both variants have the same major premiss. The one on the left is only valid when "either... or..." is interpreted as exclusive disjunction (*aut*-junction; in the sense of *exactly one of A, B, C,...*, not in the sense of an iteration of the binary exclusive "or"). And this is indeed the meaning of "*either*... or..." in philosophical language and usually also in ordinary language. In both variants of the Disjunctive Argument a description of the purified genus '*A*', which is to play the role of major premiss, must somehow be given in advance. It is unclear whether this premiss is assumed to result from an *Aufhebung*, whatever that may be, or whether it may be perceived directly by the mental eye.

But that is of less importance to us than is his view on the *composition* of "pure" entities or concepts. Hegel maintains that the logical subject '*A*' of the major premiss in the Disjunctive Argument, while being *either* '*B*' or '*C*' or '*D*', is at the same time *both* '*B*' and '*C*' and '*D*'. For this subject is "the genus: *A*, which is *B* as well as C and D" (349). Nevertheless: "the specification is as a distinction, however, just as much the Either-Or of *B*, *C*, and *D*, negative unity, the reciprocal exclusion of characteristics" (*loc. cit.*). In his discussion of the disjunctive judgment (297f.) we learn precisely the same.

Instead of Hegel's variable "A" I shall use an "*M*", as before. A Hegelian judgment about a purified genus 'λM' then has the following verbal form in our λ-terminology:

(1) λM is (B and C and D and...).

But Hegel also holds:

(2)   [λ?] M is either B or C or D or....

This is absurd, since an *aut*-junction (exclusive disjunction) is not a logical consequence of a conjunction. From *p and q*, the conclusion *p aut q* does not follow, nor *vice versa*. The two propositional forms are even incompatible: their conjunction is a logical contradiction. Take as an example the conjunction

(1)   The Dog has a flat nose and a pointed nose and... and four
           legs and a tail and....

It does not follow that

(2)   The Dog has *either* a flat nose *or* a pointed nose or... or four
        legs or a tail or...;

had that been the case, then from a minor premiss like "Fido is
a Pekingese" we might, with a Disjunctive Argument, have drawn the
conclusion that he has no tail, or at least that he ought not to have
one.

This is to say that by taking conjunctive-*aut*-junctive descriptions of
purified genera as his point of departure, Hegel demonstrates a remarkable neglect of the important difference between mutually exclusive
characteristics: *a flat or (excl.) a pointed nose* on the one hand, and on
the other such characteristics as *four legs and a tail* which belong to all
normal dogs. In other words: the starting-point for his Logos-theory is
quite illogical. In Chapter XI we shall return to this problem, which is
not specifically Hegelian but a characteristic of traditional logic in
general.

Hegel regards judgments having purified genera 'λM' as their logical
subject as a kind of definition: "That determination of the concept, however, which is regarded as R e a l i t y, is conversely just as much something
p o s i t e d.... This B e i n g is therefore an o b j e c t ["Sache"] which is in
a n d   t o   i t s e l f, – objectivity" (352). Here Hegel's theory of inference
ends. *Real definitions* form the premisses, at least, of *philosophical*
reasoning; other uses of the intellect may start at a lower level and may
require some *Aufhebung*. Fundamental arguments are *descending* arguments. For instance the conclusion of an Argument by Analogy is less
"general", less pure, at least, than the major premiss; the Argument
by Analogy is deductive, and so is the Categorical Argument.

We have come to a number of conclusions which are of the greatest
importance, not least from an ethical point of view:

1. Hegel looks to real definitions, which he considers to be logical
   identities, for the foundation for his use of Reason.

2. These real definitions are logophoric indefinite propositions and their
   grammatical subject terms are logophoric definite descriptions λM.

3. These logophoric terms designate *pure genera*.

4. A Hegelian real definition is a sentence in which monadic predicates
   B, C, D are conjunctively predicated of logophoric terms; these
   predicates designate the components of the "immanent" character of
   'λM', the essential nature ("Wesen") of the M-things.

5. Hegel's logic contains no introduction rule for logophoric articles;
   for him, generalization and induction is the same operation and a
   logico-semantic description of the *Aufhebung* is not given.

Hegel's philosophy "therefore becomes according to its form a knowledge
from mere concepts. It develops from pure logic" (L. Nelson 1962/443).

19. DESCENT TO THE MOON

Let us assume that someone thinks that he has reached a judgment 'ι
Earth has inhabitants', and let us take a second look at the problem of
the eliminability of "ι" and of "λ". Or let us assume more generally that
someone asserts a judgment '*the/an M is P*' of an indefinite degree of
purity, *i.e.* of a logical degree unknown to his listeners (including himself). What is then the logical relevance of such a judgment for an
individual, S, with the property M? Is there in the logic of Hegel in
addition to the mysterious *Aufhebung*, also an *Abhebung*?

The fact that Hegel recognizes "valid" inference forms with *S is P* as the
logical form of the verbalized conclusion guarantees nothing. For an
argument with such a form, *e.g.* the argument "the Bantu is primitive,
Saul is a Bantu, hence Saul is primitive" may, in Hegel's logic, be "super-
ficical"; it may be an instance of his Argument by Analogy.

Our first remark is that one cannot know whether it is an argument
by analogy or a categorical argument. This is unacceptable, even if it
were the case that categorical arguments are meant by Hegel to be
dependable arguments. For when some Bantu, say Saul, is pointed out
who is not primitive, then a proponent of the major premiss can always
escape criticism by saying that he did not mean this judgment to be of the
kind that may be used in a categorical argument. The major premiss
"ι Bantu is primitive" cannot, therefore, be refuted by pointing to
individual cases.

But, second, we have not yet investigated whether the Categorical Argument is supposed to be a dependable form. "Formally" this is certainly a "valid" form for Hegel, since it coincides with the Argument by Analogy. But does he hold that from *M is P*, *S is an M* the conclusion *S is P* follows with absolute certainty? Or, in other words, is $\varepsilon(\lambda, \theta) = \theta$?

Let 'λM', for instance 'λ Earth' or 'λ Bantu', be a pure – perhaps a purified – genus, and assume that λM *is P* is held to be true. Assume, furthermore, that S is (an) M, *i.e.* that *S is (an)* M is true. Do these two premisses force someone who believes in them to assert the truth of

(1)   *S is P*,

(2)   something else, with *S* as the grammatical subject term?

If *both* of these questions are answered negatively, then there would seem to be no elimination rule for "λ", in which case every logophoric proposition of the form λM *is P* is completely *gratuite* and indisputable, exactly like the corresponding word sequence *woof M is P*, and for precisely the same reason.

20. CONTEMPT FOR THE COUNTER-EXAMPLE

(i) *The situation in Hegel's logic.* The answer to (1) in Section 19 is negative. The argument form in question is therefore *not* dependable, for in Hegel's logic the conclusion *S is P* turns out to hold in this form only for such S-things as are P.

The reader may think that this way of putting it must be a mistake on the part of the present author, but such is not the case. That this is indeed the situation in his logic is explained by Hegel in unmistakable terms in a sub-section entitled *The Definition*, to be found in the section on *Synthetic Knowledge* in the chapter *The Idea of Knowledge*. In this sub-section we are told that in Hegel's terminology an M-specimen which does not look like or does not behave according to a "posited" ("gesetzte") real definition of the concrete universal '*the M*', posited we may assume, by some Hegelian thinker, is *bad* ("schlecht"). It is not what it *ought* to have been according to "the" definition. The theoretical context of this attitude is of considerable systematic and historical interest.

Hegel believes that from an investigation – perceptual, as it seems – of an M-specimen he can tell how it ought to have been, how it ought to have looked or behaved: "Something real therefore presumably ["wohl"] shows in itself what it ought to be" (456). The difference between "the" concept '*M*' and a realization of it, *i.e.* a specimen, "enters into the concrete things" (*loc. cit.*).

Veatch employs precisely the same notion. He holds that "through the agency of a what-logic it should be possible to recognize directly, and simply as a matter of fact, that the conduct or behaviour of something is not the conduct or behaviour that is *appropriate to the kind* of thing that it is" (1969/248, my italics).[29] Veatch does not offer a shred of evidence or proof that this is the case; he does not even attemptto do so.

So a negative case: an M-thing, S, which is not P, demonstrates that its reality corresponds only imperfectly to "the" pure concept '*the M*', hence "that it [its Reality] is b a d" (Hegel, *loc. cit.*).[30]

Hegel, then, reserves to himself and to the followers of his logic the right to reject such a counter-argument, with the excuse that *ergo* the specimen is bad, and that it *ought to have been as* described in the [?] real definition. A real definition, we now understand, sometimes (always?) describes not only a general and "concrete" *idea*, but at the same time an *ideal*. An individual M which does not behave according to the definiens of a definition of '*λM*' which is posited by a Hegelian thinker is a *bad realization of* the essential M-nature; it is "immature", "atrophied", "despicable", or simply "bad" (*loc. cit.*).

This is an unmistakable and clear *circulus vitiosus in definiendo*, a phenomenon which characteiizes all theories of real definitions, with the exception of the theory of De Morgan (cp. Mill 1965/93; Rescher 1964/34).

(ii) *The logical problem and its solution.* It is not difficult, however, to distil from Hegel's writings a logical problem which, together with the problem of the logic of the quantifiers (existential instantiation and universal generalization) and, of course, also various irrational motives and goals, prompted Hegel to the reflections reported above. Hegel mentions the problem of deformed creatures. He seems to have worried over what to say, *e.g.* of a puppy with only three legs: is it a dog, or is it not? It does not conform to a real definition of '*the Dog*', if that definition contains (conjunctively) the property of having four legs, but it seems absurd not to characterize the unhappy creature as a dog for that reason. Traditional logic, based as it was upon *aut*-junctions and conjunctions of terms, cannot deal with this problem. It can, however, be solved even within a theory of real definitions provided that one is willing to use inclusive disjunctions when constructing them. The deformed creature does fall under the definition, whether of the word "Dog" or of a concrete universal '*the Dog*', if the *vel*-junctively composed predicate "has four legs or has ancestors with four legs" is included as a conjunctive term in the (formulation of) the definiens.[31] The irrefutability of *all* sentences of one of the propositional forms *ιM is P* and *λM is P* cannot therefore any longer be defended by an appeal to the existence of deformities. Hegel, of course, used this logical problem – which was real enough in his day – to support an otherwise completely implausible theory, the theory that in principle a philosopher can reach absolutely true statements of such a kind that they cannot be refuted by reference to any ordinary reality (on the θ-level) whatsoever.

21. DEONTIC LOGIC AND THE LOGIC OF POTENTIALITY

After having given a negative answer to the first question (1) in Section 19, we now come to the second possibility (2): does something else follow necessarily from two judgments expressed by sentences *λM is P* and *S is an M*?

A statement *from p follows necessarily q* means precisely the same as *a listener* (be it S or somebody else!) *has the consent of the speaker to ascribe to him the opinion that q, on the ground of his assertion that p*. If this second question, (2), is also answered negatively, then an asserted or "posited" ("gesetztes") judgment '*λM is P*' bears no relevance whatsoever to the speaker's opinion about individual M's. The logophoric article would not then have any other logical function than an arbitrary sequence of signs has.

In the light of what we now know about Hegel's logic, two possibilities come to mind. One is that from the said premisses follows

(a)   *S, even if it is not P, ought to be or to become P,*

or (in German)

(a′)   *S sei P!*

The other possibility is that we may draw the conclusion

(b)   *S is potentially, but not actually P,*

that is to say

(b′)   *S has the logical gene (a logophore with) P-ness.*

The form (a) would nowadays be called a "deontic" propositional form, and (a′) is an imperative, whilst (b) and (b′) contain some kind of alethic modalities. The formulations (a) and (a′) are justified by what was said in the last section. It is also easy to justify the interpretation (b): "The general is the *free* Power", and many other statements demonstrate that Hegel thought in terms of some kind of dynamic concept. In the last part of his logic terms like "purpose", "good purpose", "realization", "germinating power" (in German: "Trieb") "becoming", and "striving" occur with great frequency. They all point in the direction of judgments which are not purely alethic, judgments, combined with the notion of the *unfolding* of a logical seed, or gene (428).

The most reasonable interpretation of Hegel's logical efforts is therefore that the conclusion in an Argument by (from) Necessity is a synthesis of (a), (a′), (b), and (b′), which we shall render by *S is essentially P.* Such a conclusion will then have (a) or (a′) as well as (b) or (b′) as its immediate consequences.

This is to say that the second question (2) in Section 19 must be answered affirmatively. In Hegel's logic, the two premisses of an argument by, or from, necessity do have logical consequences with *S* as the grammatical subject term, even in the usual sense of "logical consequence". The statement *S is essentially P* necessarily follows, and has the immediate logical consequences (a) and (b).

We can now add two more conclusions about Hegel's logic to our list in Section 18:

5. Individuals[32] of the M-kind which do not conform to a given ["gesetzte"] real definition of 'λM' or which do not behave according to a "posited" definition are bad, immature, or atrophied specimens which need not be taken into account.[33]

6. Hegelian logic contains no workable elimination-semantics for pure logophoric terms λM, nor for impure logophoric terms ιM where ι is "lower" than θ, since there is in that logic no elimination rule for "ought" or for "potentially". Every judgment *'the/an M is P'*, or in German: *'der (die, das) M ist P'*, with the exception of iunquely referring singular ones, is therefore unassailable.

Hegel's logic may therefore be characterized as a potentiality-variant of PPL. In order to understand how anybody can place his confidence in such a logic, knowledge is required of the old theory of "formally significant" concepts and other entities.

22. THE THEORY OF FORMAL SIGNS

Veatch's first book contains information about, and a recommendation of a theory of "formal" signs which is tacitly assumed by a great many modern philosophers of more or less neo-Platonic tendency.
References below are to this work, *Intentional Logic*.

(1) Concepts are assumed to be signs, and the formal signs which Veatch discusses are mostly concepts (13).
(2) A formal sign is not the same as an instrumental sign (13).
(3) To be a formal sign is to be a sign for a *forma*: "concepts are precisely intentions, or formal signs, of essences" (18). (The term for the concept which signifies must probably be "taken" in *suppositio essentialis* or *naturalis*.)
A sign, 'S', or *'this M'*, is clearly a formal sign of 'λM' if and only if it points to or represents its signification 'λM' "naturally – *i.e.* in virtue of the sign's natural properties" (12), that is to say in virtue of the natural properties of 'S' (or of S?).
(4) "The real object, in short, which the formal sign ['S', or *'this M'*] signifies is itself that sign's very *content* ['λM']" (14, my italics).
(The theory of the syllogistic *vis universalis* (Wallis *et al.*) of singular propositions should be studied in the light of this assumption. It is now also clear why some philosophers hold the difference between an "intensional" and an "extensional" approach in logic to be of such an overwhelmingly great philosophical importance.)

(5) "A thing [this M, S, or *'this M'*, *'S'*] is identified with its own "what"" (39).

(6) The [asymmetric!] relation Father Of is a relation of logical identity (38.) A father is clearly assumed to be the "what" of his child.

(7) Mathematical (*i.e.*, modern) logic neglects formal signs altogether (35; cp. 2 above).

It should be noticed that the process of "intending" the essence of something by studying (contemplating) the (concept of) a specimen, i.e. an individual entity or an individual case, is a method of *induction* in the traditional sense of "*inductio*" (cp. IX-12). As Veatch says, in his last book, of judgments about the intended *formae* or "whats":

(8) "even though they purport to be necessary truths, they are nevertheless subject to error" (1969/253).

It is not explained how the possibility of error is introduced. In the terminology we used before, one can say that according to the theory of formal signs an individual concept signifies its own logophore and all its properties and relationships. As long as the "signs" in question are assumed to be concepts, the theory of formal signs is therefore nothing but a statement of the basic tenet of essentialist or logophoric logic. This logic is *intensional* in the sense of "comprehensional", and the same time *intentional* in as much as an individual concept is assumed to intend or point to its own essence. However, when entities other than concepts are also assumed to signify essences, for instance entities with spatial co-ordinates, then the theory of formal signs is seen to be indistinguishable from *chiffre* in general, and more especially from phenomenological pharmacy, botany, and psychology. If, as in idealism, little or no heed is paid to the difference between an individual with spatial co-ordinates and somebody's concept of that individual, then the difference between concepts as so-called formal signs and formal signs in nature fades away.

On any account, Hegel's Argument by Analogy can be seen to be an application of exactly this theory of formal signs, exemplified by the earth, or by 'the Earth', as a formal sign of its essence, with which it is "analogical".

23. COMPARISON BETWEEN HEGEL, PFÄNDER, MARITAIN, AND VEATCH

Hegel, Maritain, and Pfänder, each of whom is the author of a work on logic, ascribe more than one logophoric meaning to one and the same sentence *the/an M is P*. They all assume that some uses of such a proposition are assertions about entities which are designated by expressions of the form *the absolute M, das M überhaupt, das M an und für sich, le M comme tel*. Other uses of the same proposition express only "relative generic judgments", as in Pfänder's logic, or form the starting point only for an "argument by analogy", not for a "categorical argument", as Hegel says.

If our previous analyses of the logics of Pfänder and of Maritain are correct, then the relationship between Hegel and Maritain is still closer than that between Hegel and Pfänder. For the argument form *the/an$_{abs}$ M is P, S is an M, hence S is P* seems to be valid in Pfänder's logic without any modification of the conclusion, so that the major premiss *the/an$_{abs}$ M is P* may be refuted by reference to an M which does not have the property P. Hegel and, as it seems, Maritain as well, only permit us to draw a qualified conclusion; they can therefore "defend" any categorical or absolute proposition *the/an M is P* against any opponent whatsoever.[34] *Every* proposition of that form is therefore unassailable.

Beth writes that Hegel replaced the old logic of modalities by another logical system (Beth 1948/20). This of course does not imply that Hegel's logic is purely alethic and that he does not think in modalities at all. On the contrary, it means that he does not employ modal or deontic *language forms* as premisses or as conclusions where earlier philosophers would have done so. In arguments Hegel simply uses the word "ist", whatever he means to say.

I am afraid Hegel was considerably less original in this respect than Beth seems to have thought. The similarity between Hegel and, for instance, Maritain, who draws his direct inspiration from quite different sources, is too great for the hypothesis of Hegel as a logical pivot to be upheld. The old logic of modalities does not seem to have survived the Renaissance.

The similarity between Hegel and Veatch is quite astounding. In Veatch's work not even an outline is given of that "instrument of a what-logic" (1969/253) which the humanistic sciences, in his opinion, ought to employ. It is more striking that Veatch, just like Hegel, maintains that value-judgments in general depend upon "what-statements".

NOTES

[1] Cp. V-15, VI-1.

[2] Various uses of the word "quantitative" are discussed in Section 16 of the present chapter.

[3] Traditionally there has been a problem here; cp. IX-16 and X, especially Sections 5 and 6.

[4] See van der Meulen 1958/68 on the "tetrade or quaternio terminorum of logic". See also Ujomov 1965/150. Cp. the end of note 11 in Chapter X below.

[5] Behn blames Hegel for accepting the principle that "the" concept 'M' itself has the property called M. "The concept of [a] triangle is not itself spherical, nor does it roll, the concept of ridiculousness is not comic and the concept of untruth is not false" (Behn 1925/60). "The concept is no picture, the concept of [a] triangle is not – triangular, the concept of [a] river – does not flow, the concept of the changeable – does not change, the concept of error is not – mistaken: only the concept of truth ought to be – true" (*op. cit.*/120). One feels inclined to forgive Behn this last blunder. "A stone falls fast, but the formula s$=[\frac{1}{2}]$gt^2 does not fall" (*op. cit.*/124).

[6] On *analogia entis*, see Emmett 1966, Ch. VIII. "... the thinkers who have handled the Thomist *analogia entis* have been very well aware that their analogical thinking was concerned not with literal resemblance of terms, but in showing how the same principle or structure of relations could be discerned in different media" (183). There is a fair degree of correspondence with the use and description of certain kinds of models in present-day science; cp. Hesse 1965/27f., Nagel 1961/101–117. See also, however, Section 22 below. Maritain promised in his *Petite Logique* to treat the problem of analogy in his *Grande logique* and in the *Métaphysique*; but as it turns out a *Grande logique* by Maritain was never published.

[7] This explains why Behn has to conclude: "The romantic dialecticians, who take an interest in letting the kinds of opposition become blurred, take advantage of the verbal fog [found in language]. To pin them down to an unambiguous meaning of their judgments is impossible because they seek ambiguity by every means [or: trick; "Kunst"]" (1925/111). Cp. note 11.

[8] Unless it be the word "dialectic":"... the term "dialectical" nowadays occurs very often as a merely fashionable word without a reasonably fixed usage..." (Lorenzen 1969/87).

[9] Behn remarks: "The romantic dialectic has rightly sensed that it must oppose the application of the theory of probability to its judgments. His many and extensive discussions of degree, proportion, and measure notwithstanding, HEGEL treated the theory of probability in a step-motherly way..." (1925/112). Behn relates this to the "romantic" logic of negation. "It generally appears that the problem of negation is far more complicated than the romantic dialecticians want to admit". Cp. note 44 of Chapter VII.

[10] Cp. Hesse *op. cit.*/24, 27.

[11] Mill would have nothing of "analogous language". Some theologians have held that an "analogous" use of language may lead to some insight into the properties of God.

Among them was H. Mansel, who was of the opinion that God cannot be known as he really is, but that some knowledge of him may be attained through certain descriptions which must not be understood literally, but only "analogically". Mill has no confidence in such a procedure: "I will call no being good, who is not what I mean when I apply the epithet to my fellow creatures; and if such a being can sentence me to hell for not so calling him, to hell I will go" (quoted from Passmore 1966/31).

12 Grote explains that Aristotle's difference between "problematic" and "necessary" propositions is tied up with his cosmological theories. Aristotle was of the opinion that "from the outer sidereal sphere down to the lunar sphere, celestial substance was a necessary existence and energy.... In these sublunary sequences, as to future time, *may* or *may not*, was all that could be attained, even by the highest knowledge; certainty, either of affirmation or negation, was out of the question. On the other hand, the necessary and uniform energies of the celestial substance, formed the objective correlate of the Necessary Proposition in Logic; this substance was not merely an existence, but an existence necessary and unchangeable... he considers the Problematical Proposition in Logic to be not purely subjective, as an expression of the speaker's ignorance, but something more, namely, to correlate with an objective essentially unknowable to all." (G. Grote, *Aristotle*, part I, p. 192; quoted from the chapter on *Modality* in Venn 1962/308).

13 "A beginner in geometry, when asked what follows from 'Every *A* is *B*', answers 'Every *B* is *A*, of course'" (De Morgan 1966/211). Cp. Ong 1958/160. "Hegel paid no attention to sub-contrary opposition, although in his *Naturphilosophie* he often confounds it with contradiction" (Behn 1925/71). Cp. Hamblin 1970/158.

14 Cp. Ong 1958/160.

15 This is, I think, also Sartre's opinion. Cp. I-4 and note 39 of Chapter VII.

16 Hegel defines the expression "Aufhebung" in the following manner: "Something is *aufgehoben* only in as far as it has entered into unity with its opposite" (I/94). This notion of the unity of something and "its opposite" is, I think, a precursor of the present-day notion of an ordered couple, without which "enantiomorph" asymmetric and antisymmetric relations cannot very well be discussed. Hegel, however, unfortunately restricts himself to ordered *couples*; cp. Section 16. Compare note 25 of Chapter VII.

17 Th. Litt uses the expression "Erhebung" and speaks of "*that* identity which can only be reached by *Erhebung above* the experimental situation" (quoted from Göldel 1935/311; my italics).

18 This difference between traditional metaphysics and hypothetico-deductive methodology is stressed by Schoonbrood (1961, 1968) and also by Beth (1948/19, 1965/9–12).

19 See also Pivčević 1970/55f. on Husserl. Cp. VIII-7, VII-18, X.

20 Compare with note 9. What Behn says is true, but the problem is not limited to Hegel's inability to rank degrees of probability; the notion involved is a much more general one. Cp. note 25.

21 "When one wants to grasp such a deep ["innig"] totality it is therefore completely inappropriate to want to employ numerical and spatial relations, in which all determinations fall apart; they are rather the last and the worst medium which could be used. Natural relations like, *e.g.*, magnetism, colour-relations would be infinitely higher and truer symbols for this" (II/259). It is to be regretted that Hegel did not actually use the colour-scale (the spectrum) for the expression of higher and lower ranks of logical identity, for instance "red" for the highest and "violet" for the lowest or most "external" identity. Some politicians do that, in a rudimentary fashion. Hegel prefers the

dichotomous ordering of magnetic poles above a ranking with seven or more positions as his means of expression; cp. notes 26 and 27.

[22] The obvious solution is to abolish the whole theory of internal relations; this theory is no longer required when one drops the dogma of monadic predication so that statements about asymmetries can be formed and discussed without contradiction. "The key to Plato's theory of ideas... lies in the concept 'incompatible information [s]'" (Ubbink 1962/3). Cp. Barth 1970a, 1970b.

[23] "Ockham, and the nominalists generally, followed Scotus in regarding intension as an additive increase, characterized by some kind of unity between the original reality and that which is added. Thus Buridan, Albert of Saxony, Peter d'Ailly, Marsilius of Inghen" (Wilson 1960/20). This view is probably contained in that notion of intension which Hegel and later Bergson opposed. Scotus's and Ockham's position, according to which all intensities are "additive", is not shared by the majority of modern logicians and philosophers of science. Hempel does not call the property Wilson describes "additivity", however, but takes this property as a (conjunctive) part of his definition of "extensivity" (clause 12.4 g, Hempel 1969/66). Dijksterhuis writes about this late-scholastic problem that "when, therefore, one likes to think of growth *per additionem...*, one must try to understand how the newly added quality can have fused with the original one to form a unity" (1950/206). Wilson, in the book we just quoted, goes on to remark: "In place of the ontological problem, a logical problem comes to the fore: how to *denominate* a subject in which the intensity of a quality varies from one point to another" (*loc. cit.*; my italics). Hegel did not solve that problem. Nowadays it is dealt with by means of the infinitesimal calculus (cp. Peirce on Hegel, in Peirce I/17f., 193) and, more fundamentally still, by means of asymmetric and transitive concepts and predicates: "Sometimes scientists, especially in the fields of social science and psychology, hold the view that, in cases where no way is discovered for the introduction of a quantitive concept, nothing remains but to use concepts of the simplest kind, that is, classificatory ones. Here, however, they overlook the possibility and usefulness of *comparative concepts*, which, in a sense, stand between the two other kinds. Comparative concepts (sometimes called topological or order concepts) serve for the formulation of the result of a comparison in the form of a more-less-statement without the use of numerical values" (Carnap 1951/9). This is the gist of the matter. See Hempel and Oppenheim 1936/35-43. Cp. IV-13, VII-11 above.

[24] Read: "*the* transitive-asymmetric as such"! This is an a-logical notion also in Veatch's logic which, like the philosophy of Bergson, has no place for order, for instance, series; that is to say: for asymmetric or antisymmetric transitive relations (Veatch 1969/262, Russell 1947/821). Compare Stace's description of the difference between Brahman and "the World" (Stace 1961/214). Popma, however, has no objections against "succession"; he therefore rejects Bergson's notion of a *durée* which is somehow free from succession (Popma 1965/162; cp. I-n. 5). One doubts whether Veatch has ever tried to evaluate philosophically the difference between densely and continuously ordered sets (Veatch 1969/262).

[25] And not for Leonard Nelson either; cp. Section 1 of the present chapter and also note 1. See also the discussion between A. D. de Groot and M. J. Langeveld (cp. de Groot 1964/229 n. 2).

[26] Hegel treats everything that was ever called "an opposition", in the widest sense of that expression which even includes differences in general, as a dichotomy or polarity. This is also true of the difference between individual ("einzelne") and particular ("besondere") judgments (cp. V-14), as well as of the difference between between partic-

ular and universal judgments. "Due to the monotony of his triadic schema, Hegel is forced to exaggerate the opposition of sub-alternation and to construe it as a contradiction" (Behn 1925/70). "The verbal negation as found in language does not always show which algorithmic disquantification is meant" (*op. cit.*/111; continued in note 7 above). Behn clearly has in mind the various different ways of inserting a negation sign into a proposition containing multiple quantification; in natural languages these cannot always be distinguished clearly. A good example is offered by the ambiguous sentence form: *the (one and only) M is not P*. Hegel, however, holds natural languages in high esteem: "Since man possesses language as *the means of expression which is characteristic for Reason*, it is an idle idea to look for and to torment oneself with a more imperfect means of representation" (II/259, my italics). Behn's comment is this: "Since a rational algorithmic is bound to expose the trick in dialectic, Hegel does all he can to degrade it... Hegel's talk about the absurdity of a logical calculus is (to use his own favourite expression) nothing but an *assurance*" (99). See also the use of dichotomies in Hegel's contemporary Bentham 1952, I/88f., III/319, and compare this with Ong 1958/200–202. But since Pareto, economists are not inclined to resort to dichotomous classifications.

27 "If one first disguises the [notion of a] polar distance as a logical opposition, then of course one will also interpret the positive degrees on the scale as affirmative, and the mathematically negative degrees as logically negative. This algebra, for a mathematician incomprehensible and misconceived, persists as a vicious habit of romantics" (Behn 1925/125). For "the Either of the Positive and the Or of the Negative", see Hegel, *op. cit.* I/376. Cp. the "admixture of a quality with its contrary" (Wilson 1960/20) as one of the current views in the fourteenth century in respect of intension and remission of forms. See also Weinberg 1965/66f, 82f.

28 Cp. note 16.

29 Cp. Paracelsus' doctrine of *signature*.

30 Concerning the problem of "the dialectic of the beginning" in Hegel's logic, Kierkegaard writes: "Perhaps I may be allowed to pose a question in this connection. How comes it "that at this point Hegel and all Hegelians, who are supposed to be dialecticians, get angry, as angry as Germans? Or is that a dialectical category ["Bestemmelse"]: the bad? From where does such a predicate enter into logic? How do scorn and contempt and intimidation obtain a position as permissible forces in logic, so that the absolute beginning is assumed by the individual person, because he is afraid of what his neighbour and fellow-man will say of him if he doesn't? Is not "bad" an ethical category?" (1846/81).

31 Cp. Chapter XI. See Beckner 1968/63 f.

32 "As conceived by Hegel in his Logic, the concrete does not really allow for individuation as such; as a logical principle, the individual is inessential and lacks reality. The failure to do justice to the empirical particular has long been recognized as a cardinal weakness of his point of view" (Gray 1968/71). "With Plato he maintained that truth of being is only the essence, the universal... Aristotle allows a place for individuality which Hegel on his premises cannot admit... individual uniqueness becomes merely phenomenal" (*op. cit.*/98f). Leibniz must certainly share the blame, for in a discussion of his preference for an intensional (comprehensional) approach to logic he writes: "I prefer to consider universal concepts or ideas and their composition, *for* these do not depend on the existence of individuals" (1968/238; my italics).

33 The following example, taken from *Die Transvaler*, of a quasi-syllogism whose minor premise is denied in order that the major premise may be upheld, is quoted from

Olsson (1969/144). The major premiss is: an *Afrikaner* who is a communist is an impossibility. Mr. Bram Fischer belongs to one of the leading families in the Orange
Free-state. Mr. Fischer is a communist. The conclusion of the editor of *Die Transvaler*:
Fischer is not an *Afrikaner*. In other words: if *the (purified) M is P/not P* is true and if
this M, S, is not a P/ is P, then this M is not a (real) M. Cp. IV-15.

[34] Or in ontological terms: it is a "fact that *das Du* as a logical principle and as a
starting-point for reflexion does not occur in this [the Hegelian] system at all. The same
holds of Fichte and Schelling" (Günther 1959/102). Peirce even criticizes Hegel for
overlooking a still more general ontological category, *viz.* the category of "external
Secondness" (Peirce I/193). Cp. XII-6, *sub* (iv).

CHAPTER IX

THE PROBLEM OF THE LOGIC OF RELATIONS
AND ITS CONNECTION WITH THE LOGIC
OF THE ARTICLES

1. INTRODUCTORY REMARKS

Some of the most important differences between "what-logic" and "relating-logic" come particularly clearly to the fore in the eighty year old work on logic by W. Wundt (1893), to which I have referred several times in previous chapters. We have already seen:

(i) that in the opinion of Wundt, logic is primarily concerned with *indefinite magnitudes* (entities, "Grössen"; IV-19);

(ii) that he is a supporter of a partial identity theory of the copula (VII-24).

It turns out that Wundt's theoretical logic is based upon the notion of concept comparison; this is the neo-Platonic principle in the Fichte-Hegel logic which was so severely criticized by Leonard Nelson.

Moreover his work contains a "descending" analysis of arguments by analogy, very similar to Hegel's analysis, which turned out to be highly relevant for understanding the traditional logic of the articles. For these reasons I now return to Wundt. At the end of the chapter I shall compare him with some of his contemporaries. Born in 1832, Wundt was able to take some cognizance of the logical innovations in the second half of the nineteenth century, but he seems to have known neither De Morgan nor his fellow-countryman Frege, nor Peirce.

2. WUNDT ON RELATIONS AND LANGUAGE

Wundt expresses the traditional view of the unreality of relations[1] as follows: "In this way the formation of abstract concepts always consists in the observation ["Feststellung"] of relations which our thought finds in its own ideas or in concepts that are already given. To call this activity a separation of features [characteristics, "Merkmale"] is insufficient, if only for the reason that these relations are not really features which belong to the objects at all, but such [features] as are first formed in our thinking but which thereupon admittedly become features of things for us" (113).

It is likely that Wundt, speaking about relations, assumes relations between things of all kinds to be theoretically reducible to "internal" logical relations between essences, between "second intentions", as the schoolmen said. On that assumption the view that relations, in particular, are first formed in the mind becomes understandable.

Wundt was conscious of the fact that the logic of relations of his day was inadequate. His *Logik* was published in the period when C. S. Peirce was actively developing a logic of relations. Wundt, too, made certain attempts in this direction, in as much as he wrote a section on "the dependency of concepts" and introduced a certain notation. He compares the interdependency of concepts with functions, and chooses the letter "F" in order to express that the concept '*A*' stands in a certain relation of dependency to the concept '*B*', however without defining the kind of dependency he has in mind.

It may therefore be said to be remarkable that Wundt explicitly mentions a field of discourse which in his opinion *resists a precise use of language*: "That field which most gives rise to a mixture of logical forms of language which otherwise, even grammatically, are strictly separated, is that... which is characterized by the general concept of Relation" (121). That is to say: in Wundt's opinion, *reflections about relations cannot be formulated in clear terms*. If we want to express relational thinking by means of language, then we will have to tolerate grammatical categories being mixed up and given other than their usual tasks to fulfil.[2] This view may be characterized as Hegelian.

3. The Auxiliary Rule of Concept Comparison: Unlimited Reification and Proliferation of Articles

(i) *Wundt*. Aristotle's ten logical categories can, according to Wundt, be reduced to as few as four. "The substantive or the substance ... denotes the object-concept" (118). This is his first category. He further distinguishes between *properties*, *states* (*conditions*) and *relations*. The latter, however, do not form a real category, if by categories "the most general classes of *self-supporting* categories" are understood (122). Relations, then, are not "concrete" in the sense of ontologically self-supporting or independent.

More than once Wundt emphasizes that in his opinion there can be no very definite connection between logical categories and forms of language: "Thus the logical categories are permanent, the grammatical ones changing" (120). This is hardly compatible with his subsequent statement: "The transformation of different concepts into one another or, as we shall call it for short, the *categorical displacement* is an important auxiliary device for the free mobility of thought".

He does not regard a substantial grammatical form as a sufficient reason for saying that a piece of language conceals a *Gegenstandsbegriff*. But once we have chosen a substantival grammatical form it will be much easier to treat a property in our thoughts "as an object, (to) which other objects, properties, or states can be related".[3] When a concept falls into the category of *Gegenstände* as desired, then this is not due only to grammatical form, but is caused also "partly by its own content, partly by the logical connections it is brought into" (124).

As a determination of the logical characteristics of the category of *Gegenstände* this will not do. For Wundt has in no way explained how and on what conditions we can transform a property-concept into an object-concept. Nor has he said anything about how such a concept should be referred to. It is therefore too early to settle the question of whether he adheres to the H-thesis or not. However that may be, he immediately goes on to draw this conclusion: "Thereby the activity of our thought is clearly aimed at a gradual increase of [the number of] object-concepts ["Gegenstandsbegriffe"]" (124).[4]

I shall not discuss the truth or falsity of this statement, which clearly belongs to the field of psychology of thought. A logician cannot decide whether and under which circumstances, for which cultures, and at which levels of intellectual development this statement is true.

But we do, of course, take an interest in the consequences, for practical and for theoretical logic, of such a continual multiplication of substantive concepts. It is clear that this activity entails the *continual generation of new name phrases*, and that Wundt's thesis indirectly lends credibility to the grammatical dogma that every sentence has a noun phrase and a verb phrase, as well as to the traditional logical dogma that every proposition consists of a logical subject and a logical predicate.

Wundt does seem to think not only that it is logically permissible to form new concepts of (new) objects all the time, but even that it is necessary and desirable to do so. This postulated activity of continually forming new object-concepts is made possible by what he calls "categorical displacement". Its importance becomes clear in a section called "General conditions for the comparison of concepts" (127). Here we are informed that as the first condition of concept comparison there is the rule: the concepts that are to be compared must belong to one and the same category" (128; 126). This means that a concept of an object, 'S', cannot be compared with the meaning of a predicate M if that predicate denotes a property. This might seem to imply that predication *cannot* always be analysed in terms of a comparison. But Wundt simply assumes that predication always and for all logical purposes must be analysed as based upon a comparison of two concepts, irrespective of the nature of the grammatical subject and predicate of the proposition we choose to study. And so he concludes that "categorical displacement" of concepts must be possible not only occasionally but always and without restriction. We must and we can "*lift*" a concept "to a *self-supporting* object of thought" (129; my italics), for "it is obviously suitable in a comparison of concepts which are thought of independently, that each of them is a self-supporting object of our thought" (*loc. cit.*). But he does not define his use of the word "self-supporting" ("selbständig"). Nevertheless, he feels able to conclude: "The aforesaid Auxiliary Rule of Concept Comparison can be formulated as follows: concepts of different categories become comparable when they are transformed into concepts of one and the same category, in general into object-concepts" (*loc. cit.*).[5]

The proliferation of articles – in German: of definite articles – to which this Auxiliary Rule leads may suitably be characterized as *Wilhelm Wundt's beard*.[6]

Wundt does not instruct his readers how to carry out this displacement or transformation of, say, a property like Red or Primitive into what he calls "self-supporting objects". Nor does he throw any light upon the activity of comparing a logical subject 'S' with the object resulting from the transformation of the meaning of the predicate M. The Auxiliary Rule is therefore no more than an *ad hoc* postulate with no real meaning, a simple *flatus vocis* whereby it is *said* that it is always possible to compare a logical subject (denoted by S) with a property (denoted by M or P), whichever terms S, M, and P we choose to consider.

Let us illustrate what this means by calling once more on our Bantu

Saul. Suppose that somebody pronounces the sentence: "Saul is primitive". Wundt's Auxiliary Rule says that we may compare him with a
Gegenstand which clearly must be in some degree "self-supporting" and
which results from an undefined categorical displacement of the property
of being primitive (which we may or may not assume to be operationally
defined in advance; Wundt does not speak about that.) It certainly must
have been his intention that we call this object "das Primitive", or else
"das Primitiv-sein", since he regards the substantive language form as a
device that makes it easier to treat a property as on object. According to
the Auxiliary Rule we shall therefore be *able* to compare Saul, or 'Saul',
with 'das Primitive', as we shall *have* to do in order to judge whether Saul
is primitive or not.

Or take as another example Hegel's statement "the moon is an earth"
(VIII-2). In order to settle its truth, Wundt would say, we should have to
perform a transformation on the predicate, a categorial displacement of
the property-concept '(to be an) earth', in order to obtain an object-
concept with which 'the moon' may be compared. It seems likely that this
object-concept may be characterized as 'the earth'.

(ii) *Husserl.* The said Auxiliary Rule is of no less importance in the
logic of Wundt's sometime student, the influential Edmund Husserl. In
his *Ideas on a Pure Phenomenology and Phenomenological Philosophy*
Husserl bears witness to "the Law of "Nominalization", according to
which a nominal form corresponds to every proposition and to every
partial form distinguishable within the proposition: to the sentence itself,
say "S is p", corresponds the nominal that-sentence; to "is p", for
example, the being-P ["das P-sein"] corresponds, in the subject-place of
new sentences; to the relational form "similar", Similarity; to the
plural form, Plurality, etc. The concepts which arise from "nominalizations", and which are thought as definite solely through the pure
[grammatical?] forms, constitute formal-categorial modifications
of the idea of Being-an-object ["Gegenständlichkeit"] as such
and furnish the fundamental concept-material for formal ontology including all the formal-mathematical disciplines. This thesis is of decisive
importance for an understanding of the status of formal logic as the
logic of Apophansis, and of universal formal ontology" (Husserl 1922/
248f.).

It ought to be clear that this, for what it is worth, did not originate
with Husserl.

The problem of the logic of the German definite articles that are introduced in these nominalizations, and the difference from the logic of indefinite articles, to which Husserl in principle ascribes another logical
function (II-3), is clearly of the greatest logical and philosophical importance.

This section should be compared to Frege's discussion with Kerry,
and especially to *On Concept and Object* (Frege 1960, especially p. 46).

4. ARGUMENTS BY ANALOGY

The chapter on inference in Wundt's logic contains a discussion of Arguments by Subsumption which consists of three parts: a. *The real arguments by subsumption*, b. *The probability-argument*, and c. *The argument
by analogy*.

Wundt illustrates the class of arguments by subsumption by means of
this example (333):

> The duckbill of New-Holland possesses lacteal glands;
> those animals which possess lacteal glands are mammals;
> hence the duckbill of New-Holland belongs to the mammals.

If the order of the premisses is reversed we have an argument in the first
syllogistic figure: *the M's are P, the S is an M, hence the S belongs to the
P's*.

Like Hegel, Wundt compares arguments by analogy to arguments by
subsumption or, as Hegel called them, categoric arguments. Like Hegel
he sees the former as approximations of the latter: arguments by analogy
are improper arguments by subsumption, he says. And like Hegel he
therefore casts arguments by analogy into a form with only two premisses,
much as an Aristotelian syllogism. "For we argue," says Wundt,

> "*M* has the property *P*.
> *S* resembles the *M* in the properties *a, b, c,....*
> Hence *S* probably also has the property *P*."

In Hegel's and Mill's example of the inhabited earth Wundt replaces "the
moon" by "Mars", or more precisely, by "der Mars":

The earth is inhabited.

(The) Mars resembles the earth in many properties.

Hence (the) Mars is possibly or probably also inhabited

(346). For clarity's sake I have omitted the properties which Wundt, unlike Hegel, mentions in the minor premiss.

He holds that the minor premiss of an argument by analogy, *(the) S resembles (the) M*, may always be split up into two judgments, and that this is the reason why some authors, "following a hint given by Aristotle, have ascribed *three* premisses to the argument by analogy (*M is P, M is A, S is A*)". We recognise the A-form of simple arguments by analogy which we formulated in VIII-4. But he cannot accept this form as basic: "This representation [of an argument by analogy] is inadequate, precisely because that connection between the concepts *S* and *M* by means of the analogy-term *A* which is the characteristic mark of the argument by analogy, disappears completely when we draw the conclusion: "*S is P*" directly from three simple premisses" (348).

He holds, then, that the existence of a *relation* between S and M, in our example between Mars (or the moon) and the earth, is not brought out clearly in a formulation with three premisses, and that this relation is the rational basis of the argument by analogy. When we write down *M is A, S is A* as two separate premisses, that relation is not, in his opinion, mentioned explicitly enough.

Now I have already discussed the possibility of casting simple arguments by analogy into the B-form, whose premisses are logically "equipollent" with those of the A-form (cp. VIII-5), and we can easily agree with Wundt[7] that a formulation containing a reference to the relation of similarity between the two entities, as the B-formulation does, is more adequate from an epistemological point of view than a formulation which only contains references to monadic properties.

But Wundt's logic of concepts has further systematic features which, as we shall see, lead to disastrous consequences in the theory of inference. Our suspicions are raised by his own description of the difference between an argument by analogy and an argument by subsumption. He explains that in an argument by analogy, the subject (called) *S* is "not a special case of *M*, but a case which resembles *it*" (348; my italics: "it", "demselben"). But if *S* is (a name of) an individual or an individual concept, then there can be no concept or other entity (called) M such that it makes sense to say S *is a* (*case, instance of*) M and S *resembles* M. Take for example $S =$ "the moon" (or "Mars"), and $M =$ "the/this earth", or "Tellus". The proposition "Mars resembles the earth" makes sense (and is even true, though that is irrelevant here), but "Mars is a (case, instance of) the earth" is nonsense – because of the definite article.

5. From "resembles" to "is": a transition to mysticism

Wundt, who has only one kind of variable, does not see this point. Towards the end of this discussion of arguments by analogy he says: "These deliberations also show that the argument by analogy passes over, without any sharp borderline, into the argument by subsumption, which is founded upon induction. This is in the nature of analogy. When we become acquainted with many objects belonging to the concept M, and are able to demonstrate [the presence of] properties agreeing with S in all of them, then little by little we shall approach a borderline, where the premiss "S resembles M" passes over into the other: "S is M". The argument by analogy has thereby become an exemplifying argument by subsumption" (352).

But this is quite absurd, as is easily seen when one keeps in mind that his point of departure was this (opaque?) similarity or (intensional?) analogy:

Mars resembles Tellus.

If one knows in advance that these are two different planets, the discovery of any number of additional common properties is of no avail; we shall never arrive at the judgment expressed by: "Mars is Tellus", in the sense of strong total identity, as expressed by means of the proposition "I (Mars, Tellus)".

There are in all three possible interpretations of Wundt's assertion about the limiting case, in which the meaning of "resembles" as he says "passes into" the meaning of "is". The inferential step from

(1) S *resembles* M ("Mars resembles the earth")

(2) S *is* M ("Mars is (the, an) earth")

is either:

(a)   an application of PND (cp. VII-17), because the conclusion is
            meant as strong total identity;

or

(b)   an argument with a Leibnizian complete equality ("$\underset{\mathrm{CE}}{=}$") as
            its conclusion. This presupposes that the only known differ-
            ences (negative analogies) between Mars and Tellus are their
            spatial co-ordinates (cp. VII-16);

or

(c)   an application of An-WI (cp. VII-18), in which case the con-
            clusion is meant as a kind of weak identity, which we shall call
            "$\underset{\mathrm{EI}}{=}$".

In other words: the conclusion (2) must either mean

(2a) $I(S, M)$

or

(2b) $S \underset{\mathrm{CE}}{=} M$

or

(2c) $S \underset{\mathrm{EI}}{=} M$

which may be written

(2c′) λS is M, read: *the S (as such) is M.*

The interpretations (2b) and (2c) are somewhat less embarrassing to
Wundt than the completely illogical (2a). But Wundt does not distinguish
between the three forms (2a), (2b), and (2c).[8] He has only one symbol for
every kind of "is", and nothing corresponding to the symbol "λ" in (2c′).
What is more, the proposition "Mars is the earth" does not have the form
of the minor premiss of an argument by subsumption, *S is (an) M.* Wundt
is deceived by the habit of talking about the individual Tellus by means
of a definite description *the M, i.e.* "the earth", together with the traditional
logicians' habit of paying no attention to articles. Hegel consciously exploited both of these practices, supported by an old morphological-
phenomenological belief in the epistemological power of well-chosen

(logophoric) descriptions (VIII-20, VIII-n. 29). Wundt's Auxiliary Rule and Husserl's Law of Nominalization are remnants of this belief.

In order to see that Wundt's logic of analogy and identity is unsound, no modern logical distinctions are needed. Haenssler, whom I have had occasion to mention several times already, is not trained as a mathematical logician but judges Wundt's logical slide from *S resembles (the) M* into *S is M* to be "totally false" (1927/117). This slide has the following shape:

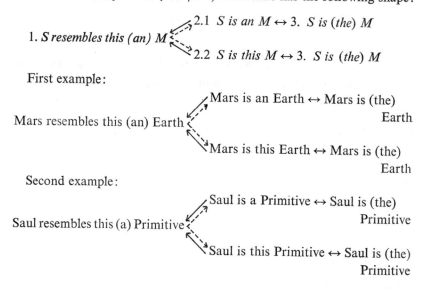

$$\text{1. } S \text{ resembles this (an) } M \begin{cases} 2.1 \ \ S \text{ is an } M \leftrightarrow 3. \ \ S \text{ is (the) } M \\ 2.2 \ \ S \text{ is this } M \leftrightarrow 3. \ \ S \text{ is (the) } M \end{cases}$$

First example:

$$\text{Mars resembles this (an) Earth} \begin{cases} \text{Mars is an Earth} \leftrightarrow \text{Mars is (the) Earth} \\ \text{Mars is this Earth} \leftrightarrow \text{Mars is (the) Earth} \end{cases}$$

Second example:

$$\text{Saul resembles this (a) Primitive} \begin{cases} \text{Saul is a Primitive} \leftrightarrow \text{Saul is (the) Primitive} \\ \text{Saul is this Primitive} \leftrightarrow \text{Saul is (the) Primitive} \end{cases}$$

6. NEGLECT OF THE ARTICLES IN DESCRIPTIONS AND THE EXACT ARGUMENT BY ANALOGY

In Wundt's logic of concepts, 2.1 and 2.2 are indistinguishable because no attention is paid to the logical role or roles of the elements of language which we call articles. Wundt's Auxiliary Rule of Concept Formation, and Husserl's Law of Nominalization as well, supply traditional logicians and philosophers with a "theoretical" basis for their habit of neglecting the logic of the articles. For if it is permissible to put the definite article before any term whatsoever, then it may as well be omitted; and Wundt frequently does omit it. He sometimes says *S* resembles *the M*, or *S is similar to the M*, and sometimes *S resembles M, S is M*.

Observe that it makes no difference at all in this connection whether logic is said to be based upon "merely subjective" concepts, upon "objective" concepts, or upon still other kinds of *Gegenstände*; such distinctions are quite ineffective here and cannot be used to meet our objections.

Faithful to traditional ideas, Wundt wanted to accommodate the logic of relations within the theory of the copula. But he wanted to do so without distinguishing different meanings of "is". Consequently, he was forced to neglect the functions of the articles. In order to make this neglect more plausible he had to invent a "device" like the categorial displacement and to formulate his Auxiliary Rule.

As the examples of his logical slide will show, Wundt cannot avoid the conclusion: "Saul is Sam" or "Coriscus is Socrates", any more than Fichte can. His systematic logic contains PND, the Principle of the Non-Distinction of the Distinct, and must therefore be subsumed under PPL.

This was most certainly not his ambition, and recently he has even been characterized as a person whose "main concern in logic was exactness in formal derivations" (Wellek 1967). In this connection his *exact argument by analogy*, or rather: his attempt to construct something worthy of that name, should not be passed over.

It turns out that Wundt understands the minor premiss of a two-premiss argument by analogy *S resembles the M*, as an expression of a four-term proportion: the relation of S to the property A is the same as the relation of M to this property. For this relation he uses the sign for algebraic division,[2] and writes

$$S:A = M:A.$$

This is to say that he interprets the minor premiss as an analogy in Aristotle's sense of "ἀναλογία" (cp. VI-11). *A* stands for the property which the entities called *S* and *M* have in common. In other words: he uses the division sign as a name for the methexis-relation (cp. VII-12). If we call that relation "δ", then the fact that S and M both participate in A may be described in the following manner: $S\delta A$ & $M\delta A$. Here "δ" refers to a binary predicate, a name for a dyadic relation. His use of the usual sign for the algebraic operation (function) of division in combination with the identity sign "$=$", wrongly suggests that the methexis-relation is a triadic functional relation. It suggests that methexis is a matter of *degrees* and that the degree of participation of S in A may be understood as a quotient, to be expressed as $S:A$. And indeed Wundt refers to *regula de tri*, upon which he wants to base his exact argument by analogy. He believes that in this manner a "factor" may be determined, which I shall call "ι", so that the conclusion of an argument by analogy may be written:

$$S = \iota P$$

(349). An example would be:

Mars $= \iota$ Inhabited.

Such an argument by analogy he calls "exact", which can only mean that a modal operator like "possibly", or "presumably", or "probably", has become redundant.

It seems likely that Wundt did not grasp the important difference between the judgment form

(ER) [aRb & cRd],

containing a second-order quantifier which *asserts the existence of a relation* between a and b which also holds between c and d, and the form

$$a:b = c:d,$$

which *presupposes the existence of a quotient* of a and b, and which is used in algebra to express the equality of this quotient with that of c and d. Wundt, and many other traditional philosophers as well, resort to the latter form when they want to express identity of structure.[9] One then easily makes the mistake of assuming that *all* properties, even the property of having (any) inhabitants, are *intensive properties* or *intensities*, *i.e.* properties susceptible of intensive magnitude, just like the properties Warm and Fast.

It seems that Wundt's efforts to construct an exact form of argument by analogy should be seen in the light of the traditional logical dimension, as an attempt to determine the terms in McTaggart's C-series (cp. VII-20), rather than as an excursion into the foundations of the theory of probabilities.

7. THE DESCENDING ARGUMENT BY ANALOGY
versus THE *dictum de exemplo*

In Wundt's *Logik* we find a discussion of the eighteenth-century logician J. H. Lambert who listed a *dictum de exemplo* among the basic principles of his logic and who insisted upon its independence of the old *dictum de omni*.

The example which Lambert gave of an expository syllogism in the third figure was:

        The earth is inhabited.
(1)   The earth is a planet.
        Consequently at least one planet is inhabited.

The validity of this argument rests, in Lambert's opinion, upon the validity of the *dictum de exemplo*. Wundt quotes this example in his discussion of Lambert's dictum (Wundt, *op. cit.* / 315).

Wundt's critique of Lambert is of great interest for us. He holds that the conclusion of an argument in the third figure either is an *example*, or else a statement with general import; in the latter case he calls the argument "an induction". What he says (315; rules 19–12 from below) becomes more easy to comprehend when expressed schematically:

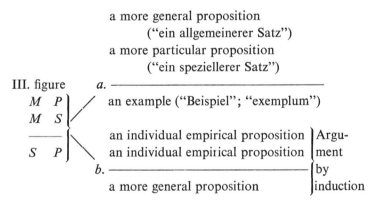

This shows that Wundt calls the expository syllogism under *b.* an "inductive" argument form. However, if its conclusion is given the foi m *some S is P*, it corresponds to Lambert's *dictum de exemplo*, and also to the simple modern rule of existential generalization, but applied to a domain of empirically observable things – which is an epistemological condition that has nothing to do with formal logic. Wundt, then, does not distinguish between induction and existential generalization in an empirical domain.

In *a*. Wundt uses the word "example", or "Beispiel". It might be thought that this a more terminological coincidence and that there is no systematic connection between this form and Lambert's *dictum de exemplo*. But in Wundt's mind there certainly was such a connection, and a very revealing one at that, as I shall now proceed to show.

As it turns out, Wundt denies that the three rules or *dicta* which Lambert assumes as fundamental in addition to the *dictum de omni* are really independent logical principles. It seems to me that what he says about this question is characteristic of German idealism as a whole even up to the present day. Wundt holds that "these rules are subject to the same *presupposition* ["Vorurtheil"] as the *Dictum de omni et nullo* itself. They presuppose that *in every argument one of the premisses is a general proposition, to which the other adds a particular case*, whereupon the conclusion produces the connection between the two" (*loc. cit.*, my italics).

He offers a technical justification for this view: "Only under this assumption is it correct that given the configuration PM, SM, one of the premisses must be negative. For it is of course only then that the transformation into a valid ["bindend"] argument in the first figure is possible" (*loc. cit.*).

We need not go into the technicalities of reduction to a syllogism in the first figure in order to recognize the argumentation which Wallis and others used in order to show that Hospinianus' new syllogistic modes were redundant. These modes, they said (cp. V-9), are redundant if and only if we may assume that an individual premiss is in reality a universal premiss of sorts; which is clearly the case (they said), since the meaning of an individual proposition *S is P* is the same as that of (*all of S*) *is P*; and on the traditional analysis of quantified propositions, that is precisely the logical form of a universal proposition (cp. IV-8).

Wundt feels that his own argument suffices to establish the existence of an underlying *Vorurtheil*, and goes on from there to conclude: "The Dictum de exemplo suffers from the same presupposition ["Vorurtheil"]. The conclusion in an argument *MP*, *MS* becomes an example only when one premiss already *has been thought as a more general one*, and the other as a more particular [special] one" (*loc. cit.*; my italics).

So the *dictum de exemplo*, though it may suffice as a basis for expository syllogisms, presupposes, according to Wundt, another and more fundamental principle or *Vorurtheil*, which also is the (semantic) basis of the *dictum de omni.* Abstracting from the syllogistic figure this is clearly the "principle" which, on the strength of Wundt's own words, I have put down as *a.* above.

Suppose we perform a *conversio simplex* upon the minor premiss in Lambert's argument (1), an operation which clearly must be valid in the article-free concept-logic of German idealism when performed upon propositions which do not overtly contain any quantifiers. After this conversion Lambert's premisses become practically the same as those of Hegel's example of the inhabited earth:

> The earth is inhabited.
> The/this planet is an earth.

And suppose, first, that the subject term of the major premiss is understood as a uniquely referring definite description; then the conclusion

> the/this planet is inhabited

can certainly be drawn validly (cp. III-7). Suppose, secondly, that the major premiss is "thought as a more general one", as Wundt recommends; then we have to do not with Lambert's example of the inhabited earth, but with Hegel's. The old theory of the *vis universalis* of singular propositions makes it plausible that the major premiss can be "thought as a more general proposition", as Wundt puts it.

Since a *conversio simplex* of the minor premiss may be assumed to be valid from an id-logical point of view, we may also consider the possibility that the causal relation sometimes worked the other way around. In other words, it may well be that the fundamental character many people have ascribed to the argument form

> *ı M is P*
> *S is an M* (or: *this M is S*)
> ——————————
> *hence S is P*

has been one of the reasons why the interpretation of a singular judgment with a *vis universalis* could triumph for so long. The two principles logically support each other.

The *dictum de exemplo* and expository syllogism are, on this traditional theory, based upon the "descending" Argument by Analogy.

8. IDEAL INSTANTIATION

In Wundt's opinion the conclusion in an argument in the mode *a.* is an example, hence a judgment about a species or about an individual. This must mean that Wundt has not grasped the intended meaning of the words "es giebt" and "ist wenigstens ein" which Lambert employs in the consequence of his *dictum de exemplo*. True, from the premisses of (1) one can also draw this conclusion: *this S* (the one which in the premisses is called or described as *M*) *is P*. But it is also possible to draw a "particular" or existential conclusion. A particular, or existential, judgment may be *introduced* on the grounds of individual premisses. Lambert felt that this ought to be expressed explicitly.

Wundt did not understand the importance of doing that. Neither in his logic nor in Hegel's is there any inference rule akin to the form of an expository syllogism in the third figure. If I have understood Hegel correctly, his Argument by Induction is an inference form with an arbitrary number of individual premisses, aimed at establishing a judgment about a genus. He rightly deems the conclusion of such an argument "problematic". But the conclusion of (1) is anything but problematic, and unless it is analysed as a subject-predicate sentence

> (at least one planet) is inhabited

it is not a judgment about a genus 'the Planet' at all.

In Hegel's and Wundt's logics the form of an argument by analogy is clearly more fundamental than the form of an expository syllogism in the third Aristotelian figure, so that the latter form as well as the *dictum de exemplo* are quite superfluous as logical principles.

This means no less than that a rule of *generalization* is reduced to a rule or principle of *instantiation*. An introduction of "there is" or "some" on the ground of individual premisses is reduced to an instantiation, namely to the introduction of a name for an individual on the grounds of a general judgment with existential import. Instead of *existential generalization* or *existence-introduction*

(2) $$\frac{(this)\ M\ is\ P}{(this)\ M\ is\ S}$$
$$\therefore\ some\ S\ is\ P$$

we find, in the logics of Hegel and Wundt, a principle of *ideal instantiation* or *id-elimination*:

$$\text{(3)} \quad \frac{\begin{array}{l} \textit{the M is P} \\ \textit{(this) S is M} \end{array}}{\textit{(this) S is P.}}$$

In Hegel's logic, the copula of the conclusion must be read "is essentially" (cp. VIII-21). In Wundt's logic, the appropriate modification may be different. Nevertheless Wundt is one of the last philosophers who wrote extensively on logical questions and who did not clearly distinguish what we today call the rule of existential generalization, which is a rule whose conclusion follows from its premiss(es) "with necessity", from inconclusive arguments by analogy and other inductive arguments (in the modern sense of "inductive", cp. Section 12). Wundt's *Logik* was published in 1880, *i.e.* the year after the publication of Frege's *Begriffschrift*, upon which rests the present-day logic of universal and existential (particular, "besondere") propositions. Peirce's analysis of the logic of statements about relations of various kinds also took place in this period.

9. The Tyranny of the First Figure and Reduction of Articles to Quantification by Means of Polyadic Predication

When an entity S_1 is used together with a relation of equality to define an equivalence class, one might if one so wished say that this entity is used as a *paradigm* of this class or property. Surely this is one of the reasons for the exaggerated confidence earlier philosophers often seem to have had in the possibilities of paradigmatic thinking? For there is indeed a valid kind of thought and of argument in which certain properties, or predicates, are transferred from S_1, to another individual, this S or S_2, "with necessity". I mean the following form, in which "R_{eq}" is a variable for names of equivalence relations:

$$\text{(4)} \quad \frac{\begin{array}{l} \textit{the } R_{eq}\textit{-ness of } S_1, \textit{ i.e. of the individual a, is P} \\ \textit{(this) S, b, possesses the } R_{eq}\textit{-ness of } S_1 \end{array}}{\textit{(this) S is P.}}$$

With "M_{eq}" for "*the R_{eq}-ness of S_1*":

(5)

>*the M_{eq} is P*
>*(this) S possesses the M_{eq}*

(this) S is P.

Here we have a "descending" valid argument in the first figure, with a prenex definite article in the major premiss. This argument form is absolutely conclusive, provided the copula of the premiss is read either as the inclusion of somebody's concept '*P*' into his concept '*the M_{eq}*', or as the inclusion of the equivalence-class under consideration in the extension of *P*. But of course it is not conclusive if the copula is read as the relation of *falling under* a concept, as Frege called it, or as the ε-relation of set theory.

The fact that this epistemologically important form is valid explains, I think, "the tyranny of the first figure" (Ross 1949/33), which we know from the logic of Wolff, Kant, Hegel, and which we found in so late a publication as the *Logik* by Wundt.

Furthermore, its epistemological importance explains why a number of thinkers have regarded the copula as symmetric. The major premiss of an argument of this form contains a symmetric dyadic predicate. The premiss itself, or in other words: its copula, is not convertible. It is easy to imagine, however, that a general philosopher who is brought up on monadic logic and who observes the symmetry of R_{eq} will be strongly tempted to think that the *copula* of the major premiss in this first-figure argument form is convertible.

More importantly, for most essentialist logicians it will be hard to distinguish (5) from the "logophoric" B-form of a simple argument by analogy in which 'λM' is one of the terms of the comparison (cp. VIII-5, 8):

(B_{log})

>λM *is P*
>$SR_{eq}\lambda M$

(perhaps, possibly) S is P.

This completely explains the subsumption, in some very influential philosophical schools, of arguments by analogy under "descending" logophoric arguments in the first figure.

The major premiss of (5), however, is not logophoric at all. For as soon as the principle of polyadic predication is admitted among the principles of syntax, then, just as in IV-13, where we discussed and modernized an argument of Ockham, the need for a prenex article in (4)-(5) disappears. Instead of the seemingly opaque and "intensional" form (5) we can now write:

$$(x) \, [xR_{eq}a \to Px]$$
$$\frac{bR_{eq}a}{\therefore \; Pb.}$$

Example:

$$\frac{\text{all individuals who in a certain "respect" } M_{eq} \text{ resemble Art, are Primitive}}{\text{Saul resembles Art in precisely this respect } M_{eq}}$$
$$\therefore \; \text{Saul is Primitive.}$$

This valid argument has nothing to do with primitive paradigmatic logic as defined in VI-2. For in order to use this argument form, a universal major premiss is required. This premiss can, however, be refuted if we are able to point out at least one individual who resembles S_1 (i.e., the individual called a) in precisely the given respect and who does not possess the property P.

Another example is:

$$\frac{\begin{array}{l}\textit{The} \text{ equilateral triangle has three equal angles} \\ \text{ABC is an equilateral triangle}\end{array}}{\therefore \; \text{ABC has three equal angles.}}$$

The equilateral triangle can be defined as a concept which is formed by virtue of an equality relation:

the equilateral triangle $\underset{\text{Df}}{=}$ the set of all triangles which are equiform with $\triangle$ DEF.

Here $\triangle$DEF is taken as paradigm. The argument has the same logical form as the previous one; again, the copula, "has", may not be read as "possesses the property of having", with the class of equilateral triangles as the logical subject.

10. SUBSTITUTION IN INDIVIDUAL PROPOSITIONS: THE ANALOGICAL DOCTRINE OF SUBSTITUTION AND THE FALLACY OF UNDISTRIBUTED MIDDLE

(i) *The analogical doctrine of substitution.* When Wundt attempts to analyse the premisses of an argument like (3) in Section 8, it is because he wants to be able to perform a *substitution*. He replaces the predicate M in the minor of (3) by the predicate P, and he does so on the strength of the major (*the*) *M is P*: "The point ["Voraussetzung"] of the major premiss $M=P$ lies precisely in the replacement of M by P..." (350). Wundt, then, indeed adheres to a neo-Platonic logic, just like Fichte and Hegel. The term M in his argument by analogy is a name or a description of an individual entity like our earth, and P is a property-name. Nevertheless he assumes that he can replace 'the earth' in the minor premiss 'Mars resembles the earth' by 'inhabited', or – in the light of his Auxiliary Rule of Concept Comparison – by 'the Inhabited'.[10]

This can be summarized in the following manner. Wundt employs the principle which we discussed in VII-16,

(12) $(K)_{-L} [Ka$ is true $\leftrightarrow Kb$ is true$] \rightarrow a = b$ is true,

in the following very strong variant:

(13) $(EK)[Ka$ is true $\leftrightarrow Kb$ is true$] \rightarrow a = b$ is true,

in combination with the principle that the consequence of (13) permits us to substitute b for a in a premiss *a is P*. I shall call this combination, which may be ascribed to the whole school which was inaugurated (or reinstated) by Fichte, "the analogical doctrine of substitution".[11] More precisely this is the doctrine that the said substitution will be in some way rationally productive and always worth making. Where this theory of substitutability is accepted there can therefore be no difference between logical reasoning and other kinds of thinking, such as day-dreaming and other forms of free association.

Barcan Marcus calls any principle of the form (12), hence also (13), "an extensionality principle" (in: Copi and Gould (eds.) 1970/280). Paradoxically it may therefore be said of great parts of the most explicitly "intensional" continental philosophical tradition that it rests upon an extremely strong principle of extensionality.

(ii) *The vis universalis of individual propositions and the fallacy of undistributed middle.* The analogical theory of substitution is equivalent to the mistake of undistributed middle in the theory of the syllogism.

In Wundt's own formulation of that principle of substitution upon which his theory of inference is for the greater part[12] based, he speaks of substitution of S for M (or of 'S' for 'M') in the major premiss M is P on the strength of the minor premiss S is P, while in his discussion of arguments by analogy he substitutes P for M in the minor S resembles M on the strength of the major M is P. He nowhere uses articles in connection with variables; the article or its meaning is clearly part and parcel of the term. In order that "the" concept 'M' may legally be replaced by "the" concept 'S' in an "identity" 'M is P' on the strength of a second "identity" 'S is M',[13] Wundt does stipulate a condition, but it is a very weak condition. He only requires that the extension of S should be contained in that of M, i.e. that $S \subset M$ be a true statement: "In the judgment "M is P" another concept S is substituted for the concept M by means of ["vermittelst"] the second judgment "S is M", a procedure which is always allowed *as long as S does not fall partly outside the extension of M*" (316); my italics).

Here Wundt refers to the *System der Logik* by F. E. Beneke published in 1842. But he overlooks the fact that Beneke formulates additional conditions that must also be satisfied if one wants to carry out a substitution logically (1842/221). For suppose the condition $S \subset M$ is indeed satisfied. If M is P is a "partial identity" in the sense of a particular proposition *some M's are P* or *a certain M is P* or another proposition with less than universal import, then the term M may still not be replaced by S, at least not if logic is to be narrower than free association. A well-known rule of syllogistic says that the middle term should be distributed in at least one of the premisses. This means that in order to obtain a valid syllogism at least one premiss must be a statement about the total extension of the middle term.

But although Wundt does not mention it, on the traditional theory of the *vis universalis* of individual propositions this requirement is satisfied in his example, since M is P is a proposition about an individual planet. This background knowledge makes Wundt's manipulation more understandable from a historical point of view.

Accoiding to present-day insights an individual proposition (*this*) M is P gives us no basis for substituting S for M, however the minor piemiss S is M is constructed, unless the former premiss is of the form $I(s, m)$; but in an argument by analogy this is not the case.

The reader might reasonably have expected Wundt to express his opinions on substitutability in terms of formulas $S = vM$, $M = vP$, or else

$S = aM$, $M = bP$, forms which he discusses elsewhere in his book (cp. VII-24 above). Especially in connection with substitutions into the premisses of an argument by analogy this seems to be a reasonable expectation. But Wundt does not do this. He may of course have had in mind, although he does not say so, that the minor be given the form of a strict (abstract) identity $S = vM$ or $I(S, vM)$, and that the factor v should be determined *before* such a premiss is used as a basis for substitution.[8] This possibility may be used to excuse him for omitting the condition that M be distributed in the major premiss. But it then becomes impossible to understand why he does mention the condition that the extension of S be included in that of M.

In any case it seems reasonable to assume that the combination of the Ockham-Wallis-Leibniz-Wolff-Kant doctrine with the lack of distinction between *at least* one "is" of identification and one "is" of predication, has been operative in generating the analogical theory of substitution.

11. DISSOCIATION OF ARGUMENTS BY ANALOGY FROM SYLLOGISTIC

Wundt tries to analyse all arguments as inference forms with three terms S, P, and M, and he tries to reduce them to syllogistic figures, if not exactly to syllogistic moods. He defends this in a logical *credo*, where his belief is so explicitly stated that it is worth quotation here:

"... nowhere is any justification to be found for making a fundamental distinction between the syllogism and other more productive forms of drawing conclusions. All argumentation may be reduced to three-term forms, with an arrangement of the concepts according to need, *i.e.* to the syllogistic forms discovered by Aristotle" (326). The arrangement of the concepts is clearly the syllogistic figure; the real Aristotelian moods seem to be of much less importance to him. Wundt's own moods are rather reminiscent of the moods of Renaissance logicians like Ramus, Melanchton and Keckermann.

Wundt turns out to be one of the last authors on logic who tried to press arguments by analogy (or rather: arguments *from* analogy) into a syllogistic strait-jacket. At the end of the *Logische Elementarlehre* by B. Erdmann there is a discussion of the relationship between arguments by analogy and (Aristotelian) syllogistic. Erdmann aptly describes that view of arguments by analogy which we have met in the logics of Hegel and Wundt, and I therefore quote him here:

"Arguments by analogy show in this form an obvious kinship with syllogisms. And not only do they seem to be shrouded in the syllogistic garment, but also to be of a *deductive* nature. For it seems to be possible to read in it the thought of a mediate predication along the lines of the *first* figure, only instead of the same ["der gleiche"] middle term we now have two which are similar to each other. The analogy between the argument by analogy and syllogisms is carried even further, into the relation of the syllogism to *induction*" (786; my italics).

As I have already said, this is not a description of Erdmann's own views on the matter. On the contrary: he quite rightly observes that "the argument by analogy, when understood as a syllogism, is an example of a *quaternio terminorum*, that fallacy which invalidates the basic requirements of the syllogism" (788, my italics). In VIII-7 and later sections I have tried to elucidate this more in detail.

Erdmann draws the conclusion that arguments by analogy should be regarded as belonging to a separate argument form, in addition to deductive and inductive arguments.

12. TWO MEANINGS OF "DEDUCTION" AND "INDUCTION" AND THE POSTULATE OF DEDUCTIVITY

(i) *What is meant by "deduction" and "induction"?* Erdmann gives us no sharp definition of what he means, or what others have meant by the expressions "deductive" and "inductive". Yet he observes, among his contemporaries or among the writers of the preceding generation, a systematic connection between their understanding of arguments by analogy and their notion of deductivity, as well as a certain inclination to associate them at the same time with a kind of inductive activity. This is to say that he observes and to a certain extent shares a feeling of uncertainty as to the range of the concepts of induction and deduction which were current up to this time.

When he says that there are two meanings of "inductive" and that the difference between them is of fundamental importance in philosophy, Veatch is completely right (1967/64f.). The traditional and the modern meanings of "deductive" and "inductive" do not only differ, but have in

fact very little in common. This important fact is, however, usually overlooked or passed over by modern authors, with the result that quite misleading interpretations of great parts of traditional philosophy are forced
upon us. A great many traditional philosophers, perhaps most, were not
trying to do what we are trying to do, and parts of their works are therefore less relevant to problems in systematic philosophy in our century than
some contemporary authors want us to think: J. N. Findlay is a good
example here. In traditional philosophy "deductio" and "inductio"
often meant the same as "descensus" and "ascensus", or at least included
these notions (cp. Risse 1964/331, on De Soto). Hunaeus, a Renaissance
logician, defines them as follows: "Descensus logicus est argumentatio
qua proceditur a voce quapiam in enuntiatione posita ad enumerationem
rerum singularum, pro quibus ea vox capiebatur.... Ascensum vocant a
singularibus ad universale progressionem" (quoted from Risse, *op. cit.*/
415n.).

Let us contrast the most important components of the old and the new
lexical definitions of "deductive" and "inductive".

Ded_{trad}. The premisses are "more general" (are true/realized at a higher
ι) than the conclusion, or at least as general ("reasoning from the general
to the particular", *e.g. ideal instantiation,* or *id*-elimination).

– Note that if generalization is regarded as a process leading to statements about an essential nucleus, then "a deductive rule of generalization"
is a self-contradictory expression.

Ded_{mod}. If the premisses are all true, then the conclusion is certainly true;
this is indicated by putting the sign "∴" before the conclusion. "Deductive" in this sense means the same as "absolutely conclusive". More often
than not this notion is nowadays fused with the notion of "(more or less)
conclusive due to verbal form", but that seems to have been different in
the time of Kant (cp. Section 19).

    a. *Instantiation.*   Here the conclusion may be an individual proposi-
                    tion, the premiss a universal or an existential (but
                    not an indefinite) proposition.

    b. *Generalization.* The conclusion is a universal or an existential
                    proposition; the premiss may be an individual prop-
                    osition, but it may also contain quantifiers.

c. *Other conclusive*
  argument forms (for instance the conclusive forms of propositional
          logic).

Ind$_{trad}$. The conclusion is "more general" (higher ι) than each of the premises ("reasoning from the particular to the general", "Aufhebung").

Ind$_{mod}$. The conclusion, which may be a universal, but never an indefinite premiss, is not completely dependable, even if the premises should all be true. The argument is not absolutely conclusive, *a fortiori* it is not absolutely conclusive due to verbal form. This is often indicated by the sign "∴" preceding the conclusion.

It will be clear that *Hegel's* Argument by Analogy is deductive in the traditional sense, and at the same time inductive in the modern sense. The same can probably be said of common (not "exact") arguments by analogy in Wundt's theory of inference, although the traditional logical dimension is not so clearly visible in his work as it is in Hegel's.

Wundt's "exact argument by analogy" which was mentioned at the end of Section 6 above is then deductive in the traditional sense but is clearly also intended to be deductive in the modern sense.

Erdmann refuses to call arguments by analogy "deductive" or "inductive". This may be explained by the assumption that he felt uncertain about these words, which in his time were about to take on a new meaning.

(ii) *The postulate of deductivity.* According to this postulate, which is quite common in traditional philosophy, "descending" arguments in the sense of Ded$_{trad}$ are logically fundamental. This postulate suggest a natural union of theoretical logic and descending metaphysics, that set of ideas which Lovejoy has aptly characterized as the Great Chain of Being and which he has described in detail in a work by that name (Lovejoy 1936). What was said in Section 9 above explains, I think, how this postulate could become so influential, since the form of "inference" we studied there seems to be a descending one while being quite fundamental from an epistemological point of view and certainly logically valid (absolutely conclusive). This inference form, if it may be called so, therefore lent to the Postulate of Deductivity a considerable amount of rational support, in addition to the support it certainly received from quite different sources and which may have been of a less natural kind.

Hegel and Wundt are clear cases of philosophers who strongly believed in this postulate and whose logics depend on its validity. Erdmann abolished the Postulate of Deductivity without abolishing essentialistic logic, but such a position cannot be maintained in the long run.

In modern logic the Postulate of Deductivity cannot be interpreted.

13. ARGUMENTS BY ANALOGY IN THE LOGICS OF LOTZE AND SIGWART

Neither Lotze nor Sigwart formulates arguments by analogy as (quasi-) syllogisms, and *a fortiori* not as first-figure syllogisms.

(i) *Lotze.* Let *S is M* be a true proposition.[14] An explanation is sought for the fact that S has the property M. Suppose it is known that the property P somehow entails the property M; then we can explain that S is M *if* we assume that S is P. What Lotze calls an argument by analogy is an argument whose conclusion is an assumption *S is P* which may be used to explain why S is M in the light of a true general statement *P entails M*. The latter statement is not listed among the premises of his so-called argument by analogy. It functions in Lotze's argument as what Toulmin calls "a warrant" (Toulmin 1964/98). If we do add this warrant to the premises, we obtain a quasi-syllogism in the second figure:

$$Px \dashv 3 Mx \quad \text{or perhaps: } the/a\ P\ is\ M$$
$$\frac{S\ is\ M}{\therefore\ S\ is\ P} \qquad\qquad \frac{S\ is\ M}{\therefore\ S\ is\ P}$$

An example of what Lotze calls "an argument by analogy" would be: "S dissolves in water, hence S is possibly sugar, since sugar dissolves in water".

The above schemas clearly differ from the A- and the B-schemas for simple arguments by analogy, which do not contain any "general" premiss. The analogy or similarity in question seems to be that between S and some P, possibly '*the P*'. Nor is the second-figure form of this argument the same as that which Hegel or Wundt ascribed to their arguments by analogy. By an argument of this kind an observed property M of S is reduced to the property P, which is ascribed to S hypothetically *if* the fact that S is M can be explained that way; that it can be so explained must be given in advance.

Lotze remarks that his so-called arguments by analogy would be conclusive if the warrant, Px–$3Mx$, were convertible (*i.e.*, if Px were strictly equivalent with Mx). However, in that case the argument could not lead to new knowledge, so Lotze says (1912/129f.). He clearly means that the first-figure argument from Mx–$3Px$, S is M, hence S is P is deductively valid in the modern sense of "deductive" (due to the validity of Barbara), and that for that very reason it cannot be the form of arguments by analogy. He does not discuss first-figure arguments with S *resembles* (*the*) M instead of S *is* M.

(ii) *Sigwart.* We know that Sigwart rejects the notion of partial identity and that he regards propositions the M *is* P with a general import as equivalent to *all M's are P*. It is therefore interesting to observe that he also describes arguments by analogy differently from Hegel and Wundt. Sigwart in no way connects arguments by analogy with conclusive arguments by subsumption. He does not even treat them in his chapter on rules of inference (1904, I/432f.) but discusses them in his methodology as a heuristic method, in a chapter *The discovery of hypotheses* (II/303f.) His analysis of arguments by analogy is that which we have rendered in schema A in Chapter VIII.

It seems that some variant of the intensive identity-theory of the copula is a necessary, though not a sufficient condition for the opinion that arguments by analogy have the form of "descending" arguments by subsumption.

14. THE FEAR OF THE COUNTER-EXAMPLE:

I. Analogical and Relative Judgments

(i) *Pfänder*, however, does formulate his arguments by analogy in the first figure:

> [*das*] Q ist P
> [*das*] S ist ähnlich [*dem*] Q
> ———————————————
> [*das*] S ist P.

I have added definite articles (in brackets) to the form given by Pfänder, because of his own example:

Der Mensch ist ein seelisches Lebewesen,
die Tiere sind *dem* Menschen *ähnlich*,
folglich sind die Tiere seelische Lebewesen

(1963/349; my italics; cp. (1) in VI-7).

Like Hegel and Wundt he compares, in the minor premiss, the logical subject '*S*' with a logophoric entity, 'der Mensch'. Erdmann did not do that. The subject of the minor premiss, 'die Tiere', is also a general term, which is not the case in Hegel's and Wundt's examples.

Pfänder points out that this argument form is not – he says: "is never" – conclusive. He says the same of arguments with a relative generic judgment among the premisses (VI-7). When the problem is to determine which conclusions do *not* follow from certain premisses, then Pfänder is always quite explicit. I add here that even Hegel did not pretend that his arguments by analogy were really conclusive in the sense of being entirely trustworthy.

In all probability Pfänder intends the major premiss of his example to be understood as an absolute generic judgment and the minor premiss does not have the form, [*das*] *S ist P*, which he normally ascribes to singular generic judgments. Since he does not treat the "ist ähnlich" in the minor and the "ist" in the conclusion as a graded copula, his discussion of arguments by analogy contributes nothing to the solution of the problem of his so-called relative generic judgments [*das*] *S ist P*. In his logic such judgments do not seem to have any consequences at all.

(ii) *Bosanquet*. The same problem is manifest in the logic of Pfänder's English contemporary, Bernard Bosanquet. The latter does not speak in terms of "relative", but of "analogical" judgments.

An analogical judgment is, so Bosanquet explains, a judgment "which treats a concrete individuality as an abstract universal" (1911, I/213).[15] He sees this kind of judgment as the "fundamental nature" of ordinary generic judgments. He tries to explain what he understands by a generic judgment in the following way: "The generic judgment is the qualification of reality *under the aspect* of a Natural Kind by attributes or relations incident to that Kind" (I/209; my italics).

Bosanquet's explanation of the role which these analogical judgments are to fill in logic is extremely revealing. He wants to avoid the word "all" in theses like the one about the sum of the angles in a triangle, and gives as his first reason that no complete *induction* is possible in this case. It is not possible to investigate whether this proposition is true or false by examining every triangle separately.

This argument for abandoning the universal propositional form does not hold water, however; for no complete induction is required here, only logical *generalization* (ug). Bosanquet does not see the difference between these two operations; and this mix-up is characteristic of the traditional way of seeing the Locke-Berkeley problem. Only recently have logicians come to realize the utility of a distinction between (1) a notion (or several notions) of complete induction, which we may call "extralogical generalization" of "non-formal generalization", and (2) notions of "logical generalization" (ug and eg).

Bosanquet, however, has yet another argument, which is perhaps even more interesting. If a generic judgment is formulated by means of the word "all" and understood as an "exhaustive judgment", then it "is helpless in the face of the most trivial exception" (I/212). Such a judgment will be falsified as soon as someone can produce one "negative case", as one often says.

For this very reason Bosanquet feels he has to reject the "all"-form as a means of expression for generic judgments. And in fact his examples of generic judgments do not contain any quantifier. They have a subject term containing a definite article, as in the "The bacillus is a septic organism", or an indefinite article, as in "A society organized on a purely commercial basis treats the working classes as little better than slaves", or else no prenex operator of any kind, as in "Man is an animal capable of social life" (I/212).

As to the question which "logical jobs" – as Geach calls it – these forms of language are called upon to fulfil, Bosanquet is a good deal clearer than Pfänder. Bosanquet chooses his point of departure as follows:

(1) *Nearly all* striped muscles can be controlled by the will. The heart muscles, however, form an exception, for though striped, they cannot be controlled by the will. In Bosanquet's opinion it nevertheless cannot be doubted that the coincidence expressed by the "nearly all"-judgment "must indicate some sort of *connection*, however circuitous" between the *appearance* of the muscle and the degree to which it can be controlled by the will. Nor can it be doubted that the exceptional case of the heart

muscle "must be accounted for by special conditions" (I/212, my italics).

This may be accepted, but he has not shown that this calls for the linguistic form.

(2) *the* striped muscle can be controlled by the will

in order to express that

(3) *there is* an as yet unknown connection (relation) between the appearance of muscles and their dependence on the will, in virtue of which one has to reckon with as yet unknown "special conditions".

Bosanquet clearly assumes this linguistic form (2), *the M is P*, to have simply the "logical" task of preventing our innermost convictions from being refuted by a counter-example, a "negative case" that may be expressed by a proposition of the simple form *this M is not P*.

Although Bosanquet took more interest in arguments by analogy (cp. Bosanquet 1911, I/92), the agreement between him and his German colleague Pfänder about the logic of generic propositions is astounding, for Pfänder says: "From the falsity of an individual judgment one may not, then, immediately infer the falsity of one of these *relative generic judgments.*[8] Human beings often make mistakes in their thinking in this respect. By taking the falsity of individual judgments as a premiss they try to show immediately [im-mediately, E.M.B.] the falsity of relative judgments which, for some reason or other, do not suit them" (259f.).

Neither Bosanquet nor Pfänder invokes statistics or probability theory. And, being unfamiliar with the possibilities offered by the combination of polyadic predication with the Fregean theory of quantification, neither of them realizes that negative cases or counter-examples to universal propositions may be of a very complex relational kind. They do not see that a counter-example may be of this kind:

$$F(a_1, a_2, ..., a_n) \text{ is false.}$$

This is the form of the description of a counter-example to a proposition of the form

$$(x_1)(x_2)...(x_n) \quad F(x_1, x_2,...x_n),$$

where $F(x_1, x_2,... x_n)$ may be a propositional function of any degree of complexity in which "special conditions" have already been included (cp. Section 9).

Bosanquet's and Pfänder's views on the function of counter-examples in logic may be explained tentatively by means of our hypothesis of a hidden parameter in traditional logic. They thought that $\theta(S$ is M & S is P) describes a counter-example to the *hic-et-nunc* proposition $\theta(M$ is P), but not to $\iota(M$ is P) for any ι higher than θ.

This does not, however, dispense them from the obligation of describing the logic of analogical or "relative" judgments *the/an M is P* in such a manner that it becomes clear how such judgments can be contested, for instance how an opponent might launch a first attack upon a judgment expressed by the sentence "the Bantu is Primitive", if such a judgment "for some reason or other does not suit him".

15. THE FEAR OF THE COUNTER-EXAMPLE:

II. *Analogical Terms*

Veatch does not explicitly talk of *arguments by* or *from* analogy, but in his *Intentional Logic* he makes a number of remarks which are of interest to the present discussion. He blames modern logic for neglecting "anything resembling either a doctrine of the categories or a doctrine of analogy", and connects this with the failure of modern logicians to deal with intentions or formal signs (137; cp. VIII-22 above). In his discussion of argument in intentional logic he also treats of "the form of argument" (301), thereby defending "the syllogism". Here he sets himself the task of demonstrating "the thoroughgoing intentional orientation of this structure". To this end he proposes "simply to disregard all the details about figure and mood and to consider directly the middle term as the key to the demonstrative power of the syllogism" (*loc. cit.*). His examples, however, if they are to be regarded as syllogisms at all, of course have to be in one or the other of the four figures; and one is not surprised to find that without exception they are arguments in the first Aristotelian figure. The major premiss is usually a universal proposition, the minor an individual or an indefinite proposition.

Veatch, as we know, belongs to the school of John of St. Thomas and Maritain. The "analogous" or "relative" judgments of idealist philosophy have a parallel in the neo-Thomist doctrine of "analogical" concepts and terms. I have already pointed out that Hegel, trying to defend his own Argument by Analogy against accusations of having committed a fallacy of four terms, probably found some comfort in this general theory of "analogical" uses of language.

In 1955 the "material" logic of John of St. Thomas was published and annotated by three American philosophers. As to "analogical" concepts and terms they express the following opinion: "Words ought to be chosen in such a way as to bring down to a minimum the risk of falsifying the analogous concept through the imposition of improper unity. But, after all possible precautions have been taken, the inclination to corrupt analogous terms into univocals is still very much alive and must be corrected with indefatigable vigilance (cp. John of St. Thomas 1965/599, n. 13).

16. PURE PHILOSOPHICAL LOGIC: ANALOGY, SYMMETRY, AND LOGICAL STABILITY

To a contemporary reader of Bosanquet's logic it is quite clear that, starting out from the old *S is P*-logic, he has struggled with problems which cannot be dealt with successfully within the old syntactic framework and the theories of meaning and reference with which this syntax can be associated. More specifically he struggled with problems belonging to the theory of functional and other relations. In the Fichtean-Hegelian tradition, to which Bosanquet belongs, all relations are expressed by means of a symmetric copula. This is also clearly brought out in his discussion of the example of striped muscles; for Bosanquet holds that the fact that

(4) the striped heart-muscle cannot be controlled by the will
is a counter-example to the "collective" judgment

(5) all unstriped muscles cannot be controlled by the will.
However, a proposition *this M is not P* can be a description of a counter-example to *all non-M are non-P*, hence *all P is (are) M*, only if this latter universal proposition allows a *conversio* so as to have the same logical implications as *all M is (are) P*. This indicates that Bosanquet is primarily interested in symmetric relations in the sense of conversely functional relations, or one-to-one-mappings. His own term for this notion is "reciprocity" (cp. I/246–249). This term is probably a translation of the German "Wechselwirkung", as used by Kant, who took this symmetric notion as a fundamental category of Reason, "derivable" from the dis-

junctive propositional form of the logic of his time, and as used after him by Hegel.

Observing his covert assumption of the symmetry of "is" and his overt reference to reciprocity it ought not to come as a surprise that, according to Bosanquet, what he calls "philosophical logic", which he opposes to the "symbolic logic" of Russell and others, "is interested in the conditions of logical stability" (II/45). For symmetry is closely related to absence of change of any kind and hence to stability in the material world of facts and things, and for an idealist that world is essentially indistinguishable from the world of judgments and concepts. Hegel, Bergson, Bosanquet, Jacoby and Veatch all reject the view that the category of asymmetric transitive relations ("series") is logically fundamental. Kant may have seen that asymmetry is logically important, but nevertheless denied that it is a logically *basic* category. Hegel's relationship to asymmetry is complex, but in propositional logic (as we would call it) he went even further than Kant in as much as he regarded the disjunctive argument form with its symmetric "either-or" as more fundamental than the hypothetical argument form with the non-symmetric "if-then".

Secondly, since the theory of negation in the school of Fichte and Hegel is an even greater muddle than in the logic of Immanuel Kant – which is to say a great deal – in the world of *positions* expressed by indisputable analogical propositions the correlation between reciprocity and stability is no less obvious. Such *Standpunkte* are indeed immune against arguments from a critical opponent, *i.e.* they are *logically* stable.

On this point, there is a complete and surprisingly explicit agreement between the Hegelian Bosanquet and the neo-Thomistically inspired thinker Veatch. Where the former speaks in terms of "philosophical logic" and von Freytag-Löringhoff of "die reine philosophische Logik", Veatch, as we know, says "what-logic". Veatch reproaches the usurper, called by Bosanquet "symbolic logic", by von Freytag "Logistik", and by Veatch "relating-logic", for being "set up entirely for the purpose of enabling us to "go on" cognitively, to get from one point to another" (1970/256). Exactly the same complaint is voiced by the Dutch Calvinist philosopher Popma.[18] Another Hegelian whom we have already met in Chapter VIII, *i.e.* J. N. Findlay, formulates it as follows: "Philosophy may well, therefore, have the task, of putting the philosopher back into the chains of necessity – chains willingly and happily accepted – after he has enjoyed the intoxication of a liberating logic – such as that of Russell in *Principia Mathematica* – which has given him too many wings". (1970/120). Needless to say, the present author sees the task of philosophy differently!

17. A PROBLEM OF MODERN LOGIC: THE LOGICAL FORM OF SCIENTIFIC LAWS

A warning should be voiced at this point lest someone should think that the opinions of Hegel, Bosanquet, Pfänder, Maritain and Veatch are forerunners of the present-day discussion about the logical form of scientific laws ("LFSL", for short) and the related problem of counterfactual conditionals, and that in the light of this lively discussion [19] the ideas of the afore-mentioned traditional logicians acquire a certain plausibility. Certainly modern philosophers agree that it will not do simply to ascribe to scientific laws the simple form *all S are P* as their logical form, without characterizing the lawlikeness in any other manner. To start with, not all true universal propositions express basic connections or laws; many such propositions are only accidentally true and not fundamental enough to merit the status of a law. If, however,

(1)   *all S are P*

rests upon a law, then given that law the counterfactual conditional

(2)   *if a* (who is not an S) *were S, then a would also have been P*

must also be true for any *a*. Now even apart from the problem of determining the truth-value of (2), it cannot be said that (2) follows from (1). Counter-examples are easily found. Modern logicians commonly require, however, of an LFSL that it should logically imply both (1) and (2). This has the result that if such a form can be found, and if a statement, *p*, has this form, then, if the corresponding statement (1) is not true, it will be necessary to make certain changes in *p*, either by imposing special new conditions, or by choosing other general terms (predicates), or by changing the stipulative nominal definitions of the terms used, or else by specifiying the *domain of discourse* and the *ranges of the various quantifiers* in a new manner, as discussed in VII-18.

The requirement that a scientific law which is in no need of revision should logically imply a proposition of the form (1) is a necessary one in order to reduce logical stability to its absolute minimum.

A common position is to give up the search for a special LFSL and to say that scientific laws are characterized not by any special logical form (they may, for instance, be of the simple form (1)), but as having a certain *status* in the scientific world.[20]

One might agree to that, but then it might well turn out to be a good idea to introduce a new logical operator by means of which this status can be expressed. The deductive part of the logic of that operator would then presumably be much more like that of the necessity-operator in modal logic than like the "logic" of a logophorically used article.

The connection between (2) and the status of scientific laws can then be explained by defining (2) – if its truth is assumed to rest upon scientific knowledge – as a meta-linguistic statement of the following kind (Hiż 1951, Walters 1967):

(2')   (En) (E$p_1, \dots p_n$)[$p_1, \dots p_n$ have the status of scientific laws &
         $p_1, \dots p_n \vdash (Sa \to Pa)$] & Sa is false.

18. ANALOGY, MORPHOLOGY, AND JUDGMENTS AND ARGUMENTS δι' αἰσθήσεως: DESCENT OR ASCENT?

Bosanquet compares his analogical judgments, which are indefinite judgments, to aesthethic judgments (I/215). It is a reasonable hypothesis that his opinion is somehow related to the view, held by Alexander of Aphrodisias and others, that proofs resting upon ἔκθεσις, were proofs δι' αἰσθήσεως and incapable of algorithmic ("formal") representation.

Bosanquet's interest in morphology, obvious from the sub-title of his work, indicates systematic connections between his logic and the mode of thought made famous by Goethe, among others. Historically and systematically this logic is also closely related to the fundamental tenets in the speculations of the medical philosopher Paracelsus and, more recently, of Samuel Hahnemann, especially their pharmaceutical principle of *similia*. A full discussion of its relation to neo-Platonic logic would go beyond the limits set by the topic of this book. Suffice it to say that Hahnemann's medical principles were accepted by Hegel (1928, IX/712, 714).

Bosanquet's preoccupation with examples from botany and other taxonomic sciences is remarkable (I/89).[21] In his opinion it would be a mistake to reduce analogical judgments to modal judgments.[22]

Maritain recognizes in addition to *raisonnement par analogie* ("argument by analogy") also a *connaissance analogique* ("analogical knowledge"; 1933/336). He warns his readers that these notions are entirely distinct, since in analogical knowledge no inference is made. It is therefore reasonable to assume that a description of a sample of analogical knowledge in his sense is an analogical judgment in Bosanquet's sense, and possibly also a relative judgment in Pfänder's sense. But it is also reasonable to assume that there is a connection, which may conceivably be discovered and described in a problem-historical investigation of the present kind, between the notion of an argument by analogy and the notion of analogical knowledge. Maritain's own description of analogical knowledge is extremely brief and gives the impression that he has two (different) things in mind. One of these is the use of equivocal, or "analogous", terms for the purpose of the description of structural equality (isomorphisms; cp. VIII-9). But he also seems to associate this with a notion of abstraction, for he ends his brief description of analogical knowledge by calling it: "... an inadequate knowledge, no doubt, but one which may be absolutely certain" (*loc. cit.*).

This, if true, points to at least one use of the word "analogy" in which an ascending, not a descending form of thought is meant.

The expression "Wesensinduktion", which may be translated as "essential induction", is used by Elzer (1967/131), who maintains – rightly, I believe – that even Francis Bacon, in his discussion of inductive methods, had this (unclear) Aristotelian notion in mind. Leonard Nelson uses the expression "regressive method" for every rational method which is assumed to lead "from the particular to the general", and classifies both induction (in an undefined sense) and abstraction as regressive methods. Nelson maintains that until J. F. Fries, the notion of induction and the notion of abstraction were systematically confused (1962/570). The philosophers of the past were all uncertain about what to understand by the Aristotelian term "ἐπαγογή". In terms of our hypothesis of a topological dimension in traditional logic this means, if Nelson is right, that induction, in the traditional sense of ascending along this dimension, and the act of defining a property "by abstraction" were taken to be one and the same logical operation or process. This may sound very strange to a modern ear; three remarks are necessary:

(1) In some schools, notably in the idealist school of Hegel, there is little or no difference between the notion of a judgment and the notion of a concept (cp., *e.g.*, Bosanquet I/33). This situation of course facilitates the identification of abstraction and induction (in any sense of "induction"), and *vice versa*.

(2) Concept formation "by abstraction", in the sense of the formation of equivalences, logically presupposes the truth of certain propositions with second-order quantifiers. For such quantifiers traditional logic had no theory at all. It seems that this fact has induced some traditional logicians to adopt the procedure of simply dropping them, especially second-order existential quantifiers, from the judgments in which, according to modern logicians, they are requisite. The ensuing "judgment" is traditionally understood as logophoric and as "induced".

(3) No distinction was drawn between induction in the traditional sense and logical generalization in the sense of universal generalization; to demonstrate this has been one of the main tasks of the present work. The reader might, at this point, like to look again at sub-section (ii) of Section 13 of the present chapter.

Every author necessarily accepts one of the following positions:

(i) indefinite judgments *the/an M is P* are logically and semantically *sui generis*. (This is the case, for instance, in Hegel's *Science of Logic*, but in this respect that work seems to be misleading as a description of the pattern of this philosophical thesis.) It can be said that Hegel's logic is both confused and incomplete on this point; see next section);

(ii) indefinite judgments *the/an M is P* rest upon other judgments, though this does not mean that they must be obtainable συλλογιστικῶς. A judgment which originates δι' αἰσθήσεως *from* certain material may also be said to be inferred. The problem then is, first, how to describe the material which forms the basis of an aesthetic-analogical judgment in Bosanquet's sense, or of a sample of analogical knowledge in Maritain's sense, and, secondly, how to describe the nature of the inference itself.

19. KANT'S TRANSCENDENTAL DEDUCTIONS

Kant's notion of a "transcendental deduction" includes neither Ded_{trad} nor Ded_{mod} (cp. Section 11). The word "deduction" must here be taken, to start with, in the broad and non-committal sense of "inference" or "derivation".

A particularly succint description of the original feature in Kant's logic has been given by Nicolai Hartmann. Before Kant, the "postulate of deductivity" was absolutely unchallenged: it was generally assumed, that one can only *derive* something from "general" judgments with an onto-logical import. This is to say that the concept of derivation, was identified, extensionally at least, with Ded_{trad}. Now, even when this identity is assumed,[23] "it does not follow that the general statements from which the derivations are made, also in fact tell us something "ontically" general. If they do not, then they are false statements – or, in the language of Kant, synthetic *a priori* judgments with no "objective validity". Of course derivations can just as well be carried out, in a logically valid manner, starting from such statements as from true statements; only the conclu-sions are [read: need be; E.M.B.] no more true than the premisses.... It is quite clear that the central argument of the *Critique* is directed against precisely this mistake, and that it is done in the following manner: against all such deductions – against [the principle of] ontological deduction, as it may be called – the *Critique* sets a "transcendental deduction", which deals precisely with the "objective validity" of such *a priori* principles, due to which synthetic *a priori* principles are true" (Hartmann 1942/208).

In the terminology of the present study, which rests upon the hypoth-esis of a special logical dimension in traditional logic, the corresponding logico-linguistic problem can, and should, be expressed as follows: pre-Kantian thinkers saw only the logical problem of *id-elimination*, in all likelihood because they recognized no other possible type of inference. Kant's achievement is to have pointed out convincingly the fundamental importance of the problem of *id-introduction* in traditional logic.

As Hartmann puts it: "the Kantian Critique is not really directed against the foundations of the old ontology [and logic, E.M.B.], but against the speculative-rational metaphysics that had been built upon these foundations. Above all, Kant drops the deductive schema of [all] procedure" (*loc. cit.*). In other words: the old logical-ontological dimension is retained in his philosophy, but Kant denies that philosophical procedures are necessarily *descending* procedures. In a philosophy which is not bound up with traditional logic Kant's problem has no obvious meaning.

After having identified Kant's problem as the problem of id-introduction, we are in a position to profit from the detailed scrutiny of the *Critique* presented some twenty-five years ago by Aebi; to my knowledge this work has never been translated and must therefore be assumed to be practically unknown to the English-speaking world (Aebi 1947). Aebi points out:

(i) that the possibility of carrying out a valid introduction – a valid "metaphysical deduction" – of a category or pure concept rests upon the validity of Kant's "transcendental deduction";

(ii) that the transcendental deduction is an argument in the first syllogistic figure,

(iii) which according to Aebi is in the mood Barbara (but which in the opinion of the present author might more aptly be described as an argument in the mood Burburu);

(iv) that this argument (like Hegel's Argument by Analogy) contains a *quaternio terminorum*, the fallacy of four terms, the middle term being used in two different senses.[24]

As Aebi shows (317–324), Kant offers more than one version of his transcendental deduction, but what was said under (i)–(iv) holds for each of them. One version may be schematized as follows (Aebi *op. cit.*/320, 322):

> Every objective unit of apperception is a unit according to a
>   rule (category in the wider sense).
> What makes the given-ness of a multiplicity possible is a
>   transcendental unit of apperception.
> _____
> What makes the given-ness of a multiplicity possible is a unit
>   according to a rule (category in the wider sense).

This argument is supposed to show that id-introduction (of the very special and meagre set of categories or pure concepts which Kant recognized) can be carried out validly, but is not itself an (instance of an) id-introduction. The distinction objective-transcendental is not a distinction along the θ-λ-dimension, but if Aebi's analysis of Kant's argument is correct, the fact remains that even Kant erects his philosophy on a logic which consists in a Barbara-like argument with a *quaternio terminorum.* It makes it comprehensible why his attempt to break with the traditionally deductive pre-"critical" philosophy was only moderately successful and could even be taken by some (though not by Kant himself) to make such philosophies as those of Fichte and Hegel legitimate.

Recalling Kant's own high opinion of the first figure and his rejection of Lambert's *dictum de exemplo* in favour of a descending principle of subsumption (Sections 7, 8), as well as the opinions of W. Wundt about the close relationship between arguments by analogy and arguments by subsumption, it would be tempting to go on to follow Kant's discussion and use of subsumption, and Aebi's discussion of Kant's discussion (esp. Aebi *op. cit.*/194). But this would lead us too far and we therefore take leave of Kant here.

Suffice it to say that Kant raised a very serious problem in traditional logic, a problem that he did not solve, and that the false solution which he offered – his transcendental deduction and the metaphysical deductions inaugurated by it – has produced immense confusion by being neither clearly deductive or clearly inductive, in either of the meanings of these two words defined above.

In the last part of this book the problem of id-introduction will be taken up, with Aristotle, not Kant, as the natural point of departure.

NOTES

[1] Here are some examples of statements of the traditional view: "It is difficult to grasp the meaning of connection or relation. Perhaps it is not inappropriate to define it as the conscious togetherness of objects" (Erdmann 1907/97). "Ontologically, After-each-other is impossible. It is a gnoseological construction on our part" (Jacoby 1955/630). "Now a relation is an immaterial tie laid between two or more phenomena by the intellect: relations are not perceived but inferred" (Kruijer 1959/44). Cp. also Peters 1967/419, 421, 434.

[2] Nevertheless Wundt believes that he can distiguish between the meanings of "Ver-hältniss" ("relatio") and "Bezichung" ("connexio"). He uses "Verhältniss" for "the comparison of independently thought concepts", and "connexio", "Beziehung", or "Verbindung" whenever "from two concepts or acts of thought which are related to each other, a new concept or act of thought arises" (122). A *Beziehung* is therefore a triadic (ternary) relation, and it is probable that he has in mind what is nowadays called a binary operation or function of two arguments. It becomes clear how fatal was the lack of a suitable relational (dyadic, triadic, etc.) notation, when we read: "If we would express the concepts by means of letters, just as in the algebra of magnitudes, then

such [letters] could only be chosen for the concepts for objects, properties, and states [of objects], whereas all kinds of thought belonging to the forms of connection ["Beziehung"] would have to be represented by symbols with a meaning similar to that of the signs for algebraic operations $+$, $-$, :, $\times$, and so forth" (*loc. cit.*). Jevons is much more careful (22f.). See also note 9 below.

3 Compare this with the opinion expressed by C. S. Peirce: "If a language has a verb meaning "is a man", a noun "man" becomes a superfluity" (1967, III/290).

4 Cp. also Ong 1958/68f. However, we are not dealing here with the psychology of thinking or with heuristic, but with the logic of expressed judgments, *i.e.* of statements.

5 This principle was also employed by Hegel, and Leonard Nelson says: "... the lack of understanding of the judgment form not only threatens logic with the loss of distinction between objects and concepts, but even with the exchange of their roles.... [we] obtain here a means, unknown to Aristotelian logic, of transforming concepts into objects, or of letting objects evaporate into concepts, just as we please..." (1962/464f.). Cp. Moody 1965/70f. on the neo-Platonist logician Porphyry.

6 Angelelli anticipated us and introduced the expression "Meinong's beard" (1967/184); see I-1.

7 And with Bertrand Russell; see Russell 1964/327.

8 Wundt tries to determine a "factor" 'v' for every judgment 'S *is* M', such that the judgment may be treated as an identity '$S=vM$' (cp. VII-24). If we attempt to understand Wundt here on the basis of the traditional doctrine of identity that was sketched in VII-21, then we are led to interpret "[the] Mars resembles the earth" as an expression of a judgment of the form $\theta(S=vM)$. Here I have interpreted Wundt's "v" as the "ι" of VII-21. If, accordingly, the upper bound of 'v' is called "λ", then the limiting case of the judgment is $\theta(S=\lambda M)$, that is to say $\theta(\lambda(S=M))$, and it is not unnatural to interpret this in the same way as $\lambda(S=M)$ (cp. Rescher 1968/232, P 5.1). In Wundt's logic this is indistinguishable from $\theta(I(S, M))$.

9 A misleading algebraic notation for relational facts was used by Angelinus as late as 1941, probably due to lack of acquaintance with a more suitable notation (*Analogie* 1941). Confusion of these forms results when quantifiers of higher order are neglected; cp. Section 18, *sub* (2), and X-5.

10 Though Wundt also refers to Jevons, there is a great difference between their theories of substitution. Jevons, too, worked at a "pure" intensional logic (though his sense of "pure" was hardly that of the idealists). But he never wrote "$=$" between an individual term A and a property term B. If A *is* B is true, then Jevons does not symbolize this as $A=B$, nor does he write $A=vB$, as did Boole; he writes: $A=AB$, referring here to Leibniz (Jevons 1958/42; but see note 4 of our Introduction). That is to say, he employs the *conjunction* of A and B. His example is "Neptune is a planet", which he takes to mean: "Neptune=Neptune planet" (55). To a statement of identity with two individual terms, like his example "The Mont Blanc is *the* highest mountain in Europe", he ascribes the simple logical form $A = B$, while the predicating proposition "The Mont Blanc is deeply covered with snow" has the form $A=AB$ in his system (53). In Jevon's logic the logical role of the articles is certainly not neglected.

11 Cp. XII-5. Mates holds that Leibniz was very much aware of the pitfalls of substitution (Mates 1968/518f.). This is corroborated by Angelelli 1967b (cp. VII-17). See, however, Parkinson 1965/50 (cp. XII-5 below).

12 Cp. Wundt 1893/318.

13 Cp. N. Hartmann 1942/207.

14 Lotze writes an "M" where I have used "S", "π" where I have used "P", and in his

text "M" is a conjunction of variables "S", "P", ...; his specification of the form of an argument by analogy is the following:

$$M \text{ is } S$$
$$M \text{ is } P$$
$$\vdots$$
$$\overline{\phantom{M \text{ is } \pi,}}$$
$$M \text{ is } \pi,$$

to which should be added the clause: *because something's being π entails its being $S, P, \ldots$* .

[15] Given the traditional syllogistic treatment of individual propositions as universal propositions we arrive at the conclusion that from the point of view of Bosanquet's logic, *all* individual propositions which are discussed in a philosophy resting upon the Ockham-reduction express analogical judgments. The results of our discussion in V-16 of the traditional treatment of definite descriptions is quite in line with this conclusion.

[16] Cp. V-5.

[17] Cp. VIII n. 11.

[18] Cp. VII-13, VIII n. 26. Concerning logical stability see also I n. 13.

[19] An excellent survey is given in the *Encyclopedia of Philosophy* (Walters 1967). A Dutch discussion is found in Nuchelmans 1957.

[20] Cp. Geurts 1971.

[21] See in this connection especially Hempel and Oppenheim 1936.

[22] Cp. VIII-12, 16.

[23] Hartmann himself, though recommending a radical break with the old ontology, is still captive to the idea that it is only possible logically to deduce something from general sentences (*op. cit.*/208), but this mistake does not invalidate the rest of his argument.

[24] "Part of the trouble is that Kant's use of 'object' is viciously ambiguous" (Bennett 1966/131). "Kant sometimes, especially in A, drifts towards the conclusion that the unity of consciousness entails objectivity by expressing its converse – that objectivity entails the unity of consciousness – in ways which invite re-conversion" (*loc. cit.*).

PART 4

ASCENT

INTRODUCTION OF INDEFINITE PROPOSITIONS
BY EKTHESIS

1. PARTICULAR – INDEFINITE – PROBLEMATIC:
INDEFINITE TERMS *some S*

As explained in Chapter III, to reason *on the ground of* a particular
proposition *some S is P* is a precarious procedure in which the introduction
of the instantiated term must be done "critically" (III-2). The "formalization" or algorithmic description of the rules that justify arguments with
such premisses has finally become a reality due to the work of Frege. Altogether we have now more than enough reasons to focus our attention,
in our search for the logic of indefinite terms, upon traditional treatments
of particular propositions, and especially upon the import and use of
particular premisses. Let us start with a brief survey of some influential
opinions throughout the centuries.

(i) *The older tradition.* In the fifth chapter of the *Prior Analytics* Aristotle
points to the indefinite ("ἀδιορίστου") nature of particular propositions.
Alexander of Aphrodisias even regarded indefinite and particular propositions as equivalent (IV-2).

In the whole of the logical tradition almost all propositions which were
used as examples or analysed and given a logical form were formulated in
the grammatical singular (IV-8).

With respect to the logical theories of *the schoolmen*, this fact makes it
very difficult to ascribe a different meaning to a proposition like "homo
currit", in which the subject term was said to have *suppositio personalis
determinata*, from that ascribed to the so-called particular propositions
"aliquis homo currit" and "quidam homo currit" (IV-10).

The following principle of traditional logical syntax, noted in Chapter
IV, is expressed particularly clearly by Albert of Saxony: in a particular
judgment, and in a particular proposition *some S_1 is P*, the expression
"some" (or its meaning) is grouped with the term S_1 and the ensuing
combination *some S_1* is called "the (logical) subject (term)" of the

particular judgment or proposition (IV-8, VII-3). A particular judgment
or proposition may then be considered to have the logical form *S is P*,
where "*S*" is a variable for logical subjects or subject terms, with an
"indefinite value" in the case of particular propositions.

Peter of Spain discusses "homo currit" and "aliquis homo currit" in
the same paragraph and indeed ascribes the same truth conditions to both
propositions (IV-11).

As *conclusions* in expository syllogisms, Duns Scotus and William of
Ockham allow not only particular propositions, as would be required by
Darapti, but indefinite propositions as well (V-4).

(ii) As to the *Renaissance* it will be remembered that in the English translation of Peter Ramus' *Dialecticae libri duo*, published in 1574, particular
propositions like "Some man is learned" are plainly called "indefinite".
Elsewhere Ramus called individual and particular propositions together
"special" (IV-15). Therewith the theoretical notion of a particular proposition in fact disappeared (cp. Ashworth 1970/20).

(iii) It seems that during the *interregnum*, in Germany at least, this
theoretical notion was absent. Fichte identifies the particular judgment
form with the problematic judgment form (*the/an*) *S may be P*. Hegel
does the same, and Trendelenburg as well. This indicates a close traditional connection between the problem of the logic of the operators "all" and
"some", which was not solved until 1879, and the use of modalities.
Lotze, too, subsumes particular judgments expressed as *some S is P* under
"problematic" judgments about a "generally expressed subject-concept"
(cp. Lenk 1968/194, 212, 266, 437, 441, 447).

(iv) Proceeding now to *contemporary German idealism*, we notice that the
phenomenological logician Pfänder shows remarkably little interest in
the particular-universal distinction. As we know, he prefers a classification
of *Singularurteile* ("singular judgments", "judgments of singularity") and
Pluralurteile ("plural judgments", "judgments of plurality"). This classification makes it impossible to distinguish universal, indefinite, and
particular propositions (IV-21). His division of judgments into *Individualurteile* and *Arturteile* has the same effect.

By contrast with most other traditional logicians, Pfänder does not regard the form *S may be P* as the ideal way of rendering "the problematic judgment", but prefers to say *S is perhaps, possibly P*.

Fortunately both von Freytag and especially Maritain are clearer than Pfänder was concerning the analysis of particular propositions and their connection with indefinite propositions.

Von Freytag understands by "some..." the same as by "an indefinite kind of..." (1961/65, 68, 98). If he had been willing to substitute "class" for "kind" in the latter expression, then he might have invoked the authority of *E. Schröder*, well-known as the author of *Vorlesungen über die Algebra der Logik* (1890 and later). Angelelli refers (1967/102) to Schröder's "curious idea (certainly not only his)" that the expression "some men" is the name of a class, albeit a class with an indefinite number of men as elements. It should be noted that the *identity* of such a class is indefinite. The consequences of this view are aptly discussed by Geach (1972/57).

(v) *Neo-Thomism*. In Maritain's logic, too, particular propositions conceal indefinite entities. He uses the expression "terme-sujet particulier" ("particular subject term"; 1933/72). By this expression he means the word-sequence *quelque S_1* which occurs before the copula in a particular proposition *quelque S_1 est P*. It forms the logical subject term of a particular proposition, he says, and he refers to it by means of the variable "S".

We need only deal here with the simplest kind of particular propositions, illustrated by Maritain by means of these examples: "quelque homme est menteur" (82), "quelque triangle a ses trois angles égaux" (143), "quelque homme est injuste" (144).

Following upon the first of these examples, "some man is (a) liar", Maritain says: "In this proposition the S ["some man"] signifies *firstly and immediately* a floating or indeterminate individual (individuum vagum) which possesses human nature, and *secondly and mediately* such or such definite individuals having this nature" (89 n. 29). The subject term "quelque homme", "some man", in this proposition does not designate an individual chosen by the speaker, but human nature, "with a mode of being which is individual, although indeterminate, or more exactly *an indeterminate individual* (*individuum vagum*) which possesses this nature"

(54f.). The word sequence *some* S_1 has its own *significatio*, sense, or meaning which is precisely this indefinite individual, this *individuum vagum* which is a carrier of "the" nature corresponding to the word S_1 (cp. our H-thesis in II-5). So, too, in Maritain's logic for those propositions whose subject term has *suppositio* for *definite* individuals (*suppositio determinata*), as in each of the three examples quoted above. In *each* case the word "quelque", *i.e.* "some", is linked to S_1, the word or word sequence following it and preceding the copula (or verb), to form "the particular subject term" *some* S_1, whose meaning is an indefinite entity, an *individuum vagum*.

Maritain advisedly and consistently writes "quelque homme", "quelque triangle" in the singular and shuns the grammatical plural form "quelques hommes", "quelques triangles". He explains this by saying that the grammatical singular "met en évidence la *nature* universelle "homme"...", which cannot be said of the grammatical plural (55 n. 39). A reasonable translation of "mettre en évidence" is "to expose". We conclude that Maritain's *individus flottants*[1] are no less *general* entities than Schröder's indefinite classes or von Freytag's indefinite kinds. It is more likely, indeed, that Schröder's classes are not general enough for Maritain and that the latter chooses the word "individual" solely to indicate the unity of a general nature or of a category (143), as our conclusions in Chapter VI suggest.

2. ALBRECHT ON EKTHESIS OF THE MIDDLE CONCEPT AS A MEANS OF RECOGNIZING NON-TAUTOLOGOUS LOGICAL TRUTHS

In this chapter we finally put the question of the logical and/or the semantic conditions (I would not like to impose a distinction yet between two kinds of conditions) which in traditional logic, and especially in id-logic, must be satisfied in order that we may *introduce* a logophoric article in a proof, argument, or discussion.

In VII-23 we formulated an hypothesis about id-logic, *i.e.*, about those logical theories which rest upon an intensive intensional conception of "the" copula. That hypothesis was that id-logicians tacitly assume a principle which I called "the thesis of id-existence", being the thesis that given a certain analogy of logical degree ζ, the existence of a judgment true at a level $\iota \succ \zeta$ may be inferred:

$$\frac{\zeta(a_1 \underset{G}{=} a_2)}{\text{ergo}, (\text{E}\iota)\,(\iota \ldots)}$$

This thesis is clearly assumed in NPF-logic, for instance, which is an id-logic. We used this hypothesis in our interpretation of Hegel's Argument by Analogy (VIII-15).

The question of how the judgment following upon the quantifier in the conclusion is related to the terms in the premiss of this im-mediate inference form (1) has not yet been taken up. All we have assumed is that the premiss expresses some kind of judgment of analogy; for we were brought to our formulation of this thesis by our investigation of the connections between judgments of similarity, judgments of equality and assorted kinds of judgments of identity. Section V-16 also deserves mention as a source of inspiration.

With all its shortcomings, this thesis is still of great importance for us as students of the problem of id-introduction. In the present and in the next chapter we shall try to develop further our insight into that problem. It may be formulated as follows: what kind of material entitles us to infer, with certainty, a logophoric judgment *the/an M is P*, and how is that material to be described? For example, what kind of premisses will – in traditional logic – justify a statement *the (a) triangle is P*, or a statement *the (a) Bantu is P*?

It was said in Section 4 of our introductory chapter that the Locke-Berkeley problem of "the general triangle" is historically and systematically related to the Aristotelian *ekthesis* of terms and its philosophical interpretation. There are also more direct indications that a notion of ekthesis plays a very central role indeed in the thought of contemporary defenders of traditional logic who are hostile to recent changes in logical theory. Thus W. Albrecht, whom von Freytag considers as a supporter and who was mentioned briefly in the Introduction, ascribes an unrivalled philosophical importance to the Aristotelian logical operation of ekthesis He discusses "ekthesis between concepts" (1954/54), and maintains that "... in order to guarantee that a syllogism smoothly results, all of the concepts that are employed in it must be thought of as having come into being by ekthesis". He further points out the special importance of a correct "ekthesis... of the Middle concept" (55). In Albrecht's opinion, "concepts *qua* general concepts... are nothing else than the ekthesis of that which is common to many things". This is clearly his reason for holding that "logic is capable of stating general laws about comprehensional ["inhaltliche"] relations between concepts in judgments and arguments: if ekthesis is to be possible, things – or rather ideas of things, or again, concepts – must be given in advance and must show something common in respect of content which then can be 'exposed' " (1954/56f.).[2] "Exposed" is here a translation of "herausgestellt".

Albrecht takes Łukasiewicz' study (1958) of the syllogistic of Aristotle as his point of departure. He puts special emphasis on Łukasiewicz' discussion of ekthesis, which in Albrecht's opinion does not touch the procedure that Aristotle called by that name and which he used in proofs of the validity of certain syllogistic forms.

Albrecht has no objections, however, to Łukasiewicz' opinion that ekthesis is a process or operation in which one starts out from *particular* piopositions. From a particular proposition *A belongs to some B*, that is to say *some B is A*, one may according to Albrecht conclude that A-things and B-things have "something in common, which again may be 'exposed' and called C" (55f.). This entity C, then, is not an individual in Albrecht's logic, but rather a characteristic (mark), *i.e.*, a "general" entity. (In order to obtain agreement with the use of variables earlier in this book, for "A", "B", and "C", read "P", "S", and "M", in that order.)

This shows that Albrecht assumes every particular proposition to be an expression of a similarity or an equality between two entities, possibly an analogy in the sense of an opaque similarity as defined in VII-17. Łukasiewicz rejects or neglects, Albrecht says, "the possibility which Aristotle exploited more than once,... of determining, in [German: "an"] the concepts a form which in a certain sense is independent of its content, yet characteristic of it. All the same he believes that he has given the first satisfactory interpretation of the so-called method of ekthesis, which is identical with this possibility..." (51).

There is no doubt about it: Albrecht completely identifies the existence of a specific method of the mind, as taught in all those philosophies that are tied up with traditional logic, with the process of ekthesis or exposition. In those schools such a spiritual method is generally assumed to exist and to enable us to "see" non-tautologous truths on logical grounds alone, *i.e.* without using anything but the crudest and most everyday kind of experience.

In Kant's *Logik*, we read that an *exposition* of a concept (NB.!) consists in the idea ("Vorstellung") of an interlocking sequence of the character-istics ("Merkmale") of the concept, provided that these characteristics are the result of analysis and not of a construction; to *expose*, or expound, a concept is to describe it, when precision is not attainable: "Die be-schreibung ist die Exposition eines Begriffes, so fern sie nicht präzis ist" (Kant 1801/220; cp. Menne 1973). Taken literally and in isolation from his discussion of constructions (for which see Hintikka 1973), this sentence is certainly Hegel in a nut-shell. Albrecht refers to Kant in the preface of his essay as an important representative of that "general pure logic" which he himself defends against the re-orientation required by Łukasiewicz.

If Albrecht is right, an analysis of the Aristotelian procedure of ek-thesis is of crucial importance to our search for rules of introduction for logophoric articles. And that means that in the present chapter we shall have to look for the connexion between logophoric articles and the traditional logic of particular premisses.

3. ARISTOTLE ON ἔκθεσις

Aristotle discusses a procedure he calls "ἔκθεσις" in the second, the sixth, and also in the twenty-fourth chapter of the *Prior Analytics*. I quote the most important passages from the second and the sixth chapters in Jenkinson's translation (Aristotle 1963b). Following up W. Albrecht's suggestion about ekthesis of the middle concept I shall, however, use "*M*" as a variable for the exposed, or expounded, term (Section 1).

(1)   "If no *B* is *A*, neither can any *A* be *B*. For if some *A* (say *M*) were *B*, it would not be true that no *B* is *A*, for *M* is a *B*.... Similarly too, if the premiss is particular. For if some *B* is *A*, then some of the *A*'s [*e.g.*, the class *M*] must be *B*. For if none were, then no *B* would be *A*" (Prior Analytics I 2 25a 15–22).

(2)   "... whenever both *P* and *R* belong to all *S*, it follows that *P* will necessarily belong to some *R* [Darapti]... It is possible to demonstrate this... by exposition. For if both *P* and *R*

belong to all *S*, should one of the *S*'s, *e.g. M*, be taken, both *P* and *R* will belong to this, and thus *P* will belong to some *R*" (I 6 28a 18–26).

Aristotle's principle of ekthesis, as used by him in these chapters, is often given the following formulation:

(Ia)

$$\frac{}{\text{there is an } M \text{ such that: } \begin{array}{l} all\ M\ is/are\ P \\ and\ all\ M\ is/are\ S \end{array}} \quad some\ S\ is/are\ P$$

(Ib)

$$\frac{\text{there is an } M \text{ such that: } \begin{array}{l} all\ M\ is/are\ P \\ and\ all\ M\ is/are\ S \end{array}}{some\ S\ is/are\ P}$$

(cp. Łukasiewicz 1951/61; Patzig 1963/171f.)

Aristotle did not reckon with terms with an empty extension. He consequently had no use for a mere "existentiator", an operator expressing the non-emptiness of a property relatively to a certain domain of discourse. If the possibility of empty terms is taken into consideration, then the conclusion of (Ia) as well as the premiss of (Ib) must begin as follows:

there is a non-empty M such that......

Aristotle employs this principle or, better, these two principles, for the purpose of proving the validity of certain syllogistic moods, one of which is the mood Darapti. If the middle term (the subject term) of the premisses of Darapti is non-empty with respect to the domain under consideration, then the premiss of (Ib) is satisfied (or better, is true).

It is a much debated question whether Aristotle thought of the "exposed (expounded) term", *M*, as designating an individual or a class. Alexander of Aphrodisias held that *M* must refer to an individual, arguing that a proof which rests upon ekthesis appeals to some kind of empirical perception, or apperception (sometimes called, in English, but rather misleadingly, "intuition").[3]

Albrecht, on the other hand, understands (Ia) in the sense that *M* has to refer to a kind or *genus*. That is to say, he regards ekthesis as a principle whereby a general term *M* is introduced.

What follows in the next sections is based upon the assumption that Albrecht's view is representative of the whole of the id-logical tradition.[4]

4. EKTHESIS AS EXISTENTIAL INSTANTIATION

Given a "some…"-proposition, a modern logician will introduce a so-called individual variable (or a new individual constant), and choose a small, *i.e.* a lower case letter, thereby indicating that no kind, class, or property is meant. Let S be the universe of discourse; that some S is P may then be written: $(Ex)\, Px$. The principle (Ia) should therefore be compared to:

(i-mod)
$(Ex)\, Px$ (domain: S)
$\therefore\ Pm$ (existential instantiation;
                    m should be chosen "critically", and the
                    conclusion may not be the last statement
                    of the argument; cp. III-2)

Now let us compare the traditional and the modern views on this matter by expressing them schematically:

I. *Id-logical tradition*	I. *Modern*
Some S is/are P (premiss)	*Some S is/are P* (premiss)
ergo, ɩM is P	$\therefore\ m\ is\ S$
and ɩM is S	*and m is P* (ex. inst.)
[Obligations of these debaters unclear.]	In a discussion *m* must be chosen by the debater who asserted the premiss.
The inference step is necessary but falls outside formalized/formalizable logic (but within epistemology).	
This is an application of *ekthesis*.	This is *ekthesis₁*.

5. EKTHESIS AS THE INTRODUCTION OF THE CONJUNCTION OF THE EXTREME TERMS

(i) *Exposition of non-empty classes.* Łukasiewicz points out that in syllogistic one may choose, instead of an individual term, "the common part of A [read: "P"] and B [read: "S"] or a term included in this common part" (1958/61). The "common part" of the terms S and P (or: its extension) is the conjunction (the intersection of the extensions) of S and P. We can, then, choose as our exposed term M either this conjunction (intersection) itself:

$$M = S \cap P,$$

or else a proper part of it:

$$M = S \cap P \cap \dots$$

Let us continue with our comparison of traditional and modern analyses:

II. *Traditional*	II. *Modern*
The conclusion under I is sometimes understood in the following way:	(1) *Some S is P* (2) ∴ (EX) [X ≠ Λ & X ⊂ S & & X ⊂ P]
∴ *all M is P* (premisses *and all M is S* of Darapti)	For "*X*" one may choose M = = S ∩ P or a name for a sub-class of the intersection:
[The difference between I and II is not clearly brought out and the obligations of the debaters are unclear.]	
The converse is valid: Darapti. [False, unless empty terms are excluded *a priori*, the reason being that the second-order quantifier (*EM*) [*the extension of M ≠ Λ &...*] has been omitted.]	M (the shaded area or a part of it). The converse is valid: (1) follows from (2), so that: ⊢ (1) ≡ (2).

(ii) *Von Freytag's Principle of Forgettability ("Vergessbarkeitsprinzip")*. Our conclusion in the left-hand column, that a second-order existential quantifier and a condition of non-emptiness has been omitted, finds support in von Freytag's *Logik* (1961/95). In addition to the *dictum de omni* and the *dictum de nullo* he introduces a new "operation rule", *i.e.* a new immediate inference rule. This is the principle that if X is a species of P, one may "forget" which one it was (*op. cit.*/95f., 112):

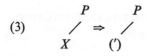

(3)

That this principle assumes the extensional non-emptiness of the said species is clear, for von Freytag uses it to prove the validity of inferences *ad subalternatam*, of Darapti, and of Felapton. His symbol for *some S's are P* is this:

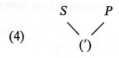

(4)

to be read: *there is a kind* (our X) *which is a sub-species of both S and P.* As Menne says, this symbol is an illustration of Darapti "with an indefinite middle term" (1957/385). For

(5)

is in von Freytag's symbolism a conjunction of two universal statements

    (a) *all M's are S*
    (b) *all M's are P.*

Von Freytag proves the validity of inferences *ad subalternatam* in the following manner (*op. cit.*/111). Starting out from his symbol

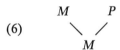

for *all M's are P*, he connects "*M*" by means of a line with another "*M*" at the upper left:

(6)

According to his own definitions of his symbolism this simply means that he adds a second premiss *all M's are M*, or *all M is M*. Von Freytag himself says about this addition of a second "*M*" that it "is not employed in syllogistic [*i.e.*, in the theory of mediate inference, E.M.B.]. It has something to do with the Law of Identity". He is of the opinion that inferences *ad subalternatam*, and Bamalip as well, though falling outside syllogistic proper, belong to pure logic (120). If (3) is applied to (6), then in the light of the symbol (4) for a particular proposition the conclusion *some M's are P* is seen to follow from *all M's are P.*

(iii) *How Plato's principle is employed.* Von Freytag makes no attempt to explain precisely what the second occurrence of "*M*" has to do with "the Law of Identity". In the opinion of the present author there can be no doubt that he has in mind the principle I have called "the H-thesis", rather than the simple tautological universal propositional form *all M's are M*. This thesis says that Plato's principle *the M is M*, understood logophorically, is well-formed and logically true for every term M (of any natural language). On the basis of this principle, which indeed "has something to do with the [or better, "a"] Law of Identity", we can justify the validity of inferences *ad subalternatam* by means of the *dictum de omni* and existential generalization. (The latter principle was not made explicit, but must have been operative in traditional minds on a less conscious level; if we do not assume this, it is hard to see how they could have mastered the simplest practical problems, which in fact they did.) Not *all M's are M*, but *the/an M is M* must be added to the premiss of this "immediate" inference form. The proof then runs as follows:

1.	*all M's are P*	(premiss)
2.	*ιM is M*	(Plato's principle; the H-thesis)
3.	*ιM is P*	(1,2, *dictum de omni*)
4.	*some M is P*	(3, existential generalization)

It seems to me that the logical truth of Plato's principle for every term M must be precisely "that new principle, which might be called the *Law of Identity* in the form in which it is important for the theory of inference" (von Freytag, *op. cit.*/111).

When Plato's principle is accepted as a general logical truth, it becomes easy to prove the validity of Darapti, Felapton, Fesapo, Bamalip, and *conversio per accidens*. For this principle allows us to drop the second-order existential quantifier and the clause $X \neq \Lambda$ in the interpretation of *some S's are P* above. No predicate, say M, can possibly be empty if there always is something, *viz.* '*the M*', of which M can be predicated truly.

In the light of the dialogical interpretation of the logic of the quantifiers it seems clearer than ever before that the chosen individual term m which we introduced under I in the last section need be identified neither with a general term M as introduced under II, nor with a subclass or species of M's reference.[5] As long as we focus our attention on monological proofs this is less obvious. But the debater who asserted the particular proposition has done his duty as soon as he mentions one individual, m, which is both S and P.

6. DEFINING *the R_{eq}-ness of S* AS A NO-RISK EKTHETIC
EPAGOGE: ANALOGICAL KNOWLEDGE AND ONTOLOGICAL ASCENT

(i) *Albrecht on the genesis of concepts*. Łukasiewicz formulates the principle of ekthesis by means of an existential quantifier: "If *B* belongs to
some *A*, then *there exists* a *C*, such that *B* belongs to all *C* and *A* belongs
to all *C*", and conversely. This quantifier is absent from von Freytag's
logic. Albrecht, who is very unhappy with Łukasiewicz's discussion.
queries the appropriateness of these quantifiers (1954/52). He believes
that Łukasiewicz has given a false interpretation because he has taken
into consideration only those parts of the *Prior Analytics* that deal with
the transformation of arguments (53). He thereupon points to the thirty-
fourth chapter, in which Aristotle discusses the "setting out", "ἐκθίθεσται",
of *the terms in the premisses* of an argument. This ekthesis, Albrecht says,
is nothing but the procedure whereby general concepts are formed.

According to Albrecht this formation of concepts by ekthesis takes
"the... immediate relationship" between *other concepts* as its point of
departure (54). And now it turns out that Albrecht draws no distinction
between this procedure and the principle which Aristotle sometimes used
in his demonstrations of the validity of certain of his syllogistic forms
(which he sometimes did by reducing, or transforming, them to other
syllogistic forms) and of certain immediate inference forms. In Albrecht's
opinion, one and the same principle, "the" ekthesis, is being consciously
employed or, in any case, is at work.

(ii) *From analogy to identity*. This view, together with what we already
know about the traditional ambition to proceed from a judgment of
analogy to a judgment of identity, makes it of paramount importance to
take the following fact into account in historical analyses. There exists a
kind of inference, which (a) has a two-term statement of an analogy as its
premiss, (b) a statement of identity as its conclusion, (c) implies an (onto-)
logical ascent in the sense of a move from individual terms to terms which
are names for general entities or concepts, (d) like existential instantiation
and like the formation of the intersection of two classes is not afflicted
with uncertainty, and which (e) may be regarded as a *definitio rei* (although
this way of speaking is uncommon in modern logic).

For suppose that we have discovered an equality between two entities (not necessarily between two individuals in the metaphysical sense, as long as they are of the same logical type). By an "equality" I mean here, as in earlier chapters, a reflexive, symmetric, and transitive relation, a so-called equivalence relation. Let us call this relation "R_{eq}". Examples are: a is just as long as b; a is just as old as b; a and b are similarly coloured; a and b are equiform; a and b are parallel, *i.e.*, a and b are directed the same way; a is identical with b, *i.e.* a is identifiable as b. A statement of the form $aR_{eq}b$ describes a fact of this sort. It is one of the principles of modern logic that the truth of a statement $aR_{eq}b$ allows us to define a "general entity" which belongs to both a and b: the class (set) of all those entities that bear the relation R_{eq} to the entity a. In our examples these general entities are commonly called: *the length of a, the age of a, the colour of a, the form of a, the direction of a, the identity of a* (cp. Frege 1884, Section 67ff., 1959/78ff.; Tarski 1964, 1941). They might also be called: *the long-as-ness of a, the old-as-ness of a, the coloured-as-ness of a, the formed-as-ness of a, the directed-as-ness of a, the identifiable-as-ness of a*; in general: *the R_{eq}-ness of a*. The inference form in question is this:

$$\frac{a \text{ bears the equality } R_{eq} \text{ to } b}{\therefore \ \textit{the } R_{eq}\textit{-ness of } a \ = \ \textit{the } R_{eq}\textit{-ness of } b,}$$

and since the sign "$=$" here should be read as strong complete identity we might as well write:

$$\text{(ii-mod)}$$
$$\frac{aR_{eq}b}{\therefore \ I \,(\textit{the } R_{eq}\textit{-ness of } a, \textit{ the } R_{eq}\textit{-ness of } b)}$$

Knowledge of the R_{eq}-ness of an individual, a, on all accounts deserves to be called "analogical knowledge" (cp. IX-16).[7] This should be compared with the inference form discussed in IX-9, which assumes the existence of a piece of analogical knowledge in this sense.

Schema (ii-mod) is not commonly called an inference form, but is said to express "the principle of abstraction", or as Russell put it: the principle that makes abstraction (in the old sense of the exclusion of certain characteristics so as to obtain a poorer concept) superfluous.

Given the fact that traditional logicians were not in the possession of a notational apparatus by which the various logically important distinctions could be expressed, we must seriously reckon with the possibility that in traditional philosophy (ii-mod) was not distinguished from the arguments by analogy which we discussed in VII-17 and VII-18.

In so far as this is right, it means that the old logic did not distinguish between abstraction on the basis of an equality relation, induction in the traditional sense, and *logical generalization*. All of these notions were assimilated into the one notion of *id-introduction*. This is in harmony with L. Nelson's view of logical history, which I mentioned previously in IX-18.

If the qualifying "perhaps" or "possibly" is omitted from the conclusion in the An-WI inference form of VII-18 and replaced by a definite or indefinite article, by the suffix "-ness", or by both, a hybrid inference form results in which the premiss is some statement of similarity while the conclusion resembles a proposition about the R_{eq}-ness of some entity a. It seems safe to assume that those logicians who were unable to keep these inference forms apart have assumed that an article sometimes expresses a position, ι, which is "lower than" the extreme logos-value λ. Conversely, belief in the appropriateness of such values may be explained by the assumption of their incapacity to see more than one inference form here.

7. PARTICULAR PROPOSITIONS AS EXPRESSIONS OF JUDGMENTS OF EQUALITY

The step from premiss to conclusion in (ii-mod) has every one of the characteristics (a)–(e) which were summed up in the last section. Certainly unary concepts like lengths, ages, colours and spatial forms are formed on the basis of equalities R_{eq}. There is no difference of opinion between Albrecht and modern logicians here. According to the traditional view, however, of which Albrecht is a particularly clear spokesman, (ii) is none other than that ekthesis which Aristotle carried out in order to derive from *some S is P* the converse proposition *some P is S*. So, if Albrecht is a trustworthy representative of the traditional logic, no distinction was made between (i), now called "existential instantiation", and (ii), *i.e.* abstraction in the sense of defining equivalence classes. Hence no distinction is made between the premiss of (i-mod) and that of (ii-mod), or in other words: no distinction was made between a particular proposition *some S is/are* P and a proposition expressing a judgment of equality, like *this S resembles this P*, or *some S's resemble some P's*. Of course the interpretation of "the copula" as identity or as "partial" identity is highly conducive to this analysis. The difference between particular propositions and identity-propositions only becomes clear when the monadic syntax is abolished. So long as that is not done, the fact that (ii) is a valid "inference form" furnishes the traditional logicians with an additional reason for taking ekthesis$_1$ to yield a general concept '$\imath M$' and to think that such a concept can *sometimes* be *determined*.

I quote below some particularly significant statements from Albrecht's monograph (the reader may first like to recall those in Section 1 above).

1. "... the validity of an argument depends among other things upon the possibility... of considering the concepts which are employed in the argument in their relationship ["Verhältniss"] to each other, even before they are so employed" (Albrecht 1954/54).

2. "... the ekthesis between concepts is capable of establishing that *immediate* relationship... which first and foremost guarantees the usefulness of the corresponding premisses in an argument" (*loc. cit*; my italics).

3. "Concepts *qua* general concepts [must] be thought of as having come into being through ekthesis" (55).

This leads us to the formulation of the following hypothesis about the traditional view of ekthesis as a procedure for producing general concepts (left columm below), which we compare with the modern view of concept formation on the basis of equalities.

III. *Traditional*

(1) Some S bear a relation of identity to some P (premiss), *i.e.*,
'*some S*' is part. id. with '*some P*'

(2) there is a general entity, '*the/an M*', of a higher (onto)logical level than the entities under consideration, which may be "looked out" (thesis of id-existence; cp. VII-23).

[It seems that in traditional logic no distinction is made between the operation under III and that under II, and that the latter is identified with the operation under I, *i.e.*:]

The step from (1) to (2) is Aristotelian *ekthesis*, the same procedure/operation as under I.

III. *Modern*

[Some S-things resemble (R_{eq}) some P-things. Assume s ε S and p ε P.]

(1) $s \, R_{eq} p$ (premiss)

(2) $I(\hat{x}(xR_{eq}s), \, \hat{x}(xR_{eq}p))$,
in other words:
the R_{eq}-*ness of s is the* R_{eq}-*ness of p*.

Let
    the $M_{eq} \underset{\text{Df}}{=}$ the R_{eq}-*ness of s*.

Then the following proposition is true:
s is/has the M_{eq} *and p is/has the* M_{eq}.

The step from (1) to (2) is *ekthesis$_2$*.

8. CONCLUSION

If we compare our own theory about traditional id-introduction with the conclusions we arrived at in IX-7, 8, then the first turns out to mirror and to complement the latter and *vice versa*, in the following sense. Like Robert Holkot's *logica fidei* the logics of Hegel, Wundt, and others contain no expository syllogism. In Chapter IX we found that instead of existential generalization these logicians offer an "ideal instantiation". In the present chapter we have seen that many traditional logicians carry out an "ideal generalization" or id-introduction under conditions which in contemporary logic prompt an existential instantiation. Or expressed more briefly: some-introduction is, in Hegel and Wundt, reduced to id-elimination; in addition, some-elimination is commonly reduced to id-introduction. The latter result makes it probable that the former is applicable to all those logicians who neither count the expository syllo-gism nor a *dictum de exemplo* among their fundamental logical principles.

In the schema below I have sketched the modern theory in traditional symbols, in order to facilitate a comparison with the assumptions under-lying traditional logic.

Traditional	*Modern*
Introduction-rule for "*id*":	Elimination-rule for "some":
some S is/are P	*some S is/are P*
ergo, ιM is S	∴ *m is S*
and ιM is P (ideal generalization by ekthesis)	*and m is P* (existential instanti-ation)
(*M* has a "philosophical" *suppositio essen-tialis/naturalis/absoluta*)	(*m* must be chosen "critically")
Elimination-rule for "*id*":	Introduction-rule for "some":
ιM is P	*m is P*
[this] S is (an) M	*m is S*
ergo, [this] S is essentially/to a certain degree/possibly M (ideal instantiation)	∴*some S is/are P* (existential generaliz-ation)
No elimination-rule for "essentially", "to a certain degree", or "possibly" is available.	

It turns out that, if our account is correct, premises with a prenex quantifier "some" (or "there is") constituted the chief stumbling-block in pre-Fregean logic. This may not be immediately clear from a study of

Beth's article of 1956. Beth discusses the traditional understanding of proofs of statements with universal import and takes the classical Locke-Berkeley problem as an example. Modern logicians who recognize no id-operator are bound to understand this general problem as the problem of universal generalization or all-introduction: under which conditions can a conclusion with a prenex universal quantifier, *e.g.*, a conclusion of the kind *all triangles are...*, be introduced?

Now it is required not only in the universal generalization rule, but in the existential instantiation rule as well, that the individual term which is eliminated (introduced), be chosen "critically" at the point where it enters into the argument for the first time (cp. III-2). The formalization of this condition is the basic achievement of Fregan logic. That is to say: the problem of giving a precise formulation of the rule of universal generalization is solved as soon as the problem of a precise formulation of the rule of existential instantiation is solved, and conversely. The latter problem, *viz.* the problem of how to get rid of a prenex occurrence of the expression "some", was traditionally solved by an *obscurium per obscurius*, namely by introducing a prenex definite or indefinite article.

Since this article was assumed to indicate an essentialistic *suppositio* of the subsequent general term and hence to refer to a general entity, say the general triangle, it will be clear that, once this article was introduced, no further generalization was felt to be required. Universal generalization, then, was regarded as a quite superfluous operation and came to be seen as irrelevant to the process of establishing truths of a deeper nature. Universal propositions were therefore commonly understood to pertain only to the θ-level of "logical space".

We are now in a position to sum up which traditional principles correspond to and which have been replaced by the introduction and elimination rules of modern logic:

1. Universal instantiation: cp. the *dictum de omni*.[10]
2. Existential instantiation: cp. the Aristotelian ekthesis; this operation was badly understood and was not distinguished from abstraction and from induction and was often assumed to result in an id-introduction (id-generalization).
3. Universal generalization: was regarded as a process of lesser importance, in the light of the views described under 2.

4. Existential generalization: here two positions must be distinguished, (a) and (b).

(a) Cp. *syllogismus expositorius* in the third figure with a particular conclusion (Durupti, cp. V-7), and also the *dictum de exemplo* (Lambert).

(b) Due to the doctrine of the *vis universalis* of singular (*e.g.*, of individual) propositions it was impossible for most philosophers to distinguish the principles mentioned under (a) from Darapti. This implies that they *must* hold Darapti to be valid. This in its turn implies the existential import of universal propositions, which leads to the assumption of the existence, for *every* general term (predicate) of natural language, of a general entity belonging to that term (the H-thesis). Hence arises the doctrine that the validity of expository syllogisms rests upon the validity of an argument containing at last one logophoric premiss $ıM$ is P. In this way there came about a widespread conviction that expository syllogisms are reducible to id-elimination (id-instantiation), namely to Burburu (cp. VIII-11), which consequently obtained a place in the nomological network of logical theory which is nowadays held by existential generalization. Id-elimination was at the same time regarded as more fundamental than the *dictum de omni*. In consequence *none* of the present four rules for the introduction and elimination of "all" and "some" were held to be philosophically important; only the two "rules" for id-elimination and id-introduction characterize the moves of "pure logic" (b1). In Hegel's *Science of Logic* even id-introduction is of no visible importance; in that famous work, logophoric judgments are, from an inferential point of view, *sui generis* (b2). This situation is explained by Kant's failure to solve the problem he had set himself (cp. IX-19).

9. Traditional logical syntax

(i) *Introductory remarks.* "When Aldrich says that 'Some B' is an ambiguous middle term he departs from the concept of a *term* by including the word 'Some'", writes Hamblin (1970/199). Aldrich, born in 1647, and Whately, born in 1787, both reckoned the quantifier as part of the expression they called a "term", and they were led by this syntactic principle to identify the Fallacy of Four Terms with the Fallacy of Undistributed Middle. Our answer to Hamblin is, first, that the word "term" itself seems to be ambiguous, and, second, that whatever the use of the word "term" is in a work on traditional logic, each of the words "some", "all", and "no" was regarded as having for its syntactic scope only the subsequent expression up to the copula (not including the latter).

I shall attempt below a first formulation of the syntactic principles of Aristotelian alethic logic, of Kantian logic, and of the logic of Sir William Hamilton and others. I do this mainly to indicate a new and highly needed approach to systematic philosophy and its history, but I cannot pretend that my formulations are final. Hamilton's logic, though not influential, has a certain systematic interest in this connection. All systems to be discussed are variants of *S-P*-logic. In order to facilitate comparison with usual text-book formulations of Fregean syntax I shall make use of the recursive form of definition which is preferred in modern logic, although there is little occasion (there are exceptions) to formulate recursive rules of the "feed-back" type.

One might wonder whether it would not be possible to say at least *something* about what does and what does not count as an atomic term according to Aristotelians, Kantians, etc. We do know that older logicians often used all kinds of complex phrases as atomic terms (as subject terms in the sense of the A-variant below). Some of them certainly allowed atomic terms containing relational predicates, external and internal quantifiers, identity expressions, and so on, but a characterization of those and only those word sequences which were permitted as atomic terms is nowhere to be found. As a consequence there is little reason for us to formulate feed-back rules.

That philosophers acquiesced for so long in an account of philosophical sentence construction which did not go seriously beyond the five systems of rules given here, is at least in part a result of the influential view of linguistic terms, simple or complex, as *names* of the atoms of *Language*, a postulated function of the human mind which it, *i.e.*, every mind, uses in its inner monologues (cp., *e.g.*, R. P. G. de Rijk 1968 on St. Augustine). This mental function or language was recently discussed by Nuchelmans (1972) under the name, borrowed from Geach, of "Mental". Mental was, by those who believed in it, assumed to be quite independent of its expression in any particular language. Many logicians and other philosophers will therefore have regarded rules like the ones formulated below as being first and foremost the rules of Mental, *to the extent to which these rules can be expressed in a (spoken or written) language at all*, and only in a secondary and derived sense as belonging to the structure of natural languages. This seems to be precisely the attitude of present-day defenders of traditional logic (cp. II-6).

(ii) *Aristotelian syntax.* The first question to be settled is the meaning of the technical terms "subject" and "predicate", or "subject term" and "predicate term". Aristotelian logic can be formulated in (at least) three different ways, here to be called A, B, and C. In the B-variant the quantifiers, or better: the words "every" ("all"), "some", and "no", are considered to be a part of the subject term, which therefore contains at least two words; in the A- and the C-variants they are not. Both in the A-variant and in the B-variant it is assumed that these quantifiers, if I may call them that, acquire their meaning in the proposition by being conjoined to the first non-syncategorematic or atomic term. The C-variant, which does not make this assumption, seems to have originated in the nineteenth century. It will be formulated under (v) below. It should be added that it is often difficult to decide whether an author bases his understanding of sentence construction and analysis upon the A-variant or upon the B-variant.

I shall use square brackets in order to indicate that the status of the rule enclosed by them is unclear, changing from one author to another within the same school, and possibly even within the work of a single author.

The A-variant

1. $A, B, C,...$ are atomic terms. Every atomic term can be both a subject term (S) and a predicate term (P).

2.   If S is a subject term, then (*every S*), [(*all S*)], (*some S*), and (*no S*) are noun phrases (name-substitutes).

[2b. If S is a subject term, then ([*the/an*]S) is a noun phrase.]

[2c. If S is a subject term, then (*not-S*) is a predicate term and a noun phrase.
        – Note that if 2c. is not assumed, a premiss or a conclusion of an obversion cannot be regarded as well-formed. However, for some purposes, such as epistemological or theological ones, this is not required.]

3.   Nothing is a noun phrase except in virtue of 1. and 2.

4.   If N is a noun phrase and P a predicate term, then $(N\ is\ P)$, $(N\ is\ not$
       $P)$ are well-formed propositions.
5.   Nothing is a well-formed proposition except in virtue of 1. to 4.

The B-variant

1. $A, B, C, \ldots$ are atomic terms.
2.   If U is an atomic term, then $(some\ U)$, $(every\ U)$, $[(all\ U)]$, $(no\ U)$
       are logical subject terms.
[2b.   If U is an atomic term, then $([the/an]U)$ is a logical subject term.]
[2c.   If U is an atomic term, then $(not\text{-}U)$ is a logical subject term and a
       logical predicate term.]
[2d.   If $U_1, U_2, \ldots, U_n$ are atomic terms, then $(U_1\ and\ U_2\ and \ldots\ and\ U_n)$
       and $(U_1\ or\ U_2\ or \ldots\ or\ U_n)$ are logical subject terms, the latter being
       also a logical predicate term provided "or" is understood in the
       exclusive sense.
       – Copulative terms are found in the logic of Maritain, and in
       Germany they occurred in logic books as late as the Wolffian
       school; Kant, however, drops them and they are unknown in the
       school of Hegel.]
3.   Nothing is a logical subject term and nothing is a logical predicate
       term except in virtue of 1. and 2.
4.   If S is a logical subject term and P an atomic term, then $(S\ is\ P)$,
       $(S\ is\ not\ P)$ are well-formed propositions.
5.   Nothing is a well-formed proposition except in virtue of 1. to 4.

Examples:

$$((All/every\ A)\ is\ B), \quad ((No\ A)\ is\ not\ B.$$
$$\underbrace{}_{S} \quad \underbrace{}_{P} \quad \underbrace{}_{S} \quad \underbrace{}_{P}$$

[In addition possibly also:

$$((A\ and\ B\ and\ C)\ is\ D), ((the/an\ A)\ is\ (either\ B\ or\ C\ or\ D\ or \ldots))]$$

The next set of syntactical rules I shall call:

(iii) *Kantian syntax: negation of the subject and of the predicate*
1.   to 2c.: as in the B-variant, without the brackets around 2b. and 2c.
[2c*. If U is a logical subject term, then $(not\text{-}U)$ is a logical subject term.

– I am not sure whether 2c*. can be ascribed to Kant personally; Hegel, however, cannot do without a rule of this kind ("Nicht ein dieses ist ein Allgemeines der Reflexion", cp. V-14; and the notion of "negation of the negation", as applied to concepts). Notice that in the present form, 2c*. is a "recursive", or feed-back, rule, and the only one. Cp. IV-18, VII-4.]

2d.   If *U* and *V* are logical predicate terms, then (*either U or V*) is a logical predicate term.

      – It is not clear whether *U* and *V* may themselves be disjunctions.

3.   Nothing is a logical subject term or a logical predicate term except in virtue of 1. and 2.

4a.   If *S* is a logical subject term and *P* a logical predicate term, then (*S is P*), (*S is-not P*) are well-formed propositions.

[4b.   If *S* is a logical subject term and *P* a logical predicate term, then (*S may-be/can-be P*), (*S must-be P*), (*S may-not-be/cannot-be P*), (*S must-not-be P*) are well-formed propositions.

      – On account of Kant's own words in the *Critique of Pure Reason*, in his discussion of the Table of Judgments, it may be doubted whether this rule should be included.]

4c.   If *p* and *q* are well-formed propositions, then so are (*If p, then q*) and (*Either p or q*) (Kant, *loc. cit.*).

5.   Nothing is a well-formed proposition except in virtue of 1. to 4.

Examples:

((*Some A*) *is-not B*), ((*The/an A*) *is-not B*). These are examples of negative (privative) (forms of) propositions (descriptions of judgments). ((*Some A*) *is not-B*), ((*The/an A*) *is not-B*). These are the simplest (forms of) limitative (infinite) propositions, the *propositiones infinitae* of Boethius (IV-18).
((*Some A*) *is-not not-B*), ((*The/an A*) *is-not not-B*). Are such propositions privative or limitative? Kant's definition does not allow for a distinction here, and cannot therefore be regarded as a classification of negative propositions (cp. Prior 1967).
((*The/an A*) *is B*), ((*The/an A*) *is* (*the/a B*)), ((*The/an A*) *is-not* (*the/a B*)), ((*The/a not-A*) *is-not* (*the/a not-B*)). All these propositions (assuming *A* and *B* to be given) are indefinite propositions of different kinds, or *proposi-*

tiones indefinitae (Kant did not use this expression).

Notice that on the strength of 2c*. the following will also be well-formed:

$$((Not\text{-}(the\ not\text{-}A))\ is\text{-}not\ (not\text{-}B),$$
$$((Not\text{-}(some\ A))\ is\ (not\text{-}the\ B)),$$
$$\underbrace{\qquad\qquad}_{S}\qquad\underbrace{\qquad\qquad}_{P}$$

etc. *This syntax is clearly quite insufficient for a serious formulation of a logophoric or id-logic* (cp. IV-9, VII-20), but Kant, however, did not realize that.

Compare this to:

(iv) *Hamiltonian syntax: "quantification of the predicate"*

1. *A, B, C,...* are atomic terms.
2. If *U* is an atomic term, then (*some U*), (*every U*), [(*all U*),] (no *U*) are terms.
2b. If *U* is an atomic term, then (*the/an U*) is a term.
3. Nothing is a term except in virtue of 1 and 2.
4. If *S* and *P* are terms, then (*S is P*), (*S is not P*) are well-formed propositions.
5. Nothing is a well-formed proposition except in virtue of 1. to 4.

Example:

$$((Every\ A)\ is\ (some\ B)),\ or\ ((All\ A)\ is\ (some\ B's)).$$
$$\underbrace{\qquad}_{S}\qquad\underbrace{\qquad}_{P}$$

In all the above variants of traditional syntax, the words "all", or "every", "some", and "no" were combined with the first non-syncategorematic term to form a noun phrase or logical subject term. The antiquated conception that these operators are names of quantities and that their use leads to a "quantificatious" philosophy has its origin here (cp. Geach 1973/57). No such combination is assumed in the last variant of Aristotelian syntax to be discussed here, which I shall call:

(v) *The C-variant: SiP-SoP syntax*

1. *A, B, C,...* are atomic terms.
2. If *U* is an atomic term, then (*not-U*) is a term.

3.   Nothing is a term except in virtue of 1. and 2.
4.   If *S* and *P* are terms, then *SiP*, *SeP*, *SaP*, and *SoP* are well-formed propositions. (Read: the extensions of *S* and *P* overlap; the extensions of *S* and *P* are disjunct; the extension of *S* is contained in the extension of *P*; part of the extension of *S* falls outside the extension of *P*.)
5.   Nothing is a well-formed proposition except in virtue of 1. to 4.

Here the "quantifier" is not combined with the next non-syncategore-matic term, but with the *copula*. Or maybe we should say: the quantifier is regarded as the expression of *a* copula, a logical relation between *S* and *P*. "The" copula has disappeared. It seems that the C-variant originated in nineteenth-century textbooks, and that there is little justification for its prominent place in logical textbooks for students of philosophy. It seems also to be of British origin. I have never found it in nineteenth-century German works on logic. The eight copulas of Christine Ladd-Franklin (1890) clearly belong in the tradition of the C-variant; Heidegger recognizes no more than two copulas (Heidegger 1914/106, Barth 1968b/285).

In order to understand continental philosophy one should start by assuming the A-variant, or, still better, the B-variant, with additions as made by Kant and others. Husserl's logic in *Erfahrung und Urteil* is obtained by taking the B-variant as point of departure, *with* the rule 2c., and by dropping from 4. the clause that a well-formed proposition (*S is not P*) may be generated from *S* and *P* (cp. Barth 1968b/283).

10. REJECTION OF INDEFINITE TERMS *some S*: BRENTANO AND FREGE

Hegel, in his *Jenenser Logik*, calls "the subject" of a particular proposition, '*some A*', "completely indefinite" (1923/84). Similarly both Maritain and von Freytag regard *some A* as a designation of an indefinite entity, '*S*', or '*M*'. It is well known that it is particularly hard to understand what it could mean to negate a proposition *an* (*indefinite*) *S is P*, or *an* (*indefinite*) *M is P*.

The expression "*an indefinite object*" and its associated meaning are rejected by Frege in his *Foundations of Arithmetic*: "I suspect that "indefinite object" is only another term for concept, and a poor one at that, being self-contradictory" (1959/60f.). He accepts no indefinite entity re-

ferred to by means of "the subject (term)" *some A*, for instance. He thus
sets himself in opposition even to such un-metaphysical contemporaries
as Schröder (1966/180) and Jevons (1958/41, 56), who still employed this
theoretical construct.[11] In Frege's syntax indefinite terms disappear completely, for in his opinion, the operators "all" and "some" must be
counted as belonging to "the predicate" and not to "the subject" of the
proposition; that is to say, if we really have the ambition to formulate a
workable theory of *negation*. It is in connection with sentences containing
a negation element that the logical role of "all" and "some" as belonging
to the predicate part of the sentence becomes clear. This is pointed out in
"On Concept and Object" (1960/48) and in Frege's discussion of Schröder's *Vorlesungen* (1960/93n.; cp. also 1971/28).[12] To regard "some numbers" as "the subject" of the proposition "some numbers are primes", or
"all bodies" as the subject of "all bodies are heavy", betrays a "superficial view as to the concept", "a mechanical or quantitative view", and
one which would force us to reject all particular judgments as *ambiguous*,
which, according to Frege, they are not. "Of course what I am here
rejecting is not the particular judgment, but only a wrong conception of
it" (1960/93 n.; 1967/194, 198f.).[13]

In his criticism and rejection of the traditional analysis of so-called
"quantified" propositional forms, Frege had a predecessor[14] in F.
Brentano (1838–1917), the man whose proclaimed ambition it was to
stop the further deterioration of a visibly ailing continental philosophy.
Brentano interprets the particular propositional form *some S is P* as the
existential form *es giebt ein SP*, which means that there exists an S which
is P ("ein *P* seiendes *S*"; 1956/118). The universal A-form is in his opinion
merely a denial of the existence of exceptions, that is to say the denial of
there is an S which is not P. In other words, according to Brentano the
A-form is logically equivalent to *es giebt nicht ein S non P* (*op. cit.*/
119).[15]

If this is accepted, then a universal proposition *all S are P* does not
logically imply the corresponding particular proposition *some S are P*.
And in fact Brentano does conclude that inferences *ad subalternatam* are
invalid (205).[16]

It follows that Brentano cannot have accepted the H-thesis, since that
thesis suffices to justify inferences *ad subalternatam* (Section 5 above).
From Brentano's interpretation of the Aristotelian propositional forms

his pupil F. Hillebrand correctly drew the further conclusion that the syllogistic forms Darapti, Felapton, Fesapo, and Bamalip are invalid too (1891/82f.).

As far as I can see, with the exception of Brentano and of those logicians who developed the forms of syntax discussed under (iv) and (v) in Section 9, all philosophers before Frege thought and reasoned in accordance with a syntactic analysis of quantified judgments which may be explained by means of the following use of parentheses:

(i)   (*some S*) *is/are P*; (*quidam S*) *est P*.

The modern description of the form of particular propositions is, as everyone knows, (Ex) $[Sx \ \& \ Px]$, or else $(Ex_S)Px_S$ with S as the domain of discourse or as the range of the variable "x_S" in a many-sorted logic (Smiley 1962). In the first case this can, by approximation, be rendered as follows:

(ii)   *some* (*are S and P*); *quidam* (*est S et P*).

This displacement of the brackets is of paramount importance for the theory of negation. It might seem that with respect to affirmative particular propositions where a term *S* is not available, *i.e.* in propositions *there are P's*, in modern symbols: $(Ex)Px$, the difference between (i) and (ii) cannot be of great importance, but this is a mistake. In the first place it should be realized that, even today, traditional continental theoretical philosophy *de facto* takes one of the variants of *S-P* syntax as its starting-point. Second, if we want to use the expressions "logical subject" and "logical predicate" with reference to form (ii), then by "the logical subject" we shall have to understand the whole domain of discourse:

(iii)   with respect to D it holds that *some are S and P* is true.

With S, the extension of *S*, as domain (range), possibly as one among several domains in a many-sorted logic, this becomes:

(iv)   for the domain S it holds that *some are P* is true.

In (iv), which is a meta-linguistic sentence, the quantifier is connected with the predicate of the I-form, in accordance with our reference to "On Concept and Object" earlier in this section. The traditional logical subject term *some S* has disappeared.

The theory of negation of universal and existential proposition which is connected with this analysis consists essentially of De Morgan's laws.

It should be realized that when Frege says that the (prenex) quantifier in a categorical proposition should be reckoned as going with "the predicate" and not with "the subject", then this vernacular is a concession to his readers, who were used to the traditional apparatus of technical concepts and terms. Frege's theory has nothing to do with, for instance, Hamilton's quantification of the predicate, which consists in the insertion of an additional quantifier before, or in, "the predicate". Some traditional logicians seem to identify these two completely different theories (cp. von Freytag 1961/67, 151, 180f.), probably because Hillebrand in 1891 used the expression "Die Quantificationstheorie", "Theory of Quantification", for the logics of Hamilton, Boole, and Jevons (Hillebrand 1891/91f.). The technical terms by means of which Frege formulates his own theory are "Argument" and "Function". "A distinction between *subject* and *predicate* finds no place in my representation of the judgment", so he writes at the beginning of his *Begriffsschrift* (1964/2). "... the words "relation of subject to predicate" denote two entirely different relations, according as the subject itself is an object or a concept. The best thing to do would therefore be to banish the words "subject" and "predicate" from logic altogether..." (1971/28; 58, 60f.).

Potts has recently shown that for the purposes of a logical syntax there may still be some merit in regarding the words *some S* as a syntactical unit even from the point of view of Fregean logic (Potts 1973). As Potts points out, the syntactic category of such a phrase will then be of quite another order or level than the category to which proper names belong. The mistakes the older logicians made was to regard word-sequences *some S, all S, no S*, and *the S* as syntactically on a par with proper names, while this way of grouping the words of a sentence will tell us something about language only if they are thought of as belonging to an entirely different syntactic category. Contrary to the word "Saul", a word sequence *some S* does not name or refer to anything, not even to an indefinite "thing". In the terminology of the linguists one could say that, for certain logical and/or linguistic purposes, *some S* may be called a "quantifier phrase" and even a "*noun* phrase" but not, if semantic confusion is to be prevented, "a *name* phrase" (cp. also Lewis 1970, Cresswell 1973).

11. REACTION: HUSSERL ON EXISTENTIAL PROPOSITIONS

Not all of Brentano's pupils followed him in his reform of logic. Edmund Husserl explains that he cannot accept Brentano's theory of judgments, "since I regard existential propositions as categorical propositions with an anomalously changed subject-meaning" (1929/266). His correspondence with Frege some decades before he wrote this clearly did not influence his opinions on this matter. It will be remembered from IV-8 that Albert of Saxony had already discussed modifications of the subject term as brought out by a preceding quantifier.[17] Husserl clearly accepts the position of traditional logic here, notwithstanding the fact that he *was* aware of a difference between the expression "a class contains something as an element" and "a class contains something as a sub-class" (cp. Frege 1960/92); but he does not elaborate his position with respect to the theory of inference.[18] Like his own pupil Pfänder, he seems to see the exercise of drawing distinctions as an end in itself. Thus it will be remembered from II-3 that Husserl defends the distinction between *das A* and *alle A* as logically different propositional elements ("Satzelemente"). An inferential relation between *das A* and *alle A*, or between *das A* and *einige A*, is, to my knowledge, not to be found in the works of Husserl.[19]

12. INTERPRETING "ALL" AND "SOME": UNIVERSE OF DISCOURSE OR METAPHYSICAL BEING?

The quantifiers[20] "all" and "some" can be given a clear inferential or dialogical meaning in use: their elimination- and introduction-rules. These rules, however, contain an implicit reference to a domain of discourse in the clauses pertaining to the choice of individual terms (or variables). In other words, "all" and "some" can only have a meaning relative to a domain, or universe of discourse, of entities. In logic, those entities are called "individuals", even when they are entirely abstract or fictitious from an epistemological point of view. If a domain is not "given", *i.e.*

reasonably well outlined in advance or definable through a study of the context and the pragmatic situation, then universal and existential premisses cannot be interpreted.

In modern logic, an argument form is said to be *logically valid* if, and only if, it is valid (absolutely dependable, a no-risk argument) in *all* domains (universes). Jacoby is clearly not familiar with this concept of logical validity, for he tries to refute the thesis that Darapti is invalid by means of an analysis of the following example (Jacoby 1962/134):

> All angels have wings
> All angels have human shape
> _____
>
> ∴ Some human shapes have wings.

Jacoby takes this to be an example of an argument which according to the *Logistiker* has true premisses (because there are no angels) and a false conclusion. That would make it a counter-example to the syllogistic form Darapti. Jacoby now tries to save Darapti by showing that this is not a counter-example after all. In his opinion this argument is therefore "logically" valid, since "existence here has the communicational value... of a kind of Being within the scope of the belief in angels" (*loc. cit.*).

To this one can only reply that the opinion of modern logicians is none other than Jacoby's own, with the sole difference that in modern logic the scope of belief in angels is called "the set of all angels". When this set is considered to be non-empty, we have a third premiss; there are angels; and one can then draw the conclusion that some human shapes have wings. For indeed: "Real Being is not at issue here" (Jacoby, *loc. cit.*). One could not agree more. Logic makes no assertion as to whether or not there are angels and does not prohibit arguments from the premiss that the universe of discourse contains angels. In this respect the dilemma between Domain and Being is a false dilemma.

But Darapti is invalid, for in another domain, the domain of human shapes of flesh and blood, Jacoby's example takes one from true premisses to a false conclusion.[21]

Jacoby is a good example of what Quine says in his preface to Clark's book: "The exaggerated notion of the cleavage between the old and the new logic fosters the very ignorance which engenders it" (Clark 1952/VI).

To a logical subject without real Being Maritain ascribes a *possible existence*. The term in question is then said to have *suppositio naturalis* or essential *suppositio* (1933/87f.).[22]

It seems that for many thinkers the H-thesis acquires a certain plausibility simply from the fact that it is perfectly possible to reason logically about a domain of completely fictitious entities, and also about a domain of non-fictitious entities without spatial or temporal co-ordinates.

Jacoby is certainly not the only philosopher who does not realize that the existential quantifier, the universal quantifier and the articles must be interpreted in terms of a domain or universe of discourse,[23] and that this universe of discourse need not be one of spatio-temporally located entities.[24] It should be recognized, however, that many early modern logicians contributed to this misunderstanding. Too often Russell and others wrote as if "there is..." or "there exists..." of itself meant existence in the *hic-et-nunc* sense, thereby confusing their personal ontology and formal logic.[25] However, as Bunge puts it: "Existential quantification, unless qualified, is ontologically neutral. Logic and mathematics have nothing to do with ontology except that they should be respected by the latter" (1973/59).

13. LINSKY'S OPERATORS: A DIALOGICAL DESIDERATUM

Modern logicians have continued the discussion about presuppositions of existence. With respect to presuppositions of non-emptiness of individual terms and description this discussion is now carried on in modal logic and in the investigation of so-called free logics. Presuppositions of non-emptiness of general terms have been discussed by, among others, Strawson, whose sympathy for the traditional view that general terms in use have existential import has led him to an attack upon Russell's theory of descriptions (Strawson 1950).

Criticism of Strawson's general attitude towards formal logics is voiced by Linsky. "Strawson is not consistent in his procedure. For the most part he accepts classical and modern logic as accurately presenting the logical relations holding between statements. It is only with Russell's Theory of Descriptions that he finds fault" (1967/91). As a result "he is confused in each of his criticisms of Russell" (98). Geach takes the Oxford theories

defended by Strawson and others, to be based upon distrust of the Fregean analysis of existence-statements in general. And he remarks, rightly I
think, that "it is no accident that the argument devised in Oxford against
the Frege-Russell analysis of existence statements has been eagerly
seized upon by theologians wedded to nonsensical doctrines about Being –
little as such an application would please the authors of the argument"
(Geach 1968/X).[26] It should be remembered that this analysis begins with
an existential quantifier (which may pertain to an abstract or to a concrete
or to a fictitious or to an ideal or... domain of discourse).[27]

Logics allowing for empty domains of discourse have been studied by
Hailperin and by Mostowski. The philosophical relevance of these investigations is not very clear.

In 1969, Hintikka expressed his doubts about the philosophical importance of the study of "free logics" as it is commonly pursued. "It seems
to me", he writes, "that we have to go beyond the usual syntactical and
semantical tools to get at the interesting problems here.... I have even
ventured to suggest that the game-theoretical interpretation of first-order
logic can serve as a starting-point for interesting philosophical and logical
theorizing" (1969/112f., Lambert (ed.) 1970/20f.).

The notion of a domain of discourse is, in fact, a fundamental dialectic
notion; and this is true whether the word "dialectic" is understood in the
classical sense of "pertaining to dialogical procedure" or in the completely different Marxist sense of "pertaining to the evaluation of human
knowledge".[28] That is to say, as long as one does not understand "dialectics" in the sense of neo-Thomist inquiries into Being, or in the sense of
Husserlian "transcendental phenomenology".[29]

To sum up, it seems to me that what is valuable in Strawson's criticism
can only appear to full advantage in a further dialogical development of
contemporary theoretical logic. It may be true that these extensions will
have to be pragmatic rather than syntactic or semantic (in the present
sense), but the possibility of syntactic refinement should perhaps not
be discarded too soon. It seems to me that Linsky's plea (*op. cit.*/115, 126)
for the introduction of special operators, and among them an *in-the-
novel*-operator, into logical theory has considerable interest. His idea
seems to be to enrich the syntax of formalized or, better, of formalizable,
logic with operators which are to have the function of indicating the
domain of discourse to which the quantifiers pertain.[30] In that way

Strawson's "presuppositions" may be rendered more tractable. This idea may lead to a fusion with the Hailperin-Mostowski investigation; but a pragmatically complete description of dialogical use of language may have to contain a "nest" of domains of discourse.[31]

NOTES

[1] Angelelli uses the expression "a vaguely floating entity" for the notion of a relation. There seems to be little reason for this choice of words unless he means that the logic of relations has taken over the connecting, or unifying, role which was earlier played by the *individus flottants* (as Maritain calls them). Cp. Hempel and Oppenheim 1936, Weinberger 1965.

[2] Similarly, Hegel speaks of *Denkbestimmungen* which may be *herausgestellt* (exposed, or expounded; cp. VIII-1), and according to Husserl, general entities can be *herausgeschaut* ("looked out"; Husserl 1948/392f.).

[3] Cp. Patzig 1963/169f.

[4] Wieland admits that his earlier (1958) attempt to save as much as possible of Albrecht's interpretation of Aristotle was a failure. He now (1967) likens ekthesis to existential generalization; but this, of course, is a mistake (Wieland 1967/24 n. 43, 44).

[5] "The Euler diagrams are a lame analogy for logical relations, since they do not bring out this important distinction [between an element of a class and (a sub-class of) that class]" (Frege 1960/93, 1967/198).

[6] The "is" in a proposition like "the length of this rod is 1.3 m" must be understood as the "is" of identification and not as the "is" of predication. That is to say, by means of such a proposition no property is being ascribed to any *class* (such as the class of all things that are as long as this rod).

[7] In addition to a *connaissance analogique*, as Maritain calls it, a *connaissance dysanalogique* should play an important role in an epistemology based upon modern logic; that is to say, the perceiving of the elements of non-symmetric relations: ordered n-tuples. In a *connaissance dysanalogique* one becomes aware of the fact that an entity, a, stands in a non-convertible relation to another entity, b. For an epistemologist, the problem is then how to describe this fact in epistemologically natural terms. The most reasonable solution would seem to be a Gestalt-psychological description of our perception of (at least) four entities, a, b, c, and d. The entities c and d determine the difference between the "directions", or "senses" (Russell 1964/86), of the relation of a to b and the relation of b to a. For this reason Hausdorff's explication of an ordered couple $\langle a, b \rangle$ as a set $\{\{a, c\}, \{b, d\}\}$ is preferable from an epistemological point of view to the Wiener-Kuratowski explication: $\langle a, b \rangle = \{\{a\}, \{a, b\}\}$, which has found general acceptance in mathematics because of its suggestion of the two first segments of a series: a, b, c, d, ... Perhaps it is possible to analyse a *connaissance analogique* as the perception of a set $\{\{a, b\}, \{b, c\}\}$, *i.e.*, as the special case in which c = d. Hausdorff's explication suggests that all dyadic relations, including dyadic analogies, are sets indicating at least four-termed structures or *Gestalten*. This also throws light upon the traditional confusion between equalities and four-term "proportionalities" which has been described in Haenssler (cp. IX, n. 9). The discussion of four-term proportions in the writings of Iamblichus may turn out to open interesting perspectives here for the history of ideas (cp. Haenssler 1928/24f.). – Cp. van Heijenoort 1967/224.

[8] Cp. Prantl IV/7, n. 22; Jacoby 1962/109.

[9] "The first question, whether all judgments are of the form A est B, must be denied, especially with respect to relational judgments and existential judgments" (G. Martin 1960/50).

[10] Brentano formulates (1956/206) the classical *dictum de omni et nullo* as follows: *quidquid valet de omnibus, valet etiam de quibusdam et singulis*; *quidquid valet de nullo, neque valet de quibusdam neque de singulis*. The words "de quibusdam" imply that a universal proposition has existential import. If these words are omitted from the first part of the *dictum*, then we are left with the rule of universal instantiation.

[11] "Herr Schröder gives an example (p. 180) of *quaternio terminorum* that arises because the expression 'some gentlemen' does not always designate the same part of the class of gentlemen. Accordingly such an expression would have to be rejected as ambiguous, and it must, in fact, be rejected if one regards it (like our author, p. 150) as designating a class that consists of 'some' gentlemen" (Frege 1960/93n). Cp. Hamblin 1970/199; Geach 1972/57.

[12] By fusing the quantifier, the negation element, and the subsequent word or phrase together into one "term", one can prove that one cat has three tails:

> (no cat) has two tails,
> one cat has one tail more than no cat,
> ergo one cat has three tails.

Locke's proof for his thesis that everything has a cause rests upon the same theory of judgment, *i.e.*, upon the same logical syntax. For assume something is to have no cause. It then has Nothing as its cause; which contradicts the assumption (cp. Quine 1964/133). This argument was repeated by Chr. Wolff (cp. Scholz 1967/42). Later on, Kant's own discussion of "den Begriffen von Allem, Vielem and Einem..., ... Keines" leads him to the following conclusion: "Reality is Something, negation is Nothing, namely a concept of the absence of an object, like the shadow, or coldness (*nihil privativum*)" (*Critique of Pure Reason*, I, Book II, Appendix; 1922/312, 1968/373). Using the same principle, Hegel succeeds in proving that tautological identities are self-contradictory. For if one expresses a judgment of the form $A = A$, one can easily convince oneself that nothing has been said, that is to say: "Nothing has come out of it". And this again means that "such identical talk therefore contradicts itself" (II/30). It seems that in E. von Hartmann's opinion, too, the word "Nothing" occurs in the proposition "Nothing is more worthless than metaphysics" as a name of a logical subject. It would then follow that there is, after all, something (or, as Kant would say, a condition of something) which is still more worthless than metaphysics, but I cannot see that von Hartmann draws that conclusion (nor does Hegel, as far as I know).

[13] For a logician it is interesting to observe that a contemporary linguist, S. Yotsukura, regards "some" as an article (1970/50–56).

[14] The views which we are going to discuss here were published in Brentano's *Psychologie vom empirischen Standpunkt* in 1874. They became known through Hillebrand's work of 1891. Brentano's course of lectures on logic were not published until 1956.

[15] Brentano's logic is therefore a two-valued (non-constructive) logic.

[16] As pointed out to me by Dr. T. C. Potts, it appears from p. 24 of the *Begriffsschrift* that even Frege does not seem to have realized that these inferences are invalid (Frege 1964/24).

[17] See also Abelard's *Dialectica*, ed. de Rijk 1966/184f.

[18] "Thus the doctrine of essences has in the recent past experienced a renaissance in phenomenology, seemingly without any metaphysical tendency, but not without the reappearance of very old difficulties, many of which had already been solved, and also not without leading us astray into equally old and notorious mistakes" (N. Hartmann 1942/207).

[19] "If we turn from recent 'philosophical logic' to recent grammar, things are not much better. The sophistications of a computer age overlie ideas that might come straight out of Dionysius of Thrace and Priscian; indeed Chomsky has expressly said that "by and large the traditional views are basically correct, so far as they go". Proper names and phrases like "some man" are alike called Noun Phrases – whatever virtue there may be in the capitals – and are regarded as belonging to the same substitution class" (Geach 1972/115f.; see also 55f.).

[20] The names "existential quantifier" and "universal quantifier" are exceedingly misleading expressions, and many have been misled by them into thinking that modern logicians can and will deal only with entities that are capable of quantification in a physical and/or in an arithmetical sense. Some misunderstandings as to the scope of quantifier logic (predicate logic, logic of propositional functions) could probably be avoided if other names were introduced, as, *e.g.*, "existentiator" and "universalizer" or "generalizer". It is not sufficiently known among philosophers that as universe of discourse one may, if one so wishes, choose an infinite set and even a set with a transfinite cardinal. Nor have philosophers learned to exploit the different kinds of infinity. The same goes for the difference between densely and continuously ordered series (cp. Stace 1961/270f., Veatch 1969/262). Geach tells of Frege's own aversion against "the way of thinking that Frege called mechanical or quantificatious thinking: *mechanische oder quantifizierende Auffassung*. I have used a rude made-up word "quantificatious" because Frege was being rude; "quantificational" and "quantifying" are innocent descriptive terms of modern logic, but they are innocent only because they are mere labels and have no longer any suggestion of quantity" (Geach 1972/56f.).

[21] For that reason one must reject Menne's conclusion that Darapti, Felapton, and Fesapo are "completely correct" (Menne 1954/128), if this is supposed to mean that they are logically valid.

[22] Cp. IV-14, IV-28.

[23] In the sense of the set of all "individuals" that are assumed at the beginning of the discourse, and not in the sense of a set of all individuals *and sets* (of individuals and sets); the assumption that such a set exists leads to a contradiction (see, *e.g.*, Halmos 1968/29).

[24] A recent confusion of ontology and logic is found in Crittenden 1970.

[25] This is clearly reflected in the subtitle of Eley 1969, in English: *Sensuous Certainty as the Horizon of Propositional Logic and Elementary Predicate Logic*. The main title of this book is: *Meta-criticism of Formal Logic*.

[26] See also Geach's essay "Strawson on symbolic and traditional logic", in Geach 1972. Cp. XII-3 below. For a revealing connection between *"what"* and "objective presuppositions", see Apel 1967/29f., about Skjervheim.

[27] Cp. V-18, 19.

[28] Cp. VII n. 43 for another notion which may also be called "dialectical", *i.e.* the notion of identity. A third important "dialectical" notion is that of the *vel*-junction, F, of all propositions which at the outset of a certain discussion are regarded as certainly false. This notion, which is a logical constant for the duration of the discussion in question may be used to define the word "not": $\sim p \underset{\mathrm{Df}}{\leftrightarrow} p \to \mathrm{F}$.

[29] "Habermas' concept of dialectics corresponds precisely to what Husserl meant by his concept of transcendental phenomenology" (Eley 1969/6). "The dialectician knows nothing scientifically except second intentions. Whenever the work of a logician comprises a dialectical section, we expect to find, framed in a logical system, an inquiry into real being" (from the foreword to John of St. Thomas 1965 (p. xvii)).

[30] Cp. IV-9.

[31] Modern linguists often unwittingly support traditional logical principles rather than modern ones. Thus Lakoff and, following him, McCawley, consider quantifiers as two-place predicates, relating propositional functions to sets (McCawley 1971/220). In the case of the existential quantifier it is hard to see how they can then avoid Schröder's notion of an indefinite set.

CHAPTER XI

CONJUNCTION, POTENTIALITY, AND DISJUNCTION

1. THE PROBLEM OF THE LOGIC OF DISJUNCTIONS AND THE LOSS OF DE MORGAN'S LAWS

In his *Topics*, Cicero ascribes to the dialecticians – or logicians as we would say nowadays – a remarkable inference form. It is remarkable on account of the very special conceptual connection it reveals between conjunction and negation. Cicero, however, does not reject this inference form, which is found in the fourteenth chapter of the *Topics*. After having summed up a number of well-known inference forms like the *modus ponens* and the *modus tollens* he goes on to say: "Septimus [modus dialecticorum] autem: Non et hoc et illud; non autem hoc; illud igitur" (XIV. 57). In Hubbell's translation: "This and that are not both true; this is not, therefore that is" (1949/425).[1]

This staggering piece of argumentation would have been correct if $\sim(p \ \& \ q)$ were equivalent to $(\sim p \ aut \sim q)$, and not to $(\sim p \ v \sim q)$.[2] The equivalence of $\sim(p \ \& \ q)$ and $(\sim p \ v \sim q)$ is one of the two laws of De Morgan. It is by means of these crucial laws that one determines the negation of propositions of the forms $M \ is \ (P \ \& \ Q)$ and $M \ is \ (P \ v \ Q)$; and given the close connection between "all" and "and", and that between "some" and "or", it is easy to understand that the very same laws form the foundation of the theory of how to negate the "categorical" forms *all S are P* and *some S is P*. These laws can ill be spared in a dialogically serviceable logic, at least in as far as they are valid in a constructive logic (where we will have to content ourselves with certain implications, not with De Morgan's full equivalences).

Ockham mentions these laws explicitly, six centuries before they were re-discovered by De Morgan.[3] One might have expected logicians after Ockham never to have lost sight of them, but history shows otherwise. Thus Parkinson had to conclude that the equivalence which Ockham and much later De Morgan formulated in terms only of the words "and", "not" and the inclusive "or", are nowhere to be found in the works of

Leibniz. Leibniz only approximated to them, by using the connective "nor".[4] Parkinson blames this shortcoming of Leibniz on the fact that he held the study of disjunctions of concepts to be relatively unimportant and that he was interested almost exclusively in conjunctions, so that he did not discover that *p or q* is equivalent to *not (not p and not q)* (Parkinson 1966/lxi). True, Leibniz also formulated a *calculus alternativus*, in which formulas of the form *x is abc* express that x is *either* a or b or c; but Parkinson regards this calculus as an exception with little importance in Leibniz's logic. In addition, Leibniz must probably be taken, here as elsewhere, to work with exclusive disjunction, not with the inclusive disjunction of De Morgan's laws (cp. Couturat 1961/344).

Practically all traditional logicians regarded the notion of exclusive disjunction as a more important logical tool than that of inclusive disjunction. Even the renowned George Boole wrote: "In strictness, the words "and", "or", interposed between the terms descriptive of two or more classes of objects, imply that those classes are quite distinct, so that no member of one is found in the other" (1854/32f.).[5] Wundt, too, when discussing disjunctions in logic understood them to be *aut*-junctions (1893/202–204). The person who introduced into logic the theoretical construct of an inclusive (*vel*-junctive) union of two classes was Jevons. The explanation is not far to seek: he was for a long time De Morgan's student (Jevons 1958/xxii, 69, 72). Jevons himself considers precisely this transition from exclusive to inclusive class union as the great difference between Boole's logic and his own.

Maritain admittedly also treats of inclusive disjunction (1933/128 n. 9), but makes a distinction between propositions which he regards as *proprement disjonctive*, namely, exclusive disjunctions, and such as are only *improprement disjonctive*, to wit inclusive disjunctions (131).[6] Similarly von Freytag in his first book characterizes exclusive disjunction as "the genuine logical disjunction" (1961/83, 161f.).

2. CONJUNCTIVE AND *aut*-JUNCTIVE ANALYSIS OF GENERAL CONCEPTS

In the philosophy of recent centuries one notices a general preference for conjunctions rather than inclusive disjunctions. Let us start with Leibniz. In his theoretical-logical works there is an almost exclusive stress upon a conjunctive analysis of concepts: "Leibniz had little, if any, use for a concept such as that of rational *or* animal, in other words, he concentrated on the notion of a logical product at the expense of the notion of a logical sum – that is, he tended to think in terms of entities which are both *A* and *B* rather than in terms of those which are either *A* or *B* (or both)" (Parkinson 1966/lxi).

If we now turn to Hegel, the stress upon conjunction of terms is no less conspicuous. Thus in his study *Hegel and Greek Thought*, J. Glenn Gray writes: "... for Hegel the logical idea includes not only the common genus, but all the differentiating marks of the species as well" (1968/99 n. 6).

This conception was, however, anything but a Hegelian speciality, but formed the basis of the theoretical logic of his time. This is splendidly brought out by Arndt, who in his introduction to the new edition of the *Neues Organon* of J. Lambert, whom we have had occasion to mention several times already, writes: "A general concept is not understood as a concept of a genus, in which only those characteristics occur which are common to the species that fall under it, but not their specific characteristics. The general concept is, rather, regarded as one which already *contains in itself* all the particular specifications which it may acquire by progressive determination, *as possibilities* [*"der Möglichkeit nach bereits an sich enthält"*]. In this conception of the generality of concepts and statements the old Aristotelian doctrine of forms, which was still active due to the impact of scholasticism, found an ally in that theory and methodology of mathematical knowledge which had come into fashion since the Renaissance, after the re-discovery of the mathematical writings from antiquity, especially of Pappus's commentaries on Euclid. Mathematical statements seemed to warrant the derivation [*"Herleitung"*] of all the singular cases which they contain under them, by complete enumeration,[7] *e.g.* the so-called distinction of cases" (Arndt, in Lambert 1965, I/XIf.; my italics).

Without the use of variables the old conceptual structure could not be formulated more succinctly than Arndt has done here.

Angelelli describes the situation in traditional logic as follows: a *genus* "implicitly" contains its *differentiae*, '*P*', '*Q*',..... It is therefore *potentially indeterminate*. For instance, *the* genus 'Triangle' is "implicitly" oblique *and* right-angled, equilateral *and* scalene. Thus Locke's famous, or infamous, triangle is in Angelelli's opinion (with which I fully agree),

a product of the general traditional doctrine of the "potential indetermination" of the logical *genera* (1967/122).

We see, then, that philosophers of such divergent tendencies as Locke,
Lambert, and Hegel all assume the same conjunctive composition of
concepts which Parkinson observes in the theoretical logical writings of
Leibniz. In Locke and in Hegel, this assumption is active in their epistemological and metaphysical speculations. Hegel, as we know, complements
it with an *aut*-junctive analysis which is incompatible with the conjunctive
one.[8] Again, this is no invention of Hegel, but a product of the same
conceptual structure as Locke's paradoxical triangle and Lambert's
genera. For all these authors this is the structure implied by the doctrine
of the definition of a concept ('*S*') by referring to its *genus* ('*M*') and its
specific differentia (*e.g.*, '*Q*').

Of course they differ as to the ontological status they ascribe to such
general concepts or genera. In Hegel's opinion, at least some of these
general ideas or genera are ontologically independent, or, as he calls it,
"concrete" entities, while Locke regards them all as "fictions and con-
trivances of the mind" (1959, II/274) and "marks of our imperfection"
(275) which have no claim to a place in the sun in an ontological stock-
taking of the world. Much energy has been wasted in the discussion of
this difference of opinion and its consequences. From a logical point
of view it is vastly more important that even Locke held that "the mind
has need of such ideas" (*loc. cit.*), namely as soon as we wish to reach
conclusions with a general import.

Locke's example here is, of course, "the general idea of a triangle" (274).
He says of this "idea" that it is neither oblique nor right-angled, neither
equilateral nor scalene, "but all and none of these at once". That is to say,
this general idea 'Triangle' is *both* scalene *and* right-angled, while an
individual triangle can only possess one of these properties. Expressed
in general terms, *i.e.* in variables, this means that if an (arbitrary) individual M-thing is either P or Q or ..., where P, Q,... define species of
the genus '*M*', then the following statements about this idea are both
true:

(1)   [*the*] *M is P and Q and...*,
(2)   [*the*] *M is neither P nor Q nor....*

This idea, '*M*', certainly deserves to be called "indefinite" or indeterminate". It is no wonder that Berkeley and others rejected this philosophical construction.

Hegel, too, assumes the conjunctive composition of general concepts and speaks about

(1')   *das A, welches sowohl B als C als D ist*

where B, C and D are mutually exclusive differential marks which characterize the various A-species (VIII-18).

He adds to this that (a species of?) '*A*' is *either* B *or* C *or* D:

(3)   *A ist entweder B oder C oder D.*

Common to both authors is that the "lower" concepts or sub-classes are taken up into the general concept *conjunctively*. The addition of (3) should, I think, be explained as follows: given a ready-made classification these formulas sound as if the expression "an arbitrary" should be added at the beginning:

(3')   *ein beliebiges A ist entweder B oder C oder D*

But the logic of arbitrary entities was not distinguished from the logic of abstract entities; hence the failure to distinguish between (3), (3'), and even (2). This does not, of course, suffice to explain the conjunctions in (1) and in (1').

It is quite another matter that Locke and Hegel used this contradictory traditional theory of concept composition for different ends. Hegel associated with (1) the view that the logical genus is the ("logical") *origin* of all the kinds or species that fall under it. Locke associated with (1) the view that the imperfect and merely subjective general idea '*M*' must be able to *represent* all the M-species. If one is at all of the opinion that this general idea, be it objective or subjective, plays an important role in proofs of statements with general import, then it will be natural to ascribe all the properties of all the *M*-kinds to '*an arbitrary* (individual) *M*'. The word "arbitrary" then takes on the connotations of "vagum" or "flottant".

The following sections will deal with a number of logico-semantic phenomena that are easily observed in contemporary natural languages too and which are related in some way or other to this traditional conception of the conjunctive-disjunctive composition of general "logophoric" concepts.

3. NON-TRUTH-FUNCTIONAL CONNECTIVES:
suppositio disjuncta AND *suppositio copulata*

The conjunctive-disjunctive intensional type of definition of so-called general terms resists every attempt at a truth-functional analysis of the expressions "and" and "either-or"; what is more, it is completely incomprehensible how one could characterize the meaning of these words as they occur in the formulas of Section 2 in any way whatsoever, be it axiomatically, dialogically, or in the manner of rules for natural deduction.

The connectives in the formulas taken from Locke and Hegel occur, from a traditional point of view, in the logical predicate. In Maritain's logic, however, connectives which in his opinion cannot be reduced to ordinary truth-functional propositional connectives occur between the components of complex logical subject terms. Let us take a look at these connectives and the uses to which they are put.

(i) *Suppositio disjunctiva and suppositio disjuncta.* According to Maritain, in some particular propositions, *e.g.* in the proposition "quelque homme est menteur", "some man is a liar" (cp. X-1), the particular subject term has *suppositio determinata, seu disjunctiva.* This means that the particular proposition in question, *some S is P,* may be treated as a disjunction of singular propositions: $(S_1$ *is P) or* $(S_2$ *is P) or....*

However, among the propositions which he calls particular there are some whose subject term *quelque S* both signifies a certain general entity and even stands for, or refers to, *i.e.* has *suppositio* for, this general entity.[9] In this case Maritain, following John of St. Thomas, speaks of *suppositio disjuncta [indeterminata],* or *suppléance particulière indéterminée* (1933/85). Note the presence of the notion of indetermination here! The following proposition is offered as an example:

(1a)   some instrument is necessary for playing music.

From this the following conclusion may be drawn:

(2a)   a piano or a violin or a flute or etc. (*taken together and as the subject of one proposition*) is necessary for playing music.

The modal expressions occurring in Maritain's examples ("nécessaire", "il faut") suggest that he sees 'piano', 'violin', 'flute', etc. as components of 'instrument', and that he regards the latter concept as a genus of 'music' or of 'playing music'. For as von Freytag says, in traditional logic "Species must be genus" (1961/80). If we assume, Locke-Lambert-Hegel-wise, that a (description of a) *definitio rei* of 'instrument' has the conjunctive-disjunctive form:

(3a) $\begin{cases} [\lambda] \ I \ is \ (P \ and \ V \ and \ F \ and...) \\ [\iota] \ I \ is \ (either \ P \ or \ V \ or \ F \ or...) \end{cases}$

then it becomes understandable that Maritain ascribes to the proposition "An instrument [genus] is necessary for making music [species] the following form:

(2a')   *(P or V or F or...) is necessary for M,*

where "or" may be read as *aut*-junction if and only if one assumes, as Martain clearly does, that a clear-cut classification is already available. Now Maritain says that (2a') obviously does not imply a disjunction of singular propositions *P is necessary for M or V is necessary for M or...*, as would be the case if the subject term of (1a) had *suppositio disjunctiva*. The subject term of (1a) is *immobilis*, that is to say: from (1a) to (2a) only a *descensus disjunctus* is possible. In such cases "there is no passage or inference from a more universal proposition to less universal propositions" (*loc. cit.*).

The view that a special kind of *suppositio* must be assumed here is understandable on the assumption that the proposition (1a) has the simple subject-predicate form. However, as soon as polyadic predication and internal quantification are admitted into the logical syntax, the replacement of the term-disjunction "a piano or a violin or a flute or..." by a disjunction of propositional functions becomes a possibility. To this end (1a) should first be turned around:

(1b)   for every x, if x is playing music, then x employs some/an instrument.[11]

(1b')(x)[x plays music → (Ey)(y is an instrument and x employs y)].

Given a certain knowledge of the kinds of musical instruments, this can be put as follows:

(2b)   for every x, if x is playing music, then x employs (a piano or a violin or a flute or....).

This is intuitively equivalent to

> (4b)   for every x, if x is playing music, then (x employs a flute or...).

The "or" in (4b) is a propositional connective. The choice between an inclusive or an exclusive disjunction may here be left to Maritain. The important fact is that the difference in the *suppositio* between two identically formed traditional *terms*, *i.e.* the so-called logical subject terms in "some man is a liar" and "some instrument is necessary for playing music" are now reduced to the difference in *logical form* between the two *propositions*. On the principles of Fregean grammar, neither of them is held to have a logical subject or subject term at all.[12]

The propositional function in parentheses in (4b),

> (5b)   x employs a piano or x employs a violin or x employs a
>   flute or...,

is nowadays commonly said to occur in (4b) as a *necessary condition* for the propositional function in the antecedent, "x plays music". In ordinary language this is often expressed as in Maritain's (1a). So, on all accounts a notion of necessity is involved, but in the modern analysis this special notion of necessity is consigned to the meta-language if it is used at all, and does not figure explicitly in the *analysans* of (1a). If Maritain should insist upon an explicit modal operator somewhere in (1b), (2b), and (4b), then he would land himself in a study of contemporary modal logic.

It is difficult to see how one could reduce the term-disjunction in (2a) to a propositional disjunction, and so to a truth-functional connective, without resort to polyadic predication. If, on the other hand, one is prepared to make use of this very natural device, no special assumptions of *suppositio*, no articles, and no modal operators are needed in this connection. We are completely in command of the implications and truth-conditions of (1b), and for critical listeners there is no problem how to criticize a proposition of this form.

This means that the need for logophoric terms cannot be demonstrated by reference to propositions with a term with *suppositio disjuncta*, since that *suppositio* can be defined in terms of polyadic predication and quantification. In IV-13 we saw how the same means could be employed to reduce a case of *suppositio simplex* to a case of *suppositio personalis*.

(ii) *Suppositio copulativa and suppositio copulata.* Maritain discusses the sentence "tout homme est mortel", the literal translation of which is "all man is mortal", a singular proposition with "all man" as its so-called subject term (84). That term is said to have *suppositio copulativa* which means that a *descensus copulativus* is permitted, *i.e.* that we may validly infer the propositional conjunction "this man is mortal *and* that man is mortal *and*...", and from there proceed to, *e.g.*, "this man is mortal". Clearly the premiss is assumed to be equivalent to "(this man and that man and...) is mortal".

If the subject-predicate analysis is maintained, then there are also propositions with a conjunctively composed subject term which do not permit the "descent" to a conjunction of propositions. Maritain offers two examples, which from the point of view of modern logic are widely different in logical form:

(1)   The apostles were twelve,

or:

(1′)   (Peter and Paul and...) were twelve,

(1933/85), and also (122)

(2)   Peter and Lewis are cousins.

We have discussed propositions of the form of (2) in VII-14 above. As to (1), a modern logician might say that for serious purposes it must be counted as ungrammatical; the apostles never were twelve since they *could* not be; the number of the (class of) apostles, however, was indeed twelve. We shall not here go into that analysis, which can be found in a great many modern texts. The history of the propositional forms

$$(S_1 \,\&\, S_2 \,\&...)\; is\; P$$
$$(S_1 \,v\, S_2 \,v...)\; is\; P$$

in which the subject term was held to have *suppositio copulativa* or *disjunctiva*, has, however, considerable historical interest.

Kemp-Smith reports that in Germany the Wolffian text-books of logic assumed the classification of propositional forms:

```
┌─Simple =   Categorical
│          ┌─Copulative (i.e. categorical with more than one
│          │             subject or more than one predicate)
└─Complex ─┤─Hypothetical
           └─Disjunctive
```

(quoted from Swing 1969/14). Swing points out that the third division of Kant's table of judgments, which Kant calls "Relation", should be understood against the background of this chart, but that Kant dropped the then traditional distinction between simple and complex propositions, and the so-called copulative judgment form as well (Swing, *loc. cit.*). In fact, his omission of the copulative judgment form (or forms) may, for Kant, have had the aim of getting rid of the simple-complex distinction, which he needed to dismiss in order to give free play for his triadic obsession. But, as Swing shows, Kant had to pay a high price for this; and so, in fact, do all of us, through the impact of the Hegelian school. For the copulative judgment is, as we saw in VII-14, a worthy predecessor of the more general relational judgment form

$$\langle s_1, s_2, ..., s_n \rangle \ is \ P,$$

and would have been the only obvious basis from which to abstract (or "derive") the notion ("category") of *community*. Having cut himself off from this possibility, Kant wriggles through the only opening which is now left to him, *i.e.* the *disjunctive judgment*. As Swing says, the "derivation" of 'community' from the disjunctive judgment form is not very convincing, to say the least. It would be natural to infer from this judgment form the logical need for the notion of *disunity* or *opposition* (*op. cit.*/25). For Kant, like everyone else, took disjunction in the exclusive or *aut*-junctive sense. The *only* kind of community known to post-Kantian German idealism, therefore, is *community in opposition*, to which is generally added the notion of (mutual) *causality*. The latter becomes understandable as a consequence of the assumption that '*the M*' contains all its species "potentially". As de Rijk points out, *suppositio naturalis* was in the late middle ages sometimes said to be *potencialis* (de Rijk 1973/73). Kant gave a new impetus to the latent desire to interpret logical possibility as active and dynamic by characterizing the category of community which he "derived" from the disjunctive judgment form as *interaction*: "Wechselwirkung zwischen dem Handelnden und Leidenden", reciprocal action between an active (subject) and a passive (entity; object). Kant, not Hegel, is to be blamed for the entry of this hybrid "category" into the theory of the "pure" elements of Reason.

4. ANGELELLI'S FINDINGS IN THE LIGHT OF THE TRADITIONAL INTERPRETATION OF EKTHESIS

Angelelli distinguishes three kinds of proposition with a universal as its logical subject in the traditional theory of predication, (a)–(c). He illustrates them by means of these examples (1967/123f.):

(a)   Homo est animal;
(b)   Man is white;
        Man is non-white;
        Triangle is right-angled;
        Triangle is non-right-angled;
(c)   Man is universal.

In (a) a genus is predicated of a species: (*species*) *is* (*genus*). He takes this to be the normal case. It is also possible, however, to predicate a species of its genus, as in the (b) examples, provided it is understood that the species are predicated of the genus only "potentially". This implies that the predicates "black" and "white" may be *predicated*, albeit "potentially", of the genus 'man' *simultaneously*. Angelelli points out that this conflicts with the *tertium non datur*.[13] And he remarks that in Frege's theory of predication such predicates as "right-angled" can only be predicated of individual triangles, not of a general concept or universal 'triangle'.

All this is in agreement with what we know from Locke, Lambert and Hegel. Angelelli takes his examples from Aristotle, from Boethius, from Thomas Aquinas, and from the *Systema logicae* of the Renaissance logician Keckermann (and in addition, of course, from Locke). These authors furnish him with the following examples (1967/130 n. 61):

"*Quoniam* Cicero sedet, Cicero autem homo est, homo sedet" (Boethius);
"Homo enim est albus *in quantum* hic homo est albus" (Aquinas);
"Homo est albus *propter* Socratem *et* niger *propter* Platonem" (Aquinas);
"Homo est albus, *quia* Callias est albus" (Keckermann, who refers to Aristotle's *Metaphysics*).

Notice (i) that all the connectives in these examples except the conjunction in the penultimate one are non-truth-functional, and (ii) *that a clear
definition of their meaning or use is nowhere to be found*. Earlier in this
book we saw that Ockham and Melanchton draw an indefinite conclusion "homo est animal" from two individual premisses (V-4, 6).
Melanchton's example is this: "Hoc est animal, et hoc est homo, ergo
homo est animal". Melanchton calls this an *expository syllogism* in the
third figure, but introduces a (*species*) *est* (*genus*)-proposition *ıS is P*,
going from individual statements directly to a statement that may be
understood logophorically, and so by-passing the particular conclusion
that according to the traditional view of ekthesis might, by a second
step, have yielded the same result.

In connection with (b) reference ought to be made to the seventh
chapter of *De interpretatione*, where Aristotle holds that "man is white"
and "man is not white" may both be true (7 17b 29–33). [14]

All these examples may be explained by our theory about the traditional view of ekthesis, which was explained in the last chapter. For that
term which is introduced "by ekthesis" on the strength of a paiticular
judgment, expressed as *some S is P*, I chose the variable "*M*", which
ought to suggest the "middle term": *ıM is S and P*... (cp. X-8). Of course
I did not thereby mean to imply that a traditional logician felt obliged
to introduce a word other than *S*. Just as the conclusion *ıS is P* may be
drawn from a premiss *all S are P* (cp. X-5), so, on the strength of *some
S is P*, a traditional logician may conclude that '*P*' has something to do
with '*S*', the reference of *ıS*. From "Socrates est albus" Thomas Aquinas
and Keckermann infer "Homo est albus...", since Socrates was a man.
In general: from

> *some S is/are P*

the conclusion

> *ıS is P (in quantum.../propter.../quia...)*

follows. In this way one easily arrives at conjunctions like *the Triangle is
right-angled in quantum... and the triangle is oblique in quantum... and...*.

In addition to "in quantum", "propter", and "quia" many other
auxiliary devices are found, whose task clearly is to signalize a non-
normal, philosophical "intentional"[15] use of connectives. Leibniz

employed, in a very complicated and extremely unclear manner, undefined expressions like "eo ipso", "qua", and "quatenus" in order to analyse comparative and other relational propositions; Mates speaks of "linguistic contortions" (1968/521). The next sections will deal with special classes of such "signals of intentionality".

5. SIDES AND ASPECTS: INTRODUCTION OF SPATIAL OPERATORS

The problem of how to criticize and perhaps falsify logophoric statements *the/an M is P* is very often dismissed as unimportant with the following excuse: it is always possible to "correct" a judgment by adding new "sides" or "aspects" to the subject-matter or *Sache*. For example, Lenin offers the following definition: "Dialectics is a ["die"] living many-sided knowledge (whereby the number of the sides is constantly growing)" (*Philosophische Heften*; quoted from Weinberger 1965). That makes a workable theory of negation superfluous, for, as Bradley says, "Error is truth when it is supplemented" (1883, I/173).

In Husserl's last work, published posthumously by Landgrebe (Husserl 1948), this way of reconsidering untenable positions is set out and employed. Husserl's opinion about existential propositions will be remembered from X-11. In that part of his book which comes closest to a theory of negation (94f.), spatial operators like "the thing's aspect from the side in question" replace the much needed theory of the negation of particular propositions, as the reader can easily verify (*op. cit.*/96).

The logic which is tacitly assumed there may be formulated as follows. Suppose that a proposition *some S is P* is given. This justifies the introduction of the following propositions

(1)   *on the one side* ["*einerseits*"] *the S is P,*

or:

(1′)   *P is an aspect of the subject-matter* ["*Sache*"] *S.*

If *some S are Q* is also given, that yields:

(2)   *on the other side* ["*andererseits*"] *the S is Q,*

so that

(3)   *on the one side (hand) the S is P and on the other side (hand) it is Q.*

In (3), we have "corrected" (1) by taking up the predicate Q conjunctively as a new "side" or "aspect" of '*the S*'. This procedure never leads to the cancellation of any property from the set that may be predicated of the logical subject in question, but only to an increase in their number. A predicate is never struck off the list for good. If *some S are P* and *some S are not P* are both true, then the procedure I just described will lead to this statement:

(4)   *On the one side, the S is P and on the other side it is not P,*

in German:

(4')   *das S ist einerseits P und andererseits nicht P.*

This is Husserl's language. Contrary to universal judgments, a judgment which is formulated in this manner cannot be refuted by reference to individual cases; the proponent of (4') has covered himself against attack, from all "sides".

So here is a new, "qualified" introduction rule for logophoric articles:

(5) $\dfrac{\textit{(some S) is P}}{\textit{ergo } \imath S \textit{ is, on the one side, P}}$

In this logic the sentence "On the one side the Bantu is primitive, and on the other side he is well educated" is well-formed and even true. Most contemporary logicians and epistemologists would probably consider this as a merely preliminary and sketchy form of speech.[16] Findlay, however, is of another opinion: "Instead of saying that John is tall, and Paul fat, we may say that the absolute substance is tall in its Johannine aspect, fat in its Pauline one" (1970/176).

6. SIDES, ASPECTS, AND AUT-JUNCTIONS

In all probability, the spatial vernacular found in the language of Husserl and in that of so many other philosophers has its roots in a metaphysical interpretation of the doctrine of the *loci* or τοποι as *sedes argumenti*.[17] These expressions originally belonged in the *topica* or theory of argumentation, where they seem to have been without metaphysical significance. During the Renaissance the *topica* of Cicero and Boethius became fused with formal logic, through the works of Rodolphus Agricola and Petrus Ramus. In a less explicit manner Peter of Spain also contributed to this process.[18] This must have given a strong new impetus to the already latent assumption that theoretical logic is in need of a special logical dimension.

It therefore seems that we are entitled to translate the letter "λ" in λM by "locus" as well as by "logos". A phrase λM may, then, be read *the locus generis M* as well as *the logos of M*.

In the light of what one can learn from Bird about the formalization of the *topica* in the middle ages, this spatial ('topical') terminology seems to be justified (Bird 1960).[19]

The combination of article and spatial modifier: *the locus generis*... may be condensed into one symbol "λ_z". A general concretion, an ontologically independent entity, is therefore a many-faceted entity, a pyramid in spiritual space. Since a pyramid contains all its sides, we can say:

(6) $\lambda_z M$ is *(P and Q and....)*.

One may also consider placing the spatial modifier in the copula, or perhaps in the connective itself:

(7a) λM is$_z$ *(P and Q and...)*,
(7b) λM is *(P and$_z$ Q and$_z$...)*.

Whatever one chooses, there is here a logical constant, be it an article, a copula, or a connective, which must be given some kind of definition before one can hope for critical dialogues between users of this language to ensue. In (7a) and in (7b) the copula cannot be distributed over the terms for the species, P, Q, etc., the way one can do with the "and" of class conjunction (class-product operator).

In order to explain why traditional logicians have always preferred exclusive disjunction to inclusive disjunction it is perhaps not necessary, but certainly sufficient, to assume that they have thought in terms of entities which unfold in a logical space. For in any space, only one entity at a time can occupy a certain place; on one "place" therefore, only one argument, that is to say; either P, or Q, or..., can be "seated".

This explains why many authors assume, as they clearly do, that the distribution of "is" over the terms of a conjunction, *P and Q*, results in an *aut*-junction:

(8)   *ɪM is either P or Q*.

I have here assumed that the article in (8) is "lower" than that in (7), in conformity with the "unfolding" of the "logos" of M (cp. Section 8).

7. "CAN" AND "MUST" IN VON FREYTAG'S LOGIC

Von Freytag does not employ this spatial vernacular in his books on logic. He does not say that *the* triangle is, *on the one side*, oblique, and *on the other side*, scalene as well, or that *the* mammal is, *on the one hand*, equine, while, *on the other hand*, it is feline as well. "The triangle *may* be scalene" so his formulation goes, and "the mammal *may* possess equine properties" (80).[20]

He declares that this means no more than "Genus *may* be species", "species *must* be genus" (loc. cit.). However, if that is all there is to it, then we do not need the words "may" and "must" at all. For the same may then be expressed by saying that *only some* mammals are equine; or that *some* mammals are equine and all equine animals mammals, but not conversely; or that the equine animals form a proper sub-class of the mammals. Instead of the judgment form *the class of S's is a proper sub-class of the class of P's*, von Freytag prefers *the P may be S*.

Neither von Freytag nor Jacoby will hear of a special logic of the modal word "may", in German: "kann". Such an undertaking would in his opinion be just as irrelevant to pure logic as a logic of relations (von Freytag, *op. cit.*/78, Jacoby 1962/53f.). For his rejection of modal logic, von Freytag offers the following excuse: "Not the subject matter, the S, must or may, but we ourselves must or may judge it to be so or not so" (79). The reason why we use "must" and "may" should be sought either in the mind of the judging person or in other logically irrelevant factors, or else in the "logical connections in which the S stands...". – "A modally coloured judgment... states *that there is such a connection*, and this is its content. Seen this way it is an assertoric judgment about the S, the P, and a larger logical *connection of concepts*, one *which is not further specified*. All this taken together is its logical subject..." (*loc. cit.*, italics mine).

It seems appropriate to compare this most indefinite subject with

Husserl's description of the subject of existential propositions (X-11). It should not be forgotten that in von Freytag's logic the logical subject of a particular proposition is an indefinite species (X-1). [21] Furthermore, it is revealing that a judgment *S may be P* shares with Bosanquet's analogical judgments the property of referring to a logical connection which is not further specified in the judgment (IX-14).

Von Freytag regards this logical subject as "clearly different from the verbal [subject]" (*loc. cit.*). The question why this is the case receives the following significant answer: "That must be so, for otherwise our present distinction between different forms of judgments would contradict our earlier insight that *every* judgment has the logical structure "*S is P*"" (*loc. cit.*). There seems to be a failure to distinguish object-variables from meta-variables here.

However that may be, the one word "is" is clearly assumed to suffice for the expression of the modalities of pure logic. Consequently both (*species*) *is* (*genus*) and (*genus*) *is* (*species*) hold in von Freytag's logic, just as Angelelli concluded from his investigation of other authors. In the former case we have to do with a well-defined judgment, in the latter case with an *indefinite* judgment: "Without changing the logical sense of the judgments we may, therefore, replace "must" by "it is determined that...". ... We replace "may" by "it is not determined whether..."" (80).

It seems very likely that von Freytag's ontological assumptions are far more complicated than he would have us believe. In all probability he tries to adapt his logic to that bundle of ideas which Lovejoy has characterized as the Great Chain of Being and of which he has given us an extensive description (Lovejoy 1936).

8. A SIMPLE FALLACY?

This mode of speech, *S may be P*, has been discussed by Moore (1959/236), and what he said about it has recently been taken up again and commented upon by Purtill (1971/119–122, 256–267). Moore's observations here are, I think, of a far deeper significance with respect to natural language than he himself (or Purtill) seems to have realized. [22]

(i) *Quantifiers or modalities?* Moore says that there are at least two senses of the word "possibly", or "may be" and that, if you confound them, you may turn up with some weird results. He cites the sentence "It is possible for a human being to be of the female sex", which can be paraphrased *salva veritate* in a number of ways, all containing some modal locution:

(1a)   *It is possible for a* human being *to be* of the female sex

(1b)   Human beings *may be* of the female sex

Many German authors clearly prefer the following vernacular:

(1c)   *Der* Mensch *kann* weiblich *sein,*

thus combining a modal expression "kann sein" with a definite article.

Moore first compares sentences like (1) to universal statements without modal expressions, such as

(2a)   All human beings are mortal,

or, as one often says,

(2b)   Human beings are mortal.

Again I add a German variant:

(2c)   *Der* Mensch *ist* sterblich.

He now introduces the assumption that the sentences (1), like the sentences (2), have universal import, and subsequently carries this assumption *ad absurdum*. If one of the variants of (1), in that sense which makes it a true sentence, is to have universal import, then it must be a sentence in which the modal expression comes *after* the universal quantifier: "Every human being may be of the female sex", since the opposite order would make (1) a false sentence. The most natural symbolization of this in modern logic is:

(3) $(x)[Hx \rightarrow \Diamond Fx],$

with "x is of the female sex" as the argument of the modal operator. This way of paraphrasing (1) yields, together with

(4)   I, G. E. Moore, am a human being,

the remarkable result

(5)   I, G. E. Moore, may be of the female sex.

Since Moore knew that he was not a woman this is obviously false when "may be" is understood in the epistemic sense of "it is unknown whether".

From this Moore draws the conclusion that the sentences (1) cannot be taken to express universal statements at all, and that "possible" in (1a) has a meaning which cannot even be approached by assuming it to be an operator modifying the predicate *Fx*. The truth is, Moore says, that (1) is nothing but a misleading manner of expressing what may be expressed just as well by the simple particular proposition

(6)   Some human beings are of the female sex,

or, more briefly,

(6′)   Some H are F.

And from (6) and (4), (5) does not follow. The lesson to be drawn from this is: do not read a modality where a mere particularity is "meant".

Moore refers to the argument form which takes us from (1) and (4) to (5) as "this simple fallacy". His point is that as soon as we take (1) to be of the form (3), this fallacy is forced upon us.

(ii) *This fallacy is a feature of the old NP + VP-logic (subject-predicate logic)*. Earlier in this chapter we have seen that this simple fallacy is a fundamental systematic feature of perhaps the greater part of traditional logic. In the first place, we know that practically all earlier logicians assumed the logically correct grouping of words in a particular sentence like (6) to be

(7)   (Some human beings) (are of the female sex),

or, more briefly,

(7′)   (Some H) (is F),

with "Some H" as a noun phrase or subject term. Second, a great many of them proceeded from (7) to one of the variants of (1), usually to (1c) when the language in question was German. The first is a purely syntactic principle, the second an inferential one. Together these two principles constitute the basis of *traditional* logical essentialism, and the "chains of necessity" which Findlay so dearly misses in the Frege-Russell logic are welded from precisely these principles.

We also know that due to the lack of individual variables in the pre-

Fregean variants of logical syntax, all logicians before Frege understood
(7) in such a way that if we want to describe it in modern technical terms,
then we shall have to say that (7) was taken to contain a second-order
quantifier, not a first-order one. The truth of (7) was assumed to imply or
to "presuppose" the existence of a *general* entity.

On the principles of Fregean syntax, the right analysis of (6) is, by
approximation,

(8)   Some (is H and F),

i.e.,

(8′)   There is something which (is H and F).

This sentence contains no noun phrase whatsoever and therefore does
not suggest (1c) or one of the other variants of (1) as a paraphrase.
Nobody would think of deriving or inferring (1c) from (6) when the
brackets are put as in (8) and (8′).

The reader may now want to draw the conclusion that if one starts out
from Frege's syntax and logic of the quantifiers, one does not run the
risk of landing oneself in essentialism of any kind and that modal logic,
even "quantified" (a highly misleading expression!) logic of the expressions "necessarily" and "possibly" will be all right. That may be so; I
think we do not really know that yet. The fact that one road to essentialism
has been blocked does not exclude the possibility that other roads are
still open or may be created while going along.

(iii) *Purtill on logical possibility.* Moore's argument rests upon the assumption that (5), the sentence "I, G. E. Moore, may be of the female sex",
is false, and this again rests upon the assumption that "may be" is given
a meaning which I have here called "epistemic possibility": it is unknown
whether.... Moore himself does not call it by that name. He speaks of
"a confusion of two different uses of the words "possible" or "may"".
He seems to take into account only their epistemic use and, in addition, the use of these terms to paraphrase sentences beginning with
"some".

Moore's argument was recently taken up by Purtill, who points out
that "possibly" and "may" can be given a meaning which differs from

either of those considered by Moore. Purtill holds that although (1) may be misleading if "possibly" is used in one sense (in the above sense of "epistemic possibility"), it is not misleading but simply true if you use it in a weaker or at least different sense, even if we cast (1) into the form of (3), as Moore was apparently not willing to do. This other sense he calls "logical possibility", to which belongs a notion of "logical necessity".

Purtill, who takes (3) to be well-formed and true, has also to accept (5) as true, since, as he points out, the argument

$$(9) \qquad (x)\,[Hx \rightarrow \Diamond\,Fx]$$
$$\underline{\qquad Hm \qquad}$$
$$\therefore \; \Diamond\,Fm$$

is logically valid, at least as long as one accepts the rules of *modus ponens* and the *dictum de omni*, or universal instantiation (the latter rule is in fact rejected in some modal systems, but we shall not go into that). By ending his discussion right here, Purtill, like so many others, probably succeeds in making many readers believe that there already exists in human communication a clear and simple notion of logical necessity, and connected with it a clear and simple and not unimportant notion of logical possibility, which simply lie there awaiting the right logician (or linguist, see below) to come along and provide the correct description of what every competent user of "natural" language means and always has meant.

(iv) *Lakoff's example.* For those who consider the features of language illustrated in Moore's fallacy as too superficial and as too obviously misleading to be worthy of scholarly attention, I have a neat surprise. On the first page of the very article (1970) in which he expresses his belief in a natural logic as that (one) logic which is somehow embedded in (any) natural language, Lakoff offers the following example of what he regards as a valid argument:

(1)   The members of the royal family are visiting dignitaries.
(2)   Visiting dignitaries can be boring.
(3a)   Therefore, the members of the royal family can be boring.

"Thus if 'visiting' is assumed to be a modifier of the head noun 'dignitaries'", Lakoff says, "then (3a) follows as a logical consequence."

In a logic for which that holds, the following argument is another piece of valid inference:

(1′)   The members of the presidential (Heinemann) family are German visiting dignitaries.

(2′)   German visiting dignitaries can be ferocious.

(3′)   Therefore, the members of the presidential family can be ferocious.

As P. T. Geach in the title of his latest book says: logic matters. The second argument is a clear counter-example to the logical validity of the English form of Lakoff's argument. These are better examples of Moore's simple fallacy than Moore's own example: Lakoff says "can be", not "may be", even in the conclusion, and cannot therefore be rescued by Purtill. Besides, if (2′) has the form $(x)[Gx \rightarrow \Diamond Fx]$, then it would have no more informative value than $(x)[\text{Wombat } x \rightarrow \Diamond \text{Ferocious } x]$, or than $(x)[\text{Non-ferocious animal } x \rightarrow \Diamond \text{Ferocious } x]$, which is also true but which only says that any completely non-ferocious animal like a wombat is (say) a tiger in some other conceivable world in an unspecified set of such "worlds".[23] So that would be a very unreasonable interpretation of (2′) when uttered as a statement of natural language. Lakoff's example therefore does not express a modal argument in the sense of modern modal logic, and thus *a fortiori* not a valid one. It may be used to express, but then only incompletely, an inductive argument of some kind, with hidden time references and variables and with a suppressed premiss or condition to the effect that nothing is known about the members of the royal/presidential family that can be used to refute the conclusion. The logical status of the latter assumption has often been an object of discussion. We need not go into that discussion here, for since he says that the conclusion "follows as a logical consequence", Lakoff has in all likelihood not had this kind of inductive argument in mind eithei. I think we should rather say that, as it stands, Lakoff's example is simply invalid. Since man is a prejudiced animal his example may well belong to a natural logic, but hardly to a normative one. Has the time not come to take a more critical attitude towards natural languages, and to ask from which natural un-logic this kind of locution derives?

9. LOGICAL SPACE AND LOGICAL POTENTIAL IN
PRE-KANTIAN LOGIC: THE LOGICAL FIELD OF FORCE

Pfänder, the last German author but one of a systematic work on pure logic, takes as the foundation of this science the "unfolding[24] of the judgments which are implied by a [certain] judgment" (cp. VI-6). His elucidation of this process of unfolding ("Entfaltung") contains a number of interesting auxiliary expressions:

"The judgments implied by a judgment… are merely not yet unfolded, that is to say, in these judgments the *logical movement* which goes from the subject-concept through the copula and to the predicate-concept is not *actually* in effect ["im Vollzug"], but they are effected and are in a state of *restful tension*. They are potential judgments, as it were folded up, and in an unfolded judgment they *lie at this or that place*. They can, however, be unfolded and be independently *lifted up* ["aufgehoben"] from the judgment in which, *folded up*, they are *lying in repose*. From the *potential, static state of tension* they can *awaken to actual logical life* and come into an *internal logical motion*" (287; all italics mine).

This passage contains a selection of spatial phrases, *i.e.* "(to) unfold", "motion", "place", "(to) lie", "(to be at) rest", and expressions referring to some sort of logical powers as well, *i.e.* "potential", "tension". It is impossible not to draw the conclusion that Pfänder, like Hegel and many others before him, is assuming the existence of a *logical field of force*. He offers no description, however embryonic, of the properties of such a field and does not even try to demonstrate its existence.

Maritain's terminology in connection with "the logical totality", '*ıM*', is very similar to Pfänder's. "The logical totality ,"Maritain says, "is called potential because it only contains its parts potentially." He offers this example: ""Man" and "beast" are only potentially in the concept "animal"" (1933/41). He clearly means to say that

(1)   The Animal is potentially Man and Beast.[25]

Pfänder gives examples of a different kind. He maintains that the judgment 'This object is an animal' immediately follows from the judgment 'This is an eagle', by *unfolding* of the latter. It seems reasonable, however, to say that the implication can only hold in virtue of a lexical nominal definition, *i.e.* of a definition of the word "eagle". Pfänder, however, makes no allusion to lexical definitions. The example suggests the potential containment of higher genera in a concept or object. That he speaks in terms of "lifting", "herausheben", also indicates that the unfolding is one of "higher" concepts, like 'animal', starting out from "lower" concepts, like 'eagle'. In Maritain's logic, on the other hand, the (judgments about) concepts which are potentially shut up in (a judgment about) a genus, *e.g.* in 'animal', are its various species. At least in a metaphysical interpretation of the logical dimension, these cannot but be *lower* than the genus.

This difference between the German idealist Pfänder and the French neo-Thomist Maritain calls for an explanation. The solution to the problem is suggested by the nationalities involved. What this difference reflects is nothing less than the impact of Kant upon German philosophy. Maritain here represents pre-Kantian logico-metaphysical strivings, Pfänder the post-Kantian ambition. The former assumes a field of force in which the gradient points from higher to lower, the latter assumes that forces are at work from the bottom of logical space up, which, if true, would enable us to carry out id-introductions with a certain expediency. Maritain's potentialities impose *deductive* processes, in the traditional sense, while Pfänder assumes logical forces in the direction of traditionally understood *induction*, *i.e.* of Albrecht's *ekthetic* processes.

The fact that our hypothesis about Kant's logical ambitions (cp. IX-19) enables us to give an explanation of this difference between Pfänder and Maritain must count as an important corroboration of that hypothesis.

Pfänder does not mention ekthesis. He does not propound his own theory in clear terms, nor does he inform us on which points his own logic of potentially implied judgments coincides with or differs from the traditional id-logical views about a potential logical totality. Does he, like Maritain, accept reduplicative propositions as philosophically important and as logically clear?[26] Does he assume *two* kinds of logical potential, super-imposed upon one another? These questions receive no answer, and other idealist authors on logic do not help us either.

10. Two Introduction-Rules for Potentiality-Operators

Below I shall use the letter "*M*" as a variable for names of genera, and "*P*" and "*Q*" as variables for names of the *differentiae* of the genus under discussion. *Differentiae* are mutually exclusive characteristics of the several species of a genus. The traditional theory assumes that

(1)   *the M is potentially P and Q and...*

is a well-formed proposition. Let us use "π" as an abbreviation of "potentially". Now what is the argument of this operator? The following formulations both seem to be reasonable interpretations of (1):

(1a)   *πM is (P and Q and...),*
(1b)   *λM is$_\pi$ (P and Q and...).*

In (1a) the subject is described as a "potentially indeterminate genus", in agreement with Angelelli's characterization of the traditional vernacular and intentions. In (1b) the copula – the *Aussagekopula* – is modified. I am inclined to think that most id-logical philosophers make no distinction between (1a) and (1b). A third possibility is to let π operate upon a complete judgment:

(1c)   *π (λM is (P and Q and...)).*

Since *P, Q,... ex hypothesi*[27] indicate *differentiae* of the genus 'λM', it goes without saying that from (1a), (1b) or (1c) one can, for low values of ι at least, deduce:

(2)   *ιM is (either P or Q or...).*

The π-operator is therefore an operator which forms an *aut*-junction from a conjunction, a property which it shares with the z-operator of Section 6. On what grounds is the introduction of a proposition *ιS is$_\pi$ P* justified? This form may be read: *the/an S may be P*; here *P* is the name of a property which characterizes one, and only one, of the S-species.

(i) One possibility is that one has to take a *definitio rei*[28] as point of departure, *i.e.*, a proposition of the form (1) above:

(3)
$$\frac{the\ M\ is\ potentially\ (P\ and\ Q\ and...)}{\therefore\ the/an\ M\ may\ be\ P}$$

It will then be necessary to assume that the premiss is partly or perhaps completely accessible to the mind, by direct inspection.[29] This view may be ascribed to the school of Hegel. I shall illustrate it by means of Goethe's

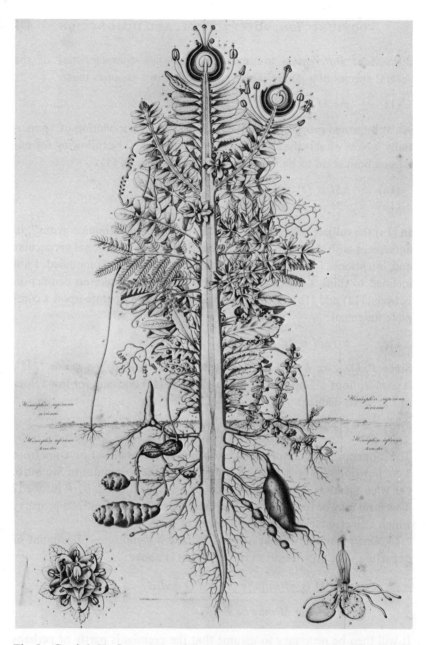

Fig. 5.   Goethe's Urpflanze. 'Vegetal Type Ideal Appendicule' from Goethe, *Oeuvres D'Histoire Naturelle Atlas* (1837), reproduced by kind permission of the Trustees of the British Museum (Natural History).

Urpflanze (Figure 5):

$$\frac{\lambda\text{Plant (the }Urpflanze\text{) is potentially (rose and carnation and cactus and ...)}}{\therefore \text{ a plant is either rose, or carnation, or cactus, or....}}$$

Now Schiller was not the only person to doubt the existence of such a spiritual entity visible to the inner eye. It would be a mistake, however, to think that those who were sceptical about the *Urpflanze* felt obliged to reject the general logic of potentialities or of powers. Another id-logical possibility would seem to be

(ii) that (1) may be reached on the ground of a conjunction of judgments of the kind that form the conclusion of (3):

(4)
$$\frac{\textit{(the S may be P) and (the S may be Q) and...}}{\therefore \textit{ the S is potentially (P and Q and...)}}$$

which, however, presupposes another possibility for arriving at judgments of the form *the/an S may be P*. Now many philosophers obviously interpret particular propositions as the lowest kind of logophoric propositions:

(5)   *some S is P* $\underset{\text{Df}}{\leftrightarrow} \theta S$ *is P*.

From what was said in Sections 7 and 8 we surmise that from a particular judgment many philosophers infer a "potentialized" indefinite higher-level judgment with the same general term S:

(6) $\dfrac{\textit{some S is/are P}}{\therefore \; \iota S \; is_\pi \; P}$, or perhaps even $\dfrac{\textit{some S is/are P}}{\therefore \; \lambda S \; is_\pi \; P}$

In order that these conclusions may be used to compose the premiss of (4), it must be certain that P is the name of a differential mark in a pre-fabricated classification. It is hard to see how one could ascertain such a thing *a priori*.

Suppose we were to accept (6) as a valid inference rule. We might then just as well take an individual proposition for our premiss:

(7) $\dfrac{\textit{this S is P}}{\therefore \; \textit{the S may be P}}$

It is quite obvious that the interpretation of individual propositions which was given by Ockham, Wallis, Leibniz, Wolff, and Kant makes an argument form like (6) or (7) seem plausible.

We still do not know on what grounds one may introduce a proposition of the form *the S as such (qua talis) is P*, which we discussed in IV-30. In VII-15, I suggested that this judgment form, too, is connected with the notion of active logical genes or nuclei, but we were not able to settle the question of the right introduction of such propositions. I think we are entitled to the hypothesis that there is no logical difference at all between *the S as such is P* and $\lambda S \ is_\pi \ P$.

Let us compare the traditional steps leading to id-introduction with the corresponding steps in modern logic. Schemas (6) and (7) above will be said to express rules for *potential id-introduction,* or "pot. id-intr." for short.

IV. *Traditional id-logic*
Id-introduction with modal "is":
  (1) *this S is P* (prem. no. 1)
or
  (1') *some S is/are P* (prem.)
This entails:
  (2a) *the/an S may be P,*
  (2b) *the/an S is potentially P,*
  (2c) *on the one side, the/an S is P.*
More briefly:
  (2) $\iota S \ is_\pi \ P$ (pot. id-intr.)
From
  (3) *that (other) S is Q* (prem. no. 2)
we obtain
  (4) $\iota S \ is_\pi \ Q.$
From (2), (4), and perhaps still other similar propositions which may not be known to us yet, we arrive at
  (5) $\iota S \ is_\pi \ (P \ and \ Q \ and...).$
At the same time we have:
  (6) *[an arbitrary] S is either P or Q or R or....*

IV. *Modern*

(1) *this S is P* (prem. no. 1)
From (1) follows:
(2) $(Ex) \ [Sx \ \& \ Px]$ *(ex. gen.)*

From
(3) *that (other) S is Q* (prem. no. 2)
we obtain
(4) $(Ey) \ [Sy \ \& \ Qy]$ *(ex. gen.)*
From (2) and (4) together we obtain:

(5) $(Ex) \ [Sx \ \& \ Px] \ \& \ (Ey) \ [Sy \ \& \ Qy].$
At the same time we have:
(6) *an arbitrarily chosen S is P or Q or R or...* (inclusive *or....*)
and
(7) $S \cap (P \cup Q) \neq \Lambda.$

11. THE LOGIC OF POTENTIALITIES IN PRACTICE

(i) It is not hard to see that the logic of potentialities, which is the most

influential variety of traditional logic today, leads to the greatest ab-
surdities. It contains no device whereby properties P which belong to all
M-things may be distinguished from properties which belong only to
some M-things. The rule of potential id-introduction leads us from the
premisses

(1) $\begin{cases} \text{some human beings are black} \\ \text{some human beings are white} \\ \text{some human beings are mortal} \\ \text{some human beings have red hair} \end{cases}$

to the conclusion that

(2)   Man is$_\pi$ (Black and White and Mortal and Red-haired and...).

The worst thing about this mode of thought is not that Man, and perhaps
even Saul, is regarded as somehow having more than one skin colour (see
further under (ii) below). The real problem is that together with the
minor "Saul is a Man", (2) implies

(3)   Saul is either Black or White or Mortal or Red-haired or....

Hence, whoever is black (or white) is immortal and cannot have red hair.
There may be, in the minds of many id-logicians, an intention to derive
(2) from (1) only when the predicates in (1) are in fact mutually exclusive,
at least on the θ-level, so that they can be taken to define species of M
(or: of 'M'). But the literature on traditional logic contains not a single
attempt to describe or to construct a criterion by which sets of mutually
θ-exclusive predicates can be distinguished from other sets.[30] So this
would-be ascending mode of thought can be seen to presuppose the
possibility of "seeing" what the right property-sets are after all,
exactly as if one were to start out consciously from a real definition
and to proceed by descent or "deduction".

Clearly, this problem is closely tied up with the problem of distinguishing
essential from merely accidental predicates. Aristotle developed no
device for distinguishing one kind from the other. In the Renaissance,
Petrus Ramus and others, whose influence upon continental philosophy,
especially upon later German idealism, has not yet been assessed, took
the not unreasonable step of dropping the essential-accidental distinction
altogether. Since they retained the stress upon exclusive disjunctions and

developed no new logical devices in their place, the net result of their activities was that in philosophy, if not in science, this circular id-logic was employed with even greater recklessness than before.

(ii) Our study of a great number of authors on traditional logic, from different philosophical schools, has made it clear that the following natural-language argument form

(I)
> *the (general) M is P*
> *S is an M*
> ───────────────
> *ergo, S is P*

is not valid in traditional id-logic. In fact both Hegel and Pfänder quite explicitly say so (cp. VI-3, VIII-20). However, a close scrutiny of the logics of Maritain, Pfänder, and Hegel led us to the conclusion that by inserting a number of qualifying phrases into the premisses and into the conclusion of (I), an argument-form results which the said authors regard as valid. Various such modifying expressions were pointed out and discussed in IV-30, VI-4, VII-15, and in VIII-21. If we combine these qualifications with the idea that a genus contains its *differentiae* only "potentially", then we come to the conclusion that in "pure" logic the following argument-form is valid:

(II)
> *the M ist P*
> *S is (an) M*
> ──────────────────────────────
> *ergo, S, in so far as it is M, is essentially P*

That is to say: the logical gene, the dressed-up logophore, of P is in S "in so far" as S is (an) M.

A logic which contains both rule (II) and the introduction rule for potential "id" (Section 9) as well, is a potentiality-version of PPL:

(1)	*this M is P*	(prem. no. 1)
(2)	$\imath M$ is_π P	(1, pot. id-intr.)
(3)	*S is (another) M*	(prem. no. 2)
(4)	*S, in so far as it is M,*	(2, 3, rule II)
	is essentially P, at least	
	up to a point.	

12. FROM LOGICAL POTENTIALITY TO INCLUSIVE DISJUNCTION

Let us now take up the problem of the "logical jobs", as Geach puts it (1968/viii), which the notion of *logical potentialities* has to do. In the preceding sections I have tried to prepare the way for a definite theory in answer to that problem. I share Angelelli's opinion (1967/122) that the potential genus is a theoretical construction and hence a solution to a problem in theoretical logic. Angelelli does not formulate that problem clearly, however, and offers no solution. My theory is this: *the doctrine of logical potentiality performed, above all, those logical jobs which are now carried out by the notion of inclusive disjunction, of which the notion of logical potentiality may be regarded as a primitive form.* I cannot but think that the logic of potentiality was invented as a consequence of

(a) the assumption, already made, that the "real philosophical disjunction" could be none other than exclusive disjunction, in combination with:

(b) the difficulties concerning the introduction of "all" and the elimination of "some", *i.e.*, around the logic of *an arbitrary M*.

It is a historical fact that the theory of logical potentialities is accompanied by a clear preference for exclusive to inclusive disjunction, and often, even, with a strong inclination towards dichotomies, as in the school of Hegel. One can explain the preference for exclusive disjunctions by assuming that philosophers usually thought in terms of a specific logical dimension and a logical space, since the notion of a space suggests that only one entity, concept, or judgment at a time can occupy a certain "place". But this is not enough to explain why a logic based on the conception of a given specifically logical space is further embellished with the notion and vernacular of potentialities, or potencies.

To brush this problem away just by referring to theological and metaphysical needs is to assume a false dilemma. It is more than doubtful whether metaphysical and logical needs have always been as separate as they may seem to us today. For suppose we decide to work only with exclusive disjunctions (i). In a generic logic, this results in a sharp distinction between species, and the dogma of monadic predication prevents us from describing the connectedness of things and kinds in the ways that are open to modern logicians. As a result there will be a strongly felt need for a logical device by means of which the species and individuals, the *ultimae species*, can be "brought together". We shall, in other words, be forced to invent some kind of *logical mixing-bowl*. The genus which, by definition, contains all the species conjunctively is precisely such a mixing-bowl (ii). Relations between individuals can now be understood as "caused" by a logical relation, *i.e.* by the participation (μέθεξις) of the individuals in one and the same genus. (The existence of non-symmetric relations calls for a dyadic genus, but we need not go into that here since this construction was never united to the rest of id-logic to form a theoretical unity.) The species are now brought into contact with one another after the first premature and exaggerated split. However, the resulting conceptual continuity of the universe has been bought at too high a price. The conjunctive unification is so clearly another exaggeration that a new theoretical move is called for. The theoretical move chosen by traditional logic was to retain both the original split and the conjunctive construction of the genus as well, but to weaken the latter by inserting a modality like "potentially". Originally this modality may have been no more than a reservation. Of course, the temptation to "dynamize" the notion of potentiality turned out to be irresistible for very many thinkers. Thus the generic logic evolved into what may be called a genetic logic, a logic of active logical genes.

Disregarding this last step of dynamization we can reconstruct the genesis of 'the potential genus', as a piece of theoretical problem-solving, in symbolic form:

(i)	*an A is either B or C or ...*	(fundamental hypothesis: descriptions of general entities should be disjunctive in the exclusive sense)
(ii)	*λA is (B and C and...)*	(*λA* is invoked as a logical mixing-bowl as a compensation for (i))
(1)	*this M is P, or some M's are P*	(first premiss)
(2)	*λM is P*	(hypothesis with respect to the optimal result of ekthesis)

(3)	*S is (another) M*	(second premiss)
(4)	*S is P*	(from (2) and (3), since (2) is assumed to have universal import)
(5)	*S is not P*	(third premiss)
(6)	*S is P and S is not P*	(4, 5)
(7)	*not*: λM *is P*	(2–6, *reductio ad absurdum*)
(8)	λM *is*$_\pi$ *P*	(modification of (2), retaining universal import)

This leads to revision of the mixing-bowl as a general principle:

(ii′) λA *is*$_\pi$ *(B and C and...)*.

The step from (1) to (8) is the introduction rule for potential *id* which we formulated in Section 9.

The problem could arise only because exclusive disjunctions were assumed to be useful in *a priori*, or philosophical, descriptions of empirical or non-empirical facts and phenomena.[31] If, however, the exclusive "philosophical" disjunctions are weakened to inclusive disjunctions, *i.e.*, if we modify the *first* principle:

(i′)   *an A is B or C or...*

(inclusive "or"), *then there is no need for an additional mixing-bowl, or hybridizer.* Our second principle (ii) and the assumption of the existence of a reference for expressions λA can be dropped altogether. We thereby get rid of the vicious *petitio principii* inherent in the use of exclusive disjunctions, for we no longer exclude the possibility that an observed or unobserved M-individual may have both properties P and Q at the same time; therefore the need for an additional unifying device does not arise.

A statement *M is P or Q* is weaker than each of the (binary) statements *M is either P or else Q* and *M is P and Q*. The first is implied by each of the latter two and may therefore be said to represent "the best of both". In fact the set of truth-conditions for the inclusive disjunction is simply the union of the truth-conditions of the binary exclusive disjunction and of the conjunction:

$$
\left.\begin{array}{cc}
f & t \\
t & t \\
t & f \\
f & f
\end{array}\right\} \Rightarrow
\begin{array}{c}
t \\
t \\
t \\
f
\end{array}
$$

As far as I know, no traditional logician ever worried about the fact that when $n>2$, an n-ary exclusive disjunction cannot be understood as an iteration of binary exclusive disjunctions. The traditional conjunctive-exclusive theory of genera clearly offers the philosopher a certain explanation of the role played by the individual terms which is "exposed" in a process of ekthesis. In payment for this explanation he has had to accept a theory of rational thought without any workable theory of negation, namely a potentiality-version of PPL; in short, he has had to accept that dialogical refutation is impossible, even in principle, of philosophical judgments of the most frequently used kind.

Boehner's question: "How could it happen that the so-called 'Laws of De Morgan' had to be rediscovered?"[32] should, I think, be answered in the following manner. Due to the impact of neo-Platonist ideas, post-mediaeval philosophers became more and more inclined to regard Aristotle's metaphysics and his philosophy of science as "more logical" than the logic of his *Analytics*. At the same time, and due to the same ideas, it became more and more usual to think and to talk in terms of a specifically logical field of force which, though badly described, had a reality for the philosophers in question that defeats the imagination of twentieth-century man. This logico-metaphysical space could accommodate the potential genus, but since an "argument" could not be located in more than one "commonplace" of the space at a time, the whole conception was incompatible with the notion of inclusive disjunctions as predicates.

This logic received an immensely strong new impetus from the seventeenth-century successes of the theories of fields of force applicable to naturally dichotomic and, hence, "*aut*-junctive" and dyadic phenomena, *i.e.* electricity and magnetism.[33] But whereas we have possessed, thanks to the genius of James Clark Maxwell, a highly satisfactory description of the properties of the physical fields of force since 1873, no similar development of the embryonic conception of a logical field of force has taken place. At the turn of the century Cohen alluded to "the importance of the potential in pure logic" (1014/339), but not so as to deepen our insight into the functions and properties of potentialities and potentials in idealist logic.[34]

13. A STANDPOINT REVISED: CASE STUDY

A weaker proposition than that formulated in the last section is that the modern use of inclusive disjunctions is a method of obtaining a "middle concept". Surprisingly enough, support can be found for this thesis in the later writings of von Freytag. This seems to imply a preparedness on his part to depart from certain very fundamental characteristics of traditional logic, and to introduce the device of inclusive disjunction into his norms for philosophical thought.

Von Freytag calls his second book – the one we shall deal with in this section – by the conspicuous name of *Logik II. Definitionstheorie und Kalkülwechsel* (*Logic II. Theory of Definition and Calculus Change*, 1967). He still defends arguments *ad subalternatam* (126). Surprisingly, he now omits the article from his examples of judgments of the kind which, in the terminology of his first book, have a "general concrete subject", *i.e.*: "Ellipse is [a] curve of the second order and encloses a finite region" (23), "Non-smoker is a man" (113).

Of course this will not solve the problem of the logic of logophoric terms. I have not been able to find the expressions "Logophor" and "allgemein Konkretes" in this second book; nor does he offer an improved theory of the logic of singular propositions, with or without definite descriptions. His omissions of the prenex articles may conceivably be caused by Scholz's harsh criticism, quoted in our Introduction (I-3). However, I believe that von Freytag has had reasons of a more theoretical kind, to be explained below, for making this change.

In any case von Freytag has felt the sting of Scholz's implicit reproach that traditional logicians never developed a decent theory of negation (*op. cit.*/69; cp. von Freytag 1967/59). The reader will admit that if this reproach is justified, traditional logical syntax and theory of inference must be accounted a failure. The assumption that traditional logic, in so far as it goes beyond the limits of Aristotelian syllogistic, contains no theory of negation and may not even permit the construction of one, is corroborated by a number of statements by adherents of that logic. It is corroborated by Bosanquet's desire for logical stability (cp. IX-14), a desire shared by Veatch and Popma (cp. I n. 13 and IX-16). It is also corroborated by Schmitt's observation on the so-called phenomeno-logical method: "The phenomenologists' account of their method... lacks... also a complete theory of truth, *at least as that term applies to statements in phenomenology*" (1967/149; my italics). That is to say: no description of the (or: a) phenomenological method contains a formu-lation of necessary conditions for the truth of indefinite propositions *das A ist B.*

Von Freytag does not deny that traditional logic does not contain much in the way of a theory of negation. The manner in which he tries to smooth over this fatal shortcoming of traditional logic is, in all its poverty, significant: "The logicians of the past will have had good reasons for not being interested in this problem." I shall not contradict von Freytag on this point.[35]

That von Freytag himself has come to see the importance of this problem seems clear. He now mentions the possibility of reproducing the laws of De Morgan in his own notational system for the logic of concepts and calls these laws "the quintessence of the whole [36] negation technique in the propositional calculus, of the procedure of forming, for every expression, its negation" (*loc. cit.*).

In the same book von Freytag for the first time shows an interest in inclusive disjunctions as a means by which, starting out from two given concepts, one can form a more general concept, a *Generalisat* of the former two. Given 'square' and 'rhombus' there is a ready-made expres-sion, *i.e.* "equilateral quadrangle", which stands for a concept more general than the first two. For 'cat' and 'dog' that is not the case. How can we, asks von Freytag, arrive in such a case at "the generalized concept"? Zoological nomenclature contains no suitable expression: "It would have to be one for something which is either cat or dog or (what we, as bad zoologists, may not exclude) both" (*op. cit.*/28).

That is quite right. I would, however, prefer the following formu-lation: good zoologists or bad, we can in no case *a priori* exclude that something is, or even will be,[37] both cat and dog. Human beings do not possess any infallible *Geistesauge* which will enable us to make, at the beginning of an investigation, a "faultless cut"[38] in the world of significant ideas. The assumption that it is possible to perform such a faultless cut is a necessary condition of the opinion that exclusive disjunction is "the real logical disjunction", as von Freytag called it in his first book (1961/ 161). In his second book, however, looking for the *Generalisat* of 'cat' and 'dog', von Freytag comes to the following conclusion: "The non-exclusive "or", which plays such an important role in logistics, says exactly what we need" (1967/28). In 1957 he still held that "pure logic" could confine itself to the investigation of systems of concepts which, in traditional manner, assumed a "complete disjunction" of concepts, for instance:

Mammal is either dog or else cat or else....[39]

But in 1967 he had left that assumption behind and maintained that "pure logic, however, may not disregard this mode of definition [by means of inclusive disjunctions, E.M.B.]". We must, he says, widen the traditional concept of definition, "and first of all we cancel the thesis: definitio fit per genus et differentiam" (*loc. cit.*).[40]

Since von Freytag is one of the exceedingly few European philosophers who not only defend a traditional logic, but who also discuss theoretical logical problems in some detail, these are words of considerable significance.

He still does not go into the question of the merits of polyadic predication. He dispenses with the obligation of establishing a theoretical connection between "not" and the articles simply by no longer using them in the way he did earlier, which is to say in connection with "can" and the exclusive "or". That is hardly a contribution to the further development of intensional logic. We shall have to content ourselves with the fact that he has come to see at least some of the importance of the logic of the inclusive "or". This point being reached, one's reasons for employing the articles "logophorically" would seem largely to have disappeared.

NOTES

[1] This argument form is also discussed in Kneale and Kneale 1964/179–181. I am inclined to think that this remarkable argument reveals more about traditional modes of thought and language than appears from the Kneales' discussion.

[2] No-one would like to state, at least not today, that *not (p and q)* is equivalent to *either not p, or else not q*. Such a statement would imply that the truth-conditions of p and q were the same as those of $p \leftrightarrow q$, so that the common "and" could be replaced

by a judgment of equivalence. In fact Fichte does precisely that. Lenk points out that Fichte rejects the common "and": "While the logical conjunction only combines the component sentences truth-functionally, the synthesis of the [Fichtean] *Wissenschafts-lehre* combines two different and opposite propositions or concepts by ranging both of them under a higher "generic concept"... that is to say, in respect of *content* ("in as far as they resemble each other")" (Lenk 1968/199). Cp. the end of Section 3. Cp. also Kyburg on "Conjunctivitis" (1970).

[3] Cp. Boehner 1952/67f.; Clark 1952/57f.; Kneale and Kneale 1964/294f.

[4] This is Quine's "dagger", in terms of which all other propositional connectives of two-valued logic may be defined.

[5] Note that the class K of all human beings is the union (the logical sum) of the class L of all white human beings *and* the class M of all black human beings *and*.... However, to each occurrence of the word "and" or of the sign "+" which is sometimes used, there corresponds, in a description of an individual human being or in a proposition with "an arbitrary human being" as its grammatical subject, an occurence of the inclusive "or" between the predicates "white", "black", etc. If, for certain class terms, the proposition

(a) $K = L + M + \cdots$

to be read:

(a')   *the class K is the class L and the class M and... taken together*

is true, then this also is true:

(b)   *an arbitrarily chosen K is L or M or....*

This has certainly contributed to the confusion in traditional intensional logic.

[6] Doyle criticizes Maritain for attributing this opinion to John of St. Thomas (Doyle 1953/21 n. 37). But John himself refers to "the particle *or* which does not join terms, but separates them" (1962/67). Jevons reports that Thomas Aquinas regarded a disjunction as false when the terms do not exclude one another (Jevons 1958/69).

[7] Compare this with the confusion of universal generalization with complete induction which we observed in Bosanquet's logic; cp. IX-14.

[8] "This 'and', uniting incompatibles, expresses the fundamental character of the Highest Being" (H. Zimmer in *Myths and Symbols of Indian Art and Civilization*, p. 46; quoted from Stace 1961/214). Cp. note 5. The Dutch Hegelian Hessing speaks in terms of "the *concrete* and *true* universal, which is *both* this *and* that, and at the same time *neither* this *nor* that" (1941/122; his italics).

[9] Cp. IV-3.

[10] Cp. John of St. Thomas 1962/66f. There one finds this example: "A horse is necessary for riding horseback". In John's opinion this entails: "(that horse *or* that horse *or* that horse *or*...) is necessary for riding horseback". This is a *descensus disjunctus*. From the proposition "All the Apostles are twelve" follows: "(this apostle *and* that apostle *and*...) are twelve". In this work, the inference in question is called a *descensus collectivus*. The complex subject term is said to have alternative *suppositio* (first example) or collective *suppositio* (second example). For the modern analysis of "the class of all apostles has twelve elements" and of "there are twelve apostles", see Tarski's textbook (1941), ch. IV.

[11] Cp. note 13 of Chapter X.

[12] Cp. IV-14, IX-9.

[13] The traditional form of *tertium non datur* is: *either p, or* (*else*) *not p*, and this entails: *not both p and not-p*, in whichever way the logical relation between *p* and *not-p* is construed, since *p aut q* entails *not both p and q*, for every *p* and *q*.

[14] Cp. IV-2, VI-30.

[15] See Veatch 1950/80f., and compare with Naess 1969 on Spinoza's terminology (especially pp. 66f., 86f.).

[16] The expressions "1 + 1" and "6 : 3" denote the same *Sache, the* number two, but "the different expressions correspond to different conceptions ["Auffassungen"] and aspects", says Frege in "Function and Concept" (Frege 1967/127). Frege did not, however, maintain that error becomes truth when it is supplemented, but offered the first clear and workable theory of negation of universal and existential propositions. – Cp. Angelelli 1967/80 n. 23.

[17] Cp. VII-23.

[18] Cp. VII n. 26.

[19] See especially p. 140 of Bird's article.

[20] Cp. III-6.

[21] Cp. Lenk 1968/194, 212, 437, 447 on Fichte and Lotze. For Fichte the problematic judgment-form (*S may/can be P*) and the particular judgment-form (*some S is P*) are one and the same. Lotze subsumes particular judgments about '*some S*' under problematic judgments about a "generally expressed" subject-concept.

[22] What follows in this section is taken from Barth 1974.

[23] Purtill would seem to agree with this: "Thus the alternative analysis of the crucial premise does not produce a version of the parallel argument which is a good argument and at the same time leads to the conclusion that, in any interesting sense, Moore may be of the female sex" (*op. cit.*/267). Cp. Dummett's discussion of Kripke's criticism of Frege's theory of reference (Dummett 1973).

[24] "Sed ad id totum de quo disseritur tum definitio adhibetur, quae quasi involutum evolvit id de quo quaeritur" (Cicero, *Topica* II 9).

[25] Cp. Maritain's criticism of Goblot (Maritain 1933/36f.). Here Goblot appears to defend concepts like Locke's general triangle in their crudest form; in the opinion of Maritain, Goblot forgets that the species is included in the genus only potentially ("en puissance"). See also the discussion of this topic in Coreth 1961/358.

[26] Cp. IV-30; see also Pfänder 1963/330.

[27] Of course, from the rules for id-introduction in Chapter X and in the present chapter this does not follow.

[28] Cp. III-5.

[29] This presupposes a view of language as essentially "Mental"; cp. X-9, sub (i).

[30] Cp. Lewis's "existentially generic pig" (D. Lewis 1970/52f.).

[31] In a footnote, Lloyd writes: "Aristotle himself often evades the dilemmas which had perplexed earlier Greek philosophers by suggesting a modification in the alternatives which had been assumed to be exhaustive and incompatible. Thus... he considers whether the primary units of bodies are divisible or indivisible, and points out the difficulties which each of these views holds. He attempts to solve this particular ἀπορία, however, by appealing to the distinction between potentiality and actuality and suggesting that the primary units are 'potentially divisible, but actually indivisible'" (168).

[32] Quoted from Clark 1952/68.

[33] Cp. Barth 1970b.

[34] Students of the history of ideas may at this point like to consult Tugendhat 1970,

where a notion of *truth-value potential* is introduced. This notion, and Tugendhat's attempt at a re-interpretation of Fiege's notion of reference, is sharply criticized in Dummett 1973/199–203. I want especially to draw attention to the following felicitous observation by Dummett: "But on an interpretation of Tugendhat's kind, there can be no question of an expression's lacking a referent, if it is a genuine semantic unit and has been supplied with a sense" (*op. cit.*/402). The reader is requested to compare this with our H-thesis and with Section (iii) of X-5.

[35] See, *e.g.*, note 24.

[36] The modern negation technique also rests upon other principles. I mention here only the important distinction between asymmetric relations and complementary ("contradictory") terms.

[37] I mean: that there ever will be something which is both cat and dog, in the sense of being a descendant both of a cat and of a dog. For philosophers who, like Hegel, are trapped in the logic of exclusive disjunction, the possibility of evolution by hybridization implied a *logical* (dialectical) problem; here, too, Goethe's *Urpflanze* may be invoked, though it will then be necessary to regard the coming into being of the hybrid (the tigon, or liger) as an instance of logical regress (or ascent) relative to its parents (lion and tiger) – that is to say, *if* the *Urpflanze* is assumed to be an entity on the λ-level.

[38] Cp. Nuchelmans 1969/21.

[39] For the method of definition *per genus proximum et differentiam specificam* and in general for that taxonomic, static (Hempel and Oppenheim 1936/8; cp. VIII – n. 23, XII-6) form of thought which approaches all Being as if it were one vast zoological garden, with or without evolution, I should like to suggest the name "the zoo-method".

CHAPTER XII

SUMMARY AND CONCLUSION

1. FOUR POINTS OF DEPARTURE AND FIVE THEMES

The following problem was investigated: what is, in traditional philosophy, the logic of the articles, in particular of the definite article? Points of departure for this investigation were the following four data:

(1) Definite articles (in Latin the zero article and in English often the indefinite article) are used with remarkable frequency in traditional philosophical literature and used in such a way that, from the point of view of modern logic, the logical form of the sentences in which these articles occur, cannot in most cases be recognized.

(2) In E. W. Beth's study *Über Lockes 'Allgemeines Dreieck'* it is argued that the following traditional problem has only recently been solved. What is the logic of proofs of theorems with general validity, particularly of theorems in geometry? The solution we are referring to here is one that is quite clearly independent of the traditional restriction of the problem to the objects of geometry.

(3) The dialogical formulation of the modern logic (or logics) of *syncategoremata* as described by P. Lorenzen.

(4) In 1955 a new study in traditional logic was published: *Logik, ihr System und ihr Verhältnis zur Logistik* by B. Baron von Freytag-Löringhoff. In this work the dialogical point of view is altogether absent. The use of definite articles in this *Logik* is not in accordance with the modern uses of these articles and it is not explained by von Freytag either.

Our opinion is that a dialogical account of a logical constant is possible only if introduction and elimination rules for these constants can be formulated as well. Introduction and elimination rules are also more in keeping with the traditional way of formulating logical rules and principles than is a dialogical formulation: *e.g.* the traditional *dictum de omni*, which is clearly a forerunner of the rule called "universal instantiation". The formulation of introduction and elimination rules for a logical constant (a *syncategorema*) does not enable us to consider this constant as semantically well-defined (*vide* Prior 1960, Belnap 1962); they are necessary, but not sufficient. If no introduction rule can be formulated we do not know what kind of premisses are required in order to justify a statement containing the logical constant under consideration. If no elimination rule can be formulated we do not know how to make use of a premiss which contains this logical constant (this *syncategorema*).

Therefore we first delimited, by means of a negative definition, a use, or a number of uses, of the definite article – in Latin the zero article and in English often the indefinite article – by defining a sub-set of the set of all propositions *a/an/the M is P*. The propositions in this sub-class we called "logophoric", a term taken from von Freytag. A proposition of this form is called logophoric if neither a universal nor an existential nor a uniquely referring proposition may be substituted for it nor a statement about a set, for all logical purposes. This provisional negative definition contains a condition which Beth called "the principle of Plato" and which, in our opinion, reflects what was known in traditional logic as "the law of identity". It is the principle that a proposition *the M is M* is logically true.

We then raised the question of how introduction and elimination rules for logophorically used definite articles might be formulated; we did not exclude the possibility that there are various rules of each kind which would imply that there is more than one well-determined logophoric use of the definite article. In that case some very obscure philosophical literature might be clarified by replacing such letter sequences as "the" or "a" or "an" in some places by other letter sequences with different introduction and elimination rules.

The following argument form was considered as a possible rule of elimination: *the M is P, S is a M, ergo S is P*, in which the major premiss is assumed to be an indefinite logophoric proposition, the minor premiss as well as the conclusion being propositions about an individual, named *S*; we have called such propositions "individual propositions". The example we used was: "the Bantu is primitive, Saul is a Bantu, ergo Saul is primitive". If further investigation proves this argument form to be valid in traditional logic, it will serve as a rule of elimination for one logophoric use of the definite article.

Chapters IV, V and VI treated of the history of the logic of indefinite propositions and individual propositions (propositions about individuals), with the ultimate goal of formulating introduction and elimination rules for logophoric uses of the articles. The most important results of these chapters are:

(i) Both Duns Scotus and William of Ockham seem to allow indefinite propositions as conclusions in expository syllogisms. In other words, given two singular premisses with the same subject term they introduce an indefinite proposition – or a particular proposition (Chapter V).

(ii) From the time of the schoolmen until the rise of modern logic the most common analysis of singular propositions was a reduction to universal propositions (Chapter V).

(iii) Since the time of the Renaissance many philosophers have treated some indefinite propositions as universal propositions, failing however to distinguish them from other indefinite propositions by means of a special prefix or any other device (Chapter IV). Consequently, indefinite propositions were logically indistinguishable from individual propositions, which were called singular propositions.

(iv) The interpretation of so-called singular propositions and, hence, of individual propositions, as a kind of universal propositions, and also the interpretation of some indefinite propositions (for instance Pfänder's *singulare Arturteile*) as propositions referring to *one* entity, was encouraged by the theories of *suppositio simplex* ("homo est dignissima creaturarum") and *suppositio simplex determinata* (IV-3 to IV-12) and especially by the theories of *suppositio naturalis* (IV-5, IV-14, IV-21, IV-28, 29). Von Freytag confirms that singular propositions (he says: "individuelle Urteile") and indefinite propositions are not distinguished in the theory of inference of contemporary versions of traditional logic (IV-1, VI-1).

(v) The analysis of singular propositions mentioned *sub* (ii) may be regarded as a consequence of the lack of a conceptual and notational distinction between a predicative copula and a copula of identification (V-11). In the work of Pfänder we find an indefinite, variable copula instead (VI-7).

(vi) The analysis mentioned *sub* (ii) tempts one to think that mere analogies between a and b are sufficient in cases where strict identity, expressed as $I(a, b)$, is required if fallacies are to be evaded (V-16).

(vii) In a logic which makes no distinction between a copula of predication and a copula of identification, the fact that logophoric indefinite propositions are treated as singular propositions, which are taken to be a type of universal proposition, leads to a primitive *paradig-matic* way of thinking. This can only be avoided by means of special precautions, as for instance a meticulous distinction between indefinite and definite articles in German and Dutch, and between different kinds of indefinite articles in English (VI-2).

(viii) The precautions which were actually introduced in the logics investigated by us do not yield a logic which is dialogically acceptable: either an elimination rule or an introduction rule for definite (or in-definite) articles is missing, or both are absent (IV-30, VI-7; see also VIII-21).

The problem of the logic of the articles, among which we count the zero article, was not solved by the schoolmen. In the Renaissance the situation grew still worse, since the logicians of that period did not even recognize the existence of the problem.

It appears that most traditional logicians in the nineteenth and twentieth centuries are hesitant about the logical validity of the above inference form, although they do credit the major premiss with some kind of general import. Nor do they provide a different elimination rule. All along the line we found too few, not too many introduction and elimina-tion rules.

None of the authors in whose works logophoric judgments expressed by propositions *a/an/the M is P* were studied describes these judgments or a sub-set of them in such a way that int-elim-rules for the (definite) article can be formulated without making use of other expressions for which no int-elim-rules can be found. We have in mind expressions such as "potentially", "on the one side", "under a certain aspect", "qua" ("as (such)"), "in quantum" ("in so far as"). Consequently the definite article is dialogically undefined.

At this point we did not close the investigation but tried to determine the conceptual background of this deplorable use of language among so many philosophers, as well as the logical problems, notions and principles which play or have played a systematic role in this matter. The main results and conclusions of our investigation can be found in Sections VII-20-23, VII-29, VIII-23, IX-12, X-8, XI-10–12. These results and conclusions can be arranged in accordance with five themes. To each of these themes a separate section is devoted below. The three sections immediately following, Sections 2, 3 and 4 deal, systematically if incompletely, with the main points of Chapters VII to XI. In Section 5 a connection is established between Section 4 of this summary and Chapter IV. It is our purpose to describe our findings in such a way as to lay bare the skeleton of the problems which faced logicians in the past.

Sections 2 to 5 below constitute an initial synopsis of the present work. Those sections of the preceding chapters that were mentioned above provide the necessary supplement to complete this survey.

2. INDIVIDUAL AND PARTICULAR PREMISSES AND UNIVERSAL CONCLUSIONS IN THE TRADITIONAL LOGIC OF THE QUANTIFIERS

(i) *The traditional logic of "all" and "some".* Frege's analysis of quantified propositions cannot be formulated with the apparatus of traditional logic because in traditional logic only one kind of variable is available, *i.e.* variables ranging over general concepts or predicates.

Formerly the quantifier in universal and particular premisses was related to the noun or adjective immediately following it (X-1). In this way the so-called subject terms *all S* and *some S* arose. The "subject" of a particular proposition (*some* S_1) *is P* or *a/an/the S is P*, in which (part of) the extension of '*a/an/the S*'='*some* S_1', is "ein völlig unbestimmtes", as Hegel says; the extension of the proposition expresses a categorical judgment "mit anomal geänderter Subjektbedeutung" according to Husserl; the subject "signifie *d'abord et immédiatement* un individu flottant ou indéterminé (*individuum vagum*)", a general nature "avec une manière d'être individuelle quoique indéterminée" in Maritain's words; it is "eine unbestimmte Art" according to von Freytag.

Given a particular premiss *some* S_1 *is P* many traditional logicians apparently introduce a proposition with a prenex definite or indefinite article yielding a so-called indefinite subject term *a/an/the S*. The proposition introduced is considered a necessary condition for the premiss, in spite of the fact that, according to the adherents of this logic, its subject is as yet indefinite in intension as well as in extension.

A universal proposition, analysed as (*all* S_1) *is P*, may now also be considered to express a judgment '(*a/an/the*) *S is P*', with (part of) the extension of '(*a/an/the*)*S*'=the extension of '*all* S_1'. In the opinion of most traditional logicians the extension of this subject '$(a/an/the)S$', at least, is determined more clearly than that of the subject of the necessary condition for a particular proposition. Similarly, in the case of an individual proposition S *is* P, *all* those individuals to which the word S applies belong to the extension of the subject term, S. In the traditional conception it is precisely this property which distinguishes a universal from a particular proposition. Viewed in this light it becomes understandable why, for such a long time, logicians persisted in interpreting individual propositions as universal (V-4ff.). The traditional treatment of individual propositions can be considered an historical consequence of the traditional method of connecting the quantifier with the noun or adjective S immediately following it, not with a propositional function containing both S and P. The procedure itself is a historical consequence of the fact that in Aristotle's logic variables were of one kind only.

From the two individual propositions (*this*) M *is* P, (*this*) M *is* S a particular conclusion may be inferred. Consequently, when Aristotle's theory of deductive inference is considered sufficient for the description of "all" and "some", then the procedure of treating individual propositions as universal propositions leads to accepting Darapti as valid (but not as "perfect", about which more will be said below). If so, universal and particular propositions (*i.e.*, the A- and the I-forms) *must* be granted the same kind of existential import. If individual variables are introduced, however, Darapti can be dispensed with, in which case there is no need to attach existential import to the subject or to the term S of universal propositions *all S is P*.

As quantifiers were connected to the word following them, it was commonly supposed that *all* categorical judgments are "really" of the form 'S *is* P'. This point of view implies that there is no fundamental logical difference between singular and other propositions, whether quantified or not. Important differences existed only in the kinds of subject that could occur in these propositions. For that reason traditional logic was considered by most of its practitioners to be primarily a science of "the concept" and only secondarily a science of "the judgment", mostly restricted to speculations about "the copula" and about "the inference", the theory of which was to be drawn up from the results of investigating "the concept".

(ii) *Ekthesis*. In a proof based on a particular premiss, such as a proof of the convertability of (1) *some S is P* into (2) *some P is S*, present-day logicians introduce an "arbitrary" individual term: (3) *m is S and m is P*. The step from (1) to (3) is called existential instantiation; just as in the case of universal generalization the introduction of the individual term *m* has to be performed "critically" (III-2). In this and similar cases Aristotle spoke of *ekthesis* but he said little in point of principle about it. In the more recent philosophical tradition ekthesis was taken to be the introduction of a general, albeit undetermined, "middle term" (*the*) *M*, to the extension of which belonged the extension or reference of *some S*, not the introduction of an individual term. It was generally thought that this procedure could not be described in unambiguous rules. At the present day the rule of existential instantiation – difficult to formalize – together with the rule of existential generalization serves to determine the inter-subjective meaning of the word "some". Before Frege the rule of universal generalization was not understood properly either. This gave rise to the Locke-Berkeley problem, which is also discussed by Kant. The formulation of the rule of universal generalization (or of an axiom related to this rule) which is an introduction rule for "all", together with the rule of universal instantiation (*the dictum de omni*), serves to determine the inter-subjective meaning of the word "all". Aristotle's syllogistic, although going quite a long way towards describing the meaning of both "all" and "some", does not define them completely.

In traditional logic, the expression "arbitrary" which is commonly used for the individual triangle which is drawn on a piece of paper in support of the proof, acquired the meaning "indefinite", "floating", ("flottant", "vagum"). Only in a dialogical formulation of the logic of "all" and "some" does the meaning of this expression become transparent (III-2).

We can now say the following. Because only one kind of variable was known, ekthesis was not considered an instantiation but a procedure leading to the introduction of an indefinite judgment, in Dutch and German expressed by means of a prenex definite article, in English more often by a prenex indefinite article, the "logical subject" of which was supposed to be a general concept (*Gegenstand*, idea,...). In consequence ekthesis was not distinguished from abstraction, in the sense of the definition of a general concept on the basis of one premiss expressing a similarity between two (or more) individuals, s and p, or individual concepts, '*S*' and '*P*' (X-6).

If one is convinced that a judgment '(*the*) *M is P*' can be reached by applying the ekthesis, which supposedly cannot be formalized, to a given particular proposition, no need will be felt for universal generalization (X-8), but rather for a supplementary characterization of this indefinite judgment.

(iii) *The potential genus.* The traditional procedure as described above raised serious problems. That *some S is P* is true and that *some S is not P* is true are not incompatible; in traditional logic therefore the judgments '(*some S*) *is P*' and '(*some S*) *is not P*' must not be incompatible either, even though they have apparently the same indefinite logical subject '*some S*'. In this way one is led to assume first the existence of an entity or concept '*the M*' such that both *P* and *not P* may be predicated of it. But this is found to be inconsistent not only with the principle of non-contradiction but also with the traditional principle that a *definitio rei* of a general concept has the following form: *the M is either P or Q or....* A preference for exclusive disjunctions was presumably motivated by a metaphysical interpretation of the rhetorical doctrine of the *loci*, because on one *locus*, conceived of as a position in a sort of logical space, not more than one "argument" or property (*either P, or Q, or...*) can be positioned or "placed". An answer to this problem was given by introducing *the potential genus:* according to this view the premisses *some S is P* and *some S is Q* lead to the introduction of a judgment '*the M is potentially P and Q and...*, as a necessary condition for these premisses, one particular M always being either P or Q or.... In this way an indefinite judgment can always be supplemented and there is no objection now to *P* and *Q* being contrary or even contradictory (complementary) terms.

The potential genus serves as a mixing-bowl in which incompatible properties can be brought together.

By means of an extra-syllogistic procedure, called ekthesis or induction, one can hope to arrive at a modalized indefinite judgment '*the M is potentially P*' which, in some way or other, is relevant to our conception of any one M-individual and which cannot be refuted by someone pointing at an M-thing which is not P (IX-14).

(iv) *Ekthesis and abstraction.* Statements which involve names of equivalence classes do not contain the modification "potentially". This has led many traditional logicians to suppose that the modern analysis of abstraction is a procedure in which some "aspects" (XI-5, 6) of the concept to be defined are left out.

(v) *Inclusive disjunction.* The logical need for this theoretical construction, the potential genus, is removed if one uses inclusive disjunction. The potential genus can be considered a primitive form of inclusive disjunction (XI-12). By using inclusive disjunctions one arrives at a "middle term" in such a way that dialogical criticism becomes possible.

(vi) *The traditional conception of ekthesis as paradigmatic thinking.* A fundamentally unchallengeable judgment about a potential genus '*the M*' which is introduced on the strength of a particular proposition, may also, of course, be derived from one individual premiss which is based on the observation of one individual. The individual concerned then serves as *paradigm* or *exemplum*. This view of ekthesis, which was the traditional theory of the use of particular premisses in arguments, partly explains how, in the Renaissance, the rhetorical doctrine of the *exemplum* came to be part of logic (V-5). For what follows in Section 4 of this summary it is important to know also that the meanings of "paradigm" and "analogy" are often conflated in the history of philosophy.

(vii) *The "tyranny of the first figure" and the postulate of deductivity.* As we have seen above, the traditional logic of particular premisses *some S is P*, the ekthesis, is incompatible with modern ideas on this matter. The imprecise formulation of traditional views in any book on logic must have made even earlier thinkers wonder about these views. It is not surprising, then, that particular propositions were often called "problematic".

The third syllogistic figure generates these and only these problematic propositions as conclusions. This must have led many logicians to regard the third figure with scepticism and indeed this figure was vehemently disputed by the schoolmen and also in the Renaissance.

Aristotle applied ekthesis to prove theorems of convertibility which were used, in turn, to reduce moods in the second and third figure to moods in the first figure. He considered the valid moods of the first figure, and these only, as "perfect" moods, in the sense of immediately evident. The inference form which he calls ekthesis does not belong to these. For many philosophers in the past this constituted a second reason for not considering the third and second figure, nor the fourth figure in some cases, nor the ekthesis, as fundamental tenets of logic. In this way the "tyranny of the first figure", about which Ross (1949) and later Patzig (1963) have written, arose, and with it also the postulate of deductivity. For ekthesis was regarded as *id-introduction* (*vide* Section 4 below) or *ideal generalization* and the postulate of deductivity means precisely that this conception of ekthesis was placed outside the basic principles of logic. This "inductive" ekthesis, and the monological-philosophical optimism that went with it (*vide supra*), were not, on the other hand, rejected by traditional logicians.

The mood of the first figure into which advocates of this postulate poured their most superior or deepest *Schlüsse*, was the Ramistic mood Burburu (V-7, VIII-11). The inference "the Bantu is primitive, Saul is a Bantu, *ergo* Saul is primitive" is formulated in this mood.

The Locke-Berkeley problem cannot be solved on the basis of these theories.

Wolff, Kant and Hegel all subscribe to the deductive point of view described above; more recently, but before the twentieth century, only Lambert had the courage to postulate a *dictum de exemplo*, which generates particular propositions, as being a fundamental principle that is completely independent of the first figure. However, a formulation of existential instantiation cannot be found in Lambert either. The attitude of the neo-Thomist logicians towards the postulate of deductivity and a *dictum de exemplo* did not emerge very clearly from the material collected.

(viii) *Value judgments.* In the literature of traditional logic no arguments can be found for the opinion held by H. B. Veatch and others, that a traditional logic enables its practitioners to arrive at value judgments (VIII-23). A possible exception is perhaps the argument of the miscreated, advanced by Hegel. We have refuted this argument by pointing out the alternative possibility of using inclusive disjunctions when formulating definitions (VIII-20), a possibility which is now also accepted by von Freytag (XI-13).

The proposition "the Bantu is primitive" can be considered a value judgment. We refer to the value judgment "homo est dignissima creatura-rum" as another case in point; William of Sherwood ascribed *suppositio simplex* to the subject of this proposition. According to our negative definition of "logophoric proposition" and "logophoric judgment" they both undoubtedly express logophoric indefinite judgments and they are both also undoubtedly "what-statements", according to the terminology of Veatch, who thinks it is possible to come to value judgments by means of a "what-logic". There is probably a close historical connection between Veatch's view and the theory of *suppositio simplex* (*vide* Section 5). In order to understand a thinker like Veatch, and others who hold similar views, one must furthermore assume a conflation of *suppositio simplex* and *suppositio naturalis*.

3. MONADIC PREDICATION, INTERNAL RELATIONS AND THE HIERARCHY OF ARTICLES

Aristotelian syllogistic as a whole can be accommodated within class logic or within monadic first-order predicate logic. This holds true, irrespective of the interpretation of "*S*", "*P*" and "*M*" as variables for class names or variables for names of "qualities", of substances, or of something else. Today, however, class logic and the monadic predicate logic are considered insufficient.

On the other hand, several contemporary philosophers have indicated that they regard a logic which is purposely constructed to deal also with propositions containing relational predicates as philosophically in-adequate and misleading, and they argue for a *logique de l'inhérence* (Maritain, VII-3) or what-logic (Veatch, 1-4). This is an indication that the traditional systems which are concealed behind these expressions possess some non-Aristotelian systematic traits which cannot be found in modern logic and which supporters of traditional logic are unwilling to abandon. Apparently they are not seriously worried by the fact that connections of various kinds can never be formulated clearly in their logic (VII-11). Wundt has argued expressly that a *Vermischung* of gram-matical categories in order to express relations is logically inevitable (IX-2).

One characteristic property of the old logic is the misconception of ekthesis as described above. If one assumes that ekthesis does not belong to formalizable logic, it must be based upon a special kind of insight

468 ASCENT

and may, accordingly, be more or less perfect. It may be based upon a more or less penetrating or profound *intention* (cp. I-8); the results may then possess different degrees of perfection. We have therefore to introduce at least one parameter "ι" if we want to describe the traditional logic which has fundamentally accepted this conception of ekthesis. This parameter determines a "quality space". We have characterized traditional logic by an expression borrowed from Rescher as a *topological logic* (VII-21) of genera and species which can be described by monadic predicates and which form "pyramids of concepts". The logical relation between the genera or species are called internal relations. On the lowest level of the logical dimension we find the "external" relations, observable relations between observable things.

We can now supplement our negative description of "logophoric judgment" with a positive one: a logophoric indefinite judgment *the (a, an) M is P* is a judgment which is true at a certain position ι in a specific logical "quality dimension". The definite article can be said to express an unspecified value of a logical parameter "ι"; the values of this parameter will in the traditional way of thinking usually form a continuum, in any case an infinite set with an upper boundary λ and a lower boundary θ.

In order to make a logic out of this conception a hierarchy, *i.e.* an ordered set of articles is required, and for each of these articles int-elim rules will have to be formulated. Traditional logic has not satisfied these minimal requirements.

In the monadic predicate logic, in which all Aristotle's syllogisms can be proved valid, provided the presuppositions of existence which he made are included with the premisses, dependency or functionality cannot be defined. In natural languages definite articles are used to expres functionality. Frege's logic of the quantifiers can be formulated on the basis of the principle of polyadic propositional functions and allows for the use of polyadic predicates. If therefore we shift to polyadic predication and also introduce a notational difference between predication and strict identification, we may introduce the use of definite articles to express functionality (dependency) in the following way:

(1)   *the M of s_1, s_2... and s_n is P* $\underset{\text{Df}}{\leftrightarrow}$
       $(Ex)\{M(s_1, s_2,..., s_n, x) \ \& \ P(x) \ \& \ (y)[M(s_1,... s_n, y) \rightarrow \\ \rightarrow I(y, x)]\}$

In the *definiens* M occurs as a $(n+1)$-adic predicate. For $n=0$ Russell's theory of definite descriptions is obtained (III-7; V-18, with "S" instead of "M").

Peirce ascribes to Hegel an unusually vivid interest in "Thirdness" (Peirce I/193). If we have understood Peirce properly, his "Thirdness" includes the notion of functionality (cp. also VII-3), in particular the traditional "middle term" 'M', considered as a function of a logical subject 'S' and a logical predicate 'P' (and perhaps also of time). In Hegel's time the notion of function was still a specifically mathematical notion and it could not be defined within the contemporary logic. In the above definition (1), however, no assumption is made about the nature of the predicate M; the definition is a general logical one. We may therefore use this definition in order to clarify the judgment form in which, according to our theory, the traditional attempt at solving the problem of functionality can be expressed:

(2)   *the M of S and P is* (to be found at the *locus*) $\imath$ $\underset{\mathrm{Df}}{\leftrightarrow}$

$(Ex)\{M(S, P, x) \,\&\, I(x, \imath) \,\&\, (y)[M(S, P, y) \to I(y, x)]\}$

As to Strawson's well-known criticism (1950) of Russell's theory of definite descriptions *the S is P*, it is important that the following should be borne in mind. In his 1952 publication Strawson discusses symmetrical co-ordinations (the only co-ordinations in which he shows great interest) of individual terms, but he does not refer to relational propositional forms (1964/79ff.), even though he does not deny in principle the importance of such forms (1964/202ff.). As a result, his discussion of the irreducibility of many conjunctively constructed subject terms strikingly resembles Maritain's and also John of St. Thomas's treatment of this matter, in which a primitive attempt is made to incorporate into Aristotelian logic a theory of relations (VII-14, XI-3). In his 1950 essay Strawson also seems to ignore the fact that, through the shift from monadic to polyadic predication which was made possible by Frege's new logic of the quantifiers, a definition (1) of functional connections became possible for the first time, a definition which, for $n=0$, yields Russell's theory. He does not provide an alternative for the traditional logical form *the S is P* but introduces instead a notion of *presupposition* which reminds anyone who is familiar with traditional logic of late scholastic conceptions and particularly of Maritain's and Jacoby's views on presuppositions of "possible" existence (Being), which is linked with the notion of an essential *suppositio* (X-12; cp. I-6). This resemblance need not be intentional on Strawson's part but it is not accidental, either, as may appear from what was said above. Unlike Russell's solution the notion of presupposition does not allow dialogical criticism of propositions *the S is P*, a reason why traditional logicians soon feel at ease with Strawson's logic (X–13). If one tries to formulate the dialogical consequences of presuppositions in such a way that *the S is P* becomes *dialog-definit*, one is likely to end up with Russell's theory.

The usual base of modern grammar: *Noun Phrase + Verb Phrase* has been taken from the traditional theory of monadic predication. It is striking that in 1969 Strawson, in a lecture entitled *Grammar and Philosophy*, pleaded for a revision of the principles of modern linguistic theory so that besides many other new basic elements, including a series of deictic elements, relational forms could be accepted as fundamentals of linguistic description (*vide* Strawson 1970, Nuchelmans 1970). It is not altogether clear whether he really means other than logical and semantic relations. If he does, Strawson's case illustrates how the scientific ontogeny can sometimes be a repetition of the phylogeny. When investigating how the grammar of the articles and, in connection with it, the grammar of negation words (VI-13) must be constructed out of deictic and relational elements, grammarians should, in our opinion, pay serious attention to the historical development of logic.

4. FROM ANALOGY TO ESSENTIAL IDENTITY

Most traditional logicians considered every proposition as the expression of a judgment of the "form" *'(the) S is P'*. They did not make a notational distinction between the "is" of predication and the "is" of identification; this can be understood against the background of the absence of more than one kind of variable. Great importance was attached to ἀδιόριστοι, indefinite propositions, by precisely those logicians in whose opinion "is" expresses an "identity" or else a "partial identy" between an '*S*' and a '*P*', not in the sense however of an identity between part of the extensions of two terms, as in Boole's and Schröder's algebras, but in the sense of an identity between their "intensions" (VII-4). It is difficult to distinguish this identity theory of the judgment-copula, or the variant in which the predicative "is" is said to express "partial identity", from the view of those authors who advocate an inherence theory of the copula. Most logicians who propagate either an identity theory or an inherence theory not only hold the intensional view of the copula in the sense mentioned above, but also credit the copula with degrees; in other words it is *elastic*. We have called this view of the copula "the *intensive* identity or inherence theory of the copula" and have abbreviated its name to "id-logic".

The partial identity between '*(the) S*' and '*(the) P*' may or may not be understood as a symmetrical relation. If it is so understood the word "is" expresses a similarity or analogy. This is the basis of the neo-Platonic-Fichtean logic (NPF-logic), which is one of the id-logics. It is not very clear what properties are attributed to the copula in the logic of authors not belonging to the school of Fichte and Hegel. Only von Freytag is explicit on this point: his "ist" is reflexive and transitive, but not symmetrical. However, the affinity *auf höherer Ebene* between his logic and the NPF-logic is remarkable.

If we combine the traditional treatment of "all" and "some" with the id-logical conception of "is" and the copula (VII-12, X-2), we are confronted with the following conception of universal and particular propositions: *all* $S_1 = P$ and *some* $S_1 = P$. In both cases, $S = P$, a formula without quantifiers or operators which is found very often in traditional works on logic, where " $=$ " is a copula which is interpreted as expressing a relation *Identität* of a certain degree. Independently of our investigation of the traditional conception of ekthesis, we found that many authors seem to assume a gradual transition from similarity to identity (VIII-19). A number of statements on this matter can only be understood if we suppose the said identity to be a "weak" form of identity, in the sense that criteria of identity which depend upon time and space are not taken into account. Such a conception of identity we called "internal" or "essential identity" (VII-18). The inductive problem of id-logic is the following: given a judgment of analogy, regarded as a low degree of essential identity, how can a judgment with, or about, a higher degree of essential identity be reached? That every judgment of analogy presupposes a judgment with a higher degree of essential identity is one of the principles of id-logic (thesis of id-existence, VII-23).

In the id-logical theory of the copula the difference between quantified propositions and propositions expressing an analogy between two entities disappears. We therefore interpreted the parameter "ι", discussed in the above section, as a parameter for degrees of internal essential identity, and proposed the word "id" as verbal form for this parameter. In id-logic induction and ekthesis coincide (X-7).

The literature of traditional logic contains a collection of labels and verbal characterizations of the parameter "ι". One may choose from among the following expressions: the ontological degree of μέθεξις (*participatio, Teilhabe*) (I-5, VII-3, 12, 25, 26); the degree to which a *forma* inheres in a substance (VII-14, 15); the degree of intensity of a quality (VI-8, 9, VIII n. 23, IX-6); the logical degree of (partial) identity (I-5, VII-4, 24, 30); the degree of equality (VII-19, IX-6); the degree of internal identity; the degree of essential identity; the degree of intention (I-4); the degree of penetration into a *Wesenskern* (VI-9); the degree of logical necessity or apodeictic modality (VII n. 49); the degree of problematic modality (VII-26, VIII-12, n. 12); the degree of abstraction or concretion (I-5); the degree of *logische Dämpfung* or of *logisches Behauptungsgewicht* (VI-7); the degree of logical possibility (VI-7); the potential of the concept or of the judgment (XI-9); the degree of probability in the sense of degree of belief (VIII n. 9); the degree of generality (VI-9, VII n. 24); the *locus* (commonplace) of a concept in a logical space (XI-6).

Many traditional logicians will undoubtedly want to distinguish the meanings of two or more of these phrases. If they persist in speaking of a logical theory, they will have to introduce more than one parameter, or additional operators, and they will have to formulate either int-elim-rules or other kinds of semantic rules which can also be formulated dialogically, for every value of every parameter and for every operator. One could think, for instance, of a hierarchy of articles combined with a separate hierarchy of symbols for copulas. A book on traditional logic in which the differences between the above mentioned expressions are treated has not been written.

5. FROM *suppositio* TO SUBSTITUTION

(i) *Decline of the theory of* SUPPOSITIO *in traditional philosophy.* The medieval theory of *suppositio* is extremely incomplete as a theory of the articles (zero articles) (IV-9). This theory can nevertheless be characterized as a critical – although not sufficiently critical – semantic theory which was constructed in order to prevent the production of fallacies (de Rijk 1962).

Different kinds of *suppositio*, however, were distinguished neither by Hegel nor by Wundt or von Freytag, and Pfänder is only interested in one kind of *suppositio* which he calls *logische Supposition*.

What, then, has happened to the theory of *suppositio* in traditional logic? As an answer to this question, especially with regard to German philosophy, the hypothesis presents itself that through the influence of neo-Platonic thinkers the critical theory of *suppositio* was transformed into an extremely uncritical theory of substitution. The neo-Platonic philosophers of the Renaissance often based their statements on Cicero. In his works the verb "supponere" did not have the meaning which it acquired in scholastic logic but the meaning *to substitute*, as is stated by Ong who says: "the substitution theory of supposition is the worst of semantics" (1958/70). In 1952, Boehner still accepted the interpretation of "*suppositio*" as substitution (1952/27ff.). Judging from the words of Risse, Dominic de Soto (1494–1560) supported this theory and according to Risse his *Summulae* became *wahrhaft schulbildend*. Risse writes about him: "To the numerous kinds of *suppositio*, [which he] understood as the replacement of one concept by another, belong also the procedures *inductio* and *deductio*, called *ascensus* and *descensus*" (1964/331). Leibniz also, in his attempts to prove the validity of the so-called oblique syllogisms, applied a kind of substitution which in our opinion suspiciously resembles the principle of analogical substitution which we formulated in IX-10 (cp. Parkinson 1965/50). J. P. Reusch was another important link in the development of the theory of *suppositio* into a theory of substitution (cp. Ziehen 1920/120, 733; Erdmann 1907/716). In VII-7 we discussed Fichte's derivation of *A is B*, "Coriscus is Socrates", from *A is X* and *B is X*; in IX-10 we investigated Wundt's logic and its connection with the principle of analogical substitution.

The language of the theory of *suppositio*, "suppositio est acceptio termini" etc., even survives into the 19th and 20th centuries and pervades many works on logic. Hegel speaks of "der Medius Terminus, die Erde, als ein Konkretes *genommen*" (1967, II/341), and Maritain of "un sujet vague et indéterminé, *pris* en bloc" (1933/85). Pfänder argues: "Wird z.B.

behauptet, die Frau sei die Liebe, oder der Deutsche sei das Gemüt, so ist offen*sicht*lich, dass... beide Arten von Menschen hier in ihrem Idealfall zu *nehmen* sind" (1963/122; my italics).

(ii) *From a small number of* SUPPOSITIONES *to the one "philosophical"* SUPPOSITIO NATURALIS. Although Ockham's theory of *suppositio* differs in important respects from those of Sherwood or Peter of Spain, all scholastic logicians agreed in distinguishing a finite number of different kinds of *suppositio*. This number was fairly small, usually less than ten. In IV-14 we suggested, as a possible explanation of the fact that the theory of *suppositio* had disappeared from traditional logic, that the virtual or "philosophical" *suppositio naturalis* of a *terminus per se sumptus* supplanted the interest in the other kinds of *suppositio* without which the theory cannot claim to be a critical semantic theory at all. This shift or limitation of interest must have taken place in late scholastic and Renaissance times.

Many influential logicians in the Renaissance do not mention the theory of *suppositio*, although articles very often occur in their examples of valid inferences. Maritain discusses the usual number of kinds of *suppositio*, but maintains a *suppositio naturalis* as well, which, he says, pertains to the copula. Such a *suppositio* makes dialogical criticism impossible and has the logical effect of a "way out". Pfänder's *reine Logik*, according to his own words, mainly applies to just one *logische Supposition* (IV-21).

(iii) *The "philosophical"* SUPPOSITIO *and ekthesis.* As said before, many philosophers believed that it is possible to "take" a term with, or in, a "philosophical" or "logical" or "essential" or "natural" *suppositio*. It will be clear that this belief is apt to increase confidence in the traditional conception of ekthesis and *vice versa*. The theory now runs as follows: ekthesis can be successful only if the subject term *some S* (according to Pfänder: the whole proposition) is "taken in" this philosophical *suppositio*; in that way only can we attain knowledge of the more profound features of the nature of a genus. In Maritain's opinion the *suppositio naturalis* relates to the copula; the belief in this kind of *suppositio*, indeed, supports the view that the copula has degrees.

(iv) *The dialectical continuum and the analogical doctrine of substitution.* In our opinion von Freytag's work, as well as those logical works written in the nineteenth and at the beginning of the twentieth century which he admires, can be understood solely against the background of the late scholastic logic and of Renaissance logic. Jacoby, in his attack on modern logic, deals with the theory of *suppositio* in less than one page. He deplores the fact that many 14th century logicians reduced this theory to *suppositio materialis* and *suppositio personalis* (1962/111). It is worth noting that Jacoby also characterizes *suppositio* as "the insertion of one concept for another, usually of a genus for a species". Hence his thoughts on this matter, as well as de Soto's, may justly be classified as what Ong calls "the worst of semantics". Jacoby's use of language seems to suggest the possibility of ascending, in the sense of the traditional meaning of induction, from species to genus by way of the theory of *suppositio* (cp. IX-12). This can be better understood if we assume that the "philosophical" *suppositio* is applied to the premises of the ekthesis-procedure.

Von Freytag keeps silent both about substitutions and *suppositiones*. Yet, in 1962 Jacoby (158) insisted: "Von Freytag's logic is one of the most important phenomena in contemporary scientific philosophy". Von Freytag himself regrets that "those doctrines which were cultivated by Aristotle *and, above all, in scholasticism*, and which still constituted the framework of the voluminous philosophical textbooks on logic published in the last century and at the beginning of ours, *i.e.* the doctrine of concept, judgment and inference which was expounded without the use of mathematical methods, this doctrine has now disappeared almost completely, both from instruction and from research" (1961/10, my italics). His own book apparently is meant to bring about a change in this state of affairs.

The logics of Pfänder, Jacoby and von Freytag, and probably also of Husserl, as well as the logics of Maritain, Veatch, Hegel, Wundt and Bosanquet are rooted in the period from 1450–1700, a period which some present-day historians call "the Interregnum" and which they characterize as a period of logical decline (cp. Thomas 1967). Ong observes the presence of a conceptual "dialectical continuum" in the Renaissance (1958/205). In Chapters VI, VII and VIII we have adduced evidence in support of the thesis that a great many authors who have written about logic in the past two centuries have embraced the intensive identity theory of the copula in "judgments". Such a conception of the copula requires a continuum of articles or copula-symbols or both, neither of which is available. This intensive theory of the copula, which has gained such a large and influential following in the past centuries, is in our opinion connected with the Renaissance "dialectical continuum" observed by Ong, and with the interpretation of "suppositio" as "substitution", "Ersetzung" or "Einsetzung" current since the Renaissance.

In the light of results and hypotheses summarized in Section 4, our thesis in IX-10, that Fichtean logic can be characterized as a theory of analogical substitution, is supported and made plausible.

In modern logic a theory of substitution is required in connection with the use of variables. This statement gains perspective if we remember that the combinatory logic of M. Schönfinkel and H. B. Curry, from which individual variables and quantifiers have been eliminated, originated partly in order to avoid the use of principles of substitution, which are difficult to formalize. Of course this does not mean that anyone who uses individual variables today and employs a principle of substitution employs the analogical theory of substitution or simple *Drittengleich-heit* (*regula de tri*).

6. DIALOGICAL CRITICISM OR LOGICAL STABILITY?

(i) *The traditional preference for symmetrical notions and syncategoremata.* Bosanquet regards the indefinite propositional form as important, partly because this form is well suited to his desire for logical stability; this desire is also expressed by other authors, for instance by A. Pfänder and by H. B. Veatch (IX-14, 16) and the plea of John of St. Thomas and his followers for "analogical" concepts also falls in with this desire (IX-15). Very often these opinions are found in conjunction with a preference for symmetrical notions as against the "dynamic" (*vide* Hempel and Oppenheim, 1936/8) asymmetrical or antisymmetrical notions and *syncategoremata* (IX-16). This is especially noticeable in the neo-Platonic-Fichtean (NPF) logic (VII-7, VII-11, VIII-20).

(ii) *Kant's theory of negation.* Kant's theory of negation (III-4, n. 6, n. 13, and IV-18) appears as an apotheosis of the Interregnum and, like a restriction to symmetrical notions and the use of "analogical" concepts, it satisfies the desire for logical stability. For, in spite of his general critical aims, Kant's theory of negation is not a critical instrument.

In our opinion his theory of negation can be understood only as a consequence of the traditional view described in Section 2, according to which the quantifiers belong with "the logical subject" so that all judgments can be said to have the logical form: (logical) subject – copula – (logical) predicate (cp. X n. 12). Consequently, if the logical form of a universal proposition is taken to be: (1) (all S) is P, and if, furthermore, a proposition of the form (1) is taken to presuppose a logophoric judgment (2) 'a/an/the S is P' (cp. the id-introduction thesis), then the external negation of (1), i.e. the proposition (3) (not all S) is P, presupposes some kind of negation of (2). In our opinion, in Kantian logic this is the "negating" ("verneinende", "privative") judgment (4) 'a/an/the S is-not P'. The internal negation of (1), i.e. the proposition (5) (all S) is not P, must then presuppose another kind of negation of (2); this is the "infinite" ("unendliche", "limitative") judgment (6) 'a/an/the S is not-P'. Anyone who is interested in the connection between this embryonic theory and the idea that particular judgments also are of the subject-predicate form, as discussed in Section 2, although, as Husserl says, "mit anomal geänderter Subjectbedeutung" (X-10), will, however, search in vain for a discussion of this connection in the traditional logical literature.

The grammatical phenomena concerning the use of articles and negation which have been observed by Kraak and which seem to prevent the construction of a coherent grammatical theory of the articles (VI-13), are in our view connected with these problems, which can be traced back via Boethius at least to Aristotle and which culminated in Kant's theory of negation.

Von Freytag discussed the difference between *privative* and *limitative* negation in his first book, but in his second book, that is to say after he came to recognize the importance of De Morgan's laws, he does not mention this Kantian distinction anymore.

(iii) *"The humanist assault on the oral disputation."* – "In disputatione dialectica sunt due partes, scilicet opponens et respondens", writes Walter Burleigh in his *De obligationibus* (quoted from de Rijk 1967/40). The following paraphrase by Lambert de Auxerre is etymologically incorrect, but it is characteristic of the close connection between logic and discussion at the height of scholasticism: "Dicitur autem dyalectica a dya, quod est duo, et lexis, quod est ratio, vel logos, quod est sermo, quasi ratio vel sermo duorum, scilicet opponentis et contradicentis in disputatione" (quoted from Prantl III/26 n. 102; cp. Petrus Hispanus, *Summulae logicales* par. 1.01).

According to Ong (1957/155; see title above this section) this con-nection between logic ("dialectica") and discussion was annihilated by the rhetorically oriented logicians in the Renaissance: "The Ramist arts of discourse are monologue arts" (Ong *op. cit.*/287). Ong's words are devastating but they seem to us to be justified. Not only oral but also written criticism was made impossible through the activities of these logicians. For, as we have seen, the Renaissance logicians, while rejecting or diluting the theory of *suppositio*, did not abandon the framework of monadic predication, did not formulate a workable logic of the quanti-fiers and did not develop a logic for the articles which they employed so frequently. Instead, we find in their works such an unsystematic use of indefinite propositions as to make serious discussion impossible. The impact of the Renaissance logicians was greatest in Germany; Melanchton and Ramus deserve special mention here.

We have discussed the slovenliness of the nineteenth century philosopher Hegel with respect to the articles. We have, furthermore, established the absolute undiscussability of Hegelian statements $a/an/the/M$ is P. The latter phenomenon is not an incidental feature, the absence of which would leave his philosophy intact: "He has usually overlooked external Second-ness altogether. In other words, he has committed the trifling oversight of forgetting that there is a real world with real actions and reactions. Rather a serious oversight that" (Peirce I/193). And even Günther writes that "*das Du*, as a logical principle and as a point of departure for re-flection, does not occur in this [Hegel's] system at all" (Günther 1959/102). In this respect Hegel's work must be regarded as merely a heritage of the Renaissance logicians as well as of Wolff and Kant. However, Hegel should not be blamed any more (or less) than his predecessors. Peirce's words about Kant as a logician are to the point and deserve to be quoted here: "After a series of inquiries, I came to see that Kant ought not to have confined himself to divisions of propositions, or "judgments", as the Germans confuse the subject by calling them, but ought to have taken account of all elementary and significant differences of form among signs of all sorts, and that, above all, he ought not to have left out of account fundamental forms of reasoning" (I/300f.). Kant contributed to "the tyranny of the first figure" (V-12). Both Hegel's inference "by analogy" and, it seems, his inference "by induction" are formulated in this figure. The same holds true of one of the three principles in Pfänder's doctrine of inferences, in which no operators occur, as well as for our own standard example of an (unacceptable) inference form with one "logophorical" premiss: "The Bantu is primitive, Saul is a Bantu, hence Saul is primitive".

Hermann Lotze did not formulate a logic of logically significant elementary signs or words either, but at least he showed a remarkable comprehension of the necessity for formulating such rules, in spite of the remarks in his long *Anmerkung über logischen Calcül* which follows after Section 198 of his *Logik*. Lotze stresses, in unmistakable words, the need for unequivocal logical *rules*, which will state clearly what follows from even the simplest connections of concepts. A clear and distinct *notation* is in his opinion not sufficient: "... die Bezeichnung allein ist nicht das, was wir bedürfen" (1912/255). "Diese Regeln sind das, was uns am empfindlichsten fehlt, wenn wir Begriffe, die nicht blos Grössen bedeuten, zur Erzielung eines Ergebnisses verknüpfen wollen, und ich glaube, dass man sich ganz grundlos mit der Hoffnung schmeichelt, sie würden plötzlich von selbst unzweideutig klar werden, sobald man nur die Inhalte, auf die man sie anwenden will, bis in ihre letzten Bestandtheile zergliedert hätte" (*loc. cit.*).

(iv) *Restoration of external Secondness in logic.* Frege's question: "In wessen Geiste?" disposes of the remnants of the doctrine of a "philosophical" *suppositio* of a term, which has to be "taken" according to the Being which "the subject" has in, or according to, "the mind" (IV-28). This question can serve as a motto for the whole development which he inaugurated. One of the many logicians who have continued Frege's work is P. Lorenzen, who, building on the work of E. W. Beth, has formulated a dialogical interpretation of the most fundamental logical rules. External Secondness, including *das Du*, has thereby regained its legitimate place in logic. One of the main goals of logic will be that the *syncategoremata* and also the propositional forms which will occur in future expansions of logic should be *dialogdefinit*, even if they are not *wertdefinit* (*wahrheitsdefinit*; cp. III-8). Logophoric indefinite propositions not only are not *dialogdefinit*, they are dialogically completely undetermined, in the sense of being fundamentally unchallengeable, since they do not even admit of a first critical question.

Practitioners of traditional logic have defended themselves against the ideal of *dialogdefinit* logical constants in various ways. They frequently employ the Zenonian-Ciceronian image, very popular in the sixteenth century, which compares logic or dialectic with a closed fist and rhetoric with an open hand. This image is found in Valla, Agricola, Joh. Sturm and Melanchton (Risse 1964/16). It is difficult to feel much sympathy for people who, at the present time, appeal to this sort of imagery.

In my opinion, Ch. Perelman and L. Olbrechts-Tyteca did not make a very fortunate choice when they announced their studies in the theory of argumentation as a new rhetoric. That their theory of argumentation does not purport to be a monological rhetoric antagonistic to theoretical "formal" logic will appear from certain statements in an essay by Olbrechts-Tyteca, who points out that "... in the works of Ramus, rhetoric as argumentation is given – or, rather, is deprived of – a position which is a function of his philosophical conceptions, of his quest for rational evidence." The subsequent remark, that "it is all the more tempting for a thinker to relegate all argumentation to the one and only logic if he does not personally develop a rigorous formal logic", indicates that the author does not (yet) see a fusion between formal logic and the theory of argumentation as a realistic possibility. However, as I have already said, Lorenzen's dialogical tableaus give us considerable reason for optimism in this respect.

Olbrechts-Tyteca correctly observes that "as it seems, those who maintain argumentation in rhetoric are also the ones who develop an analytical logic – be it Aristotle or Whately" (Argumentation/9). What is above all important here is, in my opinion, not whether argumentation is studied under the head of rhetoric or under that of a partly formalized logic, but the fact that those philosophers who are seriously interested in argumentational studies at all also take an active interest in the development of formalized logic (though the converse does unfortunately not always hold).

A very similar observation is expressed by Y. R. Simon, who, like Maritain and Veatch, admires the logic of the seventeenth century logician John of St. Thomas: "Wherever logic is principally regarded as an instrument of discussion and communication, material logic is likely to decline" (in: John of St. Thomas 1965/15). Simon deplores the fact that the so-called material logic or content logic (transcendental logic) is about to disappear.

Present-day practitioners of a traditional logic adopt the same attitude as Thomas Mann's Dr. Faustus: "Wenn ich vom Hören höre! sagte Adrian. Nach meiner Meinung genügt es völlig, wenn etwas *einmal* gehört worden ist, nämlich, als der Komponist es erdachte."

Confronted with him, Dick and Tom have lost the game in advance.

BIBLIOGRAPHY

Abaelardus, Petrus, 1956, *Dialectica* [Dialectics]. First complete edition of the Paris manuscript by L. M. de Rijk. Assen.

Ackrill, J. L., 1963, see: Aristotle 1963a.

Adorno, Th. W., 1956, *Zur Metakritik der Erkenntnistheorie. Studien über Husserl und die phänomenologischen Antinomien* [Meta-criticism of Epistemology. Studies of Husserl and the Phenomenological Antinomies], Stuttgart.

Aebi, M., 1947, *Kants Begründung der "Deutschen Philosophie". Kants transzendentale Logik: Kritik ihrer Begründung* [The Kantian Foundation of "German Philosophy". Kant's Transcendental Logic: a Critique of its Foundations], Basel.

Albrecht, W., 1954, *Die Logik der Logistik* [The Logic of Logistics], Berlin.

Analogie, 1941, *De analogie van het Zijn* [The Analogy of Being], Nijmegen. Proceedings of the eighth general meeting of De Vereeniging voor Thomistische Wijsbegeerte [the Association for Thomist Philosophy]. Supplement to *Studia catholica* **17**.

Angelelli, I., 1967, *Studies on Gottlob Frege and Traditional Philosophy*, Dordrecht.

Angelelli, I., 1967b, 'On Identity and Interchangeability in Leibniz and Frege', *Notre Dame Journal of Symbolic Logic* VIII, 94–100.

Angelinus, O. F. M. Cap., 1941, 'De eenheid van het analoge zijnsbegrip' [The Unity of the Analogous Concept of Being]. In: *Analogie 1941*.

Apel, K. O., 1967, *Analytic Philosophy of Language and the Geisteswissenschaften*, Dordrecht.

Argumentation, c. 1964, *La théorie de l'argumentation. Perspectives et applications.* [The Theory of Argumentation. Perspectives and Applications], Louvain. Publication in book-form of *Logique et Analyse* **12** (1963).

Aristotle, 1920, 22, *Philosophische Werke III: Organon* [Philosophical Works III: Organon]. Introduced, annotated and translated into German by E. Rolfes. 1: *Kategorien* [Categories]. Preceded by a translation of Porphyrius's Introduction]. (1920). 2: *Perihermenias oder Lehre vom Satz* [*De Interpretatione* or the Theory of Propositions]. (1920). 3: *Lehre vom Schluss oder Erste Analytik* [The Theory of Inference or Prior Analytics]. (1922). 4: *Lehre vom Beweis oder Zweite Analytik.* [The Theory of Proof or Posterior Analytics] (1922). 5: *Topik* [Topics]. Second edition. (1922). 6: *Sophistische Widerlegungen* [*De Sophisticis Elenchis*]. (1922). All from Leipzig.

Aristotle, 1949, *Aristotle's Prior and Posterior Analytics.* A revised text with an introduction and commentary of W. D. Ross. Oxford.

Aristotle, 1963a, *Aristotle's Categories and De Interpretatione.* Annotated translation by J. L. Ackrill. Oxford.

Aristotle, 1963b [1928], *The Works of Aristotle I.* Translated into English under the editorship of J. A. Smith and W. D. Ross. *Categoriae* and *De Interpretatione*, by E. M. Edghill; *Analytica Priora*, by A. J. Jenkinson; *Analytica Posteriora*, by G. R. G. Mure; *Topica* and *De Sophisticis Elenchis*, by W. A. Pickard. Cambridge, London.

Aristotelis, 1964, *Aristoteles Analytica priora et posteriora.* Critical edition with notes by W. D. Ross. Foreword and Epilogue by L. Minio-Paluello. Oxford.

Arnauld, A. and Nicole, P., 1965 [1662], *La logique ou l'art de penser contenant, outre les règles communes, plusieurs observations nouvelles, propres à former le jugement* [Logic or the Art of Thinking Containing, Besides the Usual Rules, Several New Observations Suited to the Formation of Judgements]. Edited by P. Clair and F. Girbal. Paris.

Arndt, H. W., 1965, see: Lambert, J. H., 1965.

Ashworth, E. J., 1970, 'Some Notes on Syllogistic in the Sixteenth and Seventeenth Centuries', *Notre Dame Journal of Formal Logic* 11, 17–33.

Bacon, J., 1973, 'The Semantics of Generic 'The'', *Journal of Philosophical Logic* 2, 323–339.

Bakker, D. M., 1971, 'Iets over het onderscheid tussen "bepaalde" en "onbepaalde" constituenten' [On the Distinction between "Definite" and "Indefinite" Constituents], *De Nieuwe Taalgids* 64.

Bakker, R., 1964, *De geschiedenis van het fenomenologisch denken* [The History of Phenomenological Thought], Utrecht.

Barcan, R., see: Marcus, R. Barcan.

Barth, E. M., 1968a, 'Het begrip 'tautologie' bij Wittgenstein en Beth' [Wittgenstein and Beth on the Concept of Tautology], *Algemeen Nederlands tijdschrift voor wijsbegeerte en psychologie* 60, 89–100.

Barth, E. M., 1968b 'De algemene hond en het niets' [The General Dog and Nothingness], *De Gids* 131, 275–292.

Barth, E. M., 1969 'On Natural Deduction in Modal Logic with Two Primitives', *Logique et analyse* 12, 157–166.

Barth, E. M., 1970a, 'Beweging en Bertrand Russell' [Motion and Bertrand Russell], *Hollands maandblad*, no. 271–272, 56–62.

Barth, E. M., 1970b, 'Enten-eller: de logica van licht en donker' [Enten-eller: The Logic of Light and Darkness], *Algemeen Nederlands tijdschrift voor wijsbegeerte* 61, 217–240.

Barth, E. M., 1972, *Evaluaties* [Evaluations]. Inaugural lecture. Assen.

Barth, E. M., 1974, 'Untimely Remarks on the Logic of "the Modalities" in Natural Language'. In: Heidrich, C. H. (ed.) 1974.

Baumgarten, A., Bloch, E., Harich, W., and Schröter, K. (eds.), 1953, *Protokoll der philosophischen Konferenz über Fragen der Logik am 17. und 18. November 1951 in Jena* [Proceedings of the Philosophical Conference on Problems of Logic held in Jena November 17–18, 1951]. Berlin. I. *Beiheft zur Deutschen Zeitschrift für Philosophie.*

Baumgarten, A. G., 1963 [1739], *Metaphysica* [Metaphysics]. Photomechanical reprint of the seventh edition, 1779. Hildesheim.

Baumgartner, H. M., Krings, H., and Wild, C. (eds.), 1973, *Handbuch philosophischer Grundbegriffe.* [A Manual of the Fundamental Concepts of Philosophy], Stuttgart.

Beckner, M., 1968 [1959], *The Biological Way of Thought,* Berkeley and Los Angeles.

Beerling, R.F., 1949, Review of Aebi 1947, *Algemeen Nederlands Tijdschrift voor Wijsbegeerte en Psychologie* 41.

Behn, S., 1925, *Romantische oder klassische Logik? Vergleichende Dialektik des antinomischen Widerspruches* [Romantic or Classical Logic? Comparative Dialectics of the Antinomic Contradiction], Münster in Westphalia. *Veröffentlichungen des Katolischen Institutes für Philosophie* 1, 5.

Bell, D., 1971, 'Fallacies in Predicate Logic?', *Mind* 80, 145–47.

Belnap, N. D., 1962, 'Tonk, Plonk and Plink', *Analysis* 22 (1961–1962), 130–134. Also in Strawson, P. F. (ed.) 1967.

Beneke, F. E., 1842, *System der Logik als Kuhnstlehre des Denkens I* [A System of Logic as the Theory of the Art of Thinking I], Berlin.

Bennett, J., 1966, *Kant's Analytic*. Cambridge.

Bennett, J., 1970, 'The Difference between Right and Left', *American Philosophical Quarterly* 7, 175–191.

Bentham, J., 1952, 54, *Jeremy Bentham's Economic Writings*. Critical edition by W. Stark. I: 1952, III: 1954. London.

Berg, I. J. M. van den, 1946, *Logica I: Inleiding. Begripsleer* [Logic I: Introduction. Theory of Concepts], Nijmegen.

Berg, I. J. M. van den, 1952, *In het voorportaal der wetenschappen. Compendium der logica ad usum scholarum* [The Gateway to the Sciences. Compendium of Logic *Ad Usum Scholarum*]. Place of publication unknown.

Bergson, H., 1959 [1910], *Time and Free Will. An Essay on the Immediate Data of Consciousness*. London. Authorised translation by F. L. Pogson of *Essai sur les données immédiates de la conscience*. 1889.

Beth, E. W., 1946, 'Historical Studies in Traditional Philosophy', *Synthese* 5 (1946–1947), 248–260.

Beth, E. W., 1948 [1944], *Geschiedenis der logica* [History of Logic]. Second revised edition. The Hague.

Beth, E. W., 1955, 'Semantic Entailment and Formal Derivability', *Mededelingen der Koninklijke Nederlandse Akademie van Wetenschappen, afd. Letterkunde* N.R. 18, No 13, 309–342. Reprinted in Hintikka, J. (ed.) 1969a.

Beth, E. W., 1956, Über Locke's "Allgemeines Dreieck" [On Locke's "General Triangle"], *Kant-studien* 48 (1956–1957), 361–380.

Beth, E. W., 1957, *La crise de la raison et la logique* [The Crisis of Reason and Logic] (*Collection de logique mathématique, série A, fasc. XII*). Paris.

Beth, E. W., 1962, *Formal Methods. An Introduction to Symbolic Logic and to the Study of Effective Operations in Arithmetic and Logic*, Dordrecht.

Beth, E. W., 1964, *Door wetenschap tot wijsheid. Verzamelde wijsgerige studiën*, Assen.

Beth, E. W., 1965 [1959], *The Foundations of Mathematics. A Study in the Philosophy of Science*. Second revised edition. Amsterdam.

Beth, E. W., 1965, 'Banks ab omni naevo vindicatur.' In: Tymieniecka, A. (ed.) 1965.

Beth, E. W., 1967, *Moderne Logica*, Assen.

Beth, E. W., 1968, *Science, a Road to Wisdom*, Dordrecht. Translation of Beth 1964.

Beth, E. W., 1970, *Aspects of Modern Logic*, Dordrecht. Translation of Beth 1967.

Bird, O., 1960, 'The Formalizing of the Topics in Mediaeval Logic', *Notre Dame Journal of Formal Logic* 1, 138–149.

Bloch, E., see: Baumgarten, A., Bloch, E., Harich, W., and Schröter, K.

Bocheński, I. M. (J. M.), 1947, see: Peter of Spain 1947.

Bocheński, I. M. (J. M.), 1956, *Formale Logik* [Formal Logic], Freiburg.

Bocheński, I. M. (J. M.), 1959, *Logisch-philosophische Studien*. With articles by I. M. Bocheński, P. Banks, A. Menne and I. Thomas. Translated from the English and edited by A. Menne. Freiberg.

Bocheński, I. M. (J. M.), 1961, *A History of Formal Logic*. Translated and edited by I. Thomas. A translation of Bocheński 1956. Notre Dame, Ind.

Bocheński, I. M. (J. M.), 1962, *Logico-Philosophical Studies*. Edited by Albert Menne. An English edition of Bocheński, I. M. 1959. Dordrecht.

Boehner, P., 1952, *Mediaeval Logic. An Outline of Its Development from 1250 to c. 1400*, Chicago.

Bolland, G. J. P. J., 1931 [1905], *Collegium Logicum. Stenographisch verslag van eenen cursus in zuivere rede gedurende het academisch studiejaar 1904–1905 te Leiden gegeven door Prof. G. J. P. J. Bolland, en uitgegeven door eenige leerlingen I–II*. Amsterdam.

Bogdan, R. J. and Niiniluoto, I. (eds.), 1973, *Logic, Probability and Language*, Dordrecht.

Bolzano, B., 1963, *Bernhard Bolzano's Grundlegung der Logik* [Bernhard Bolzano's Foundation of Logic]. Selected Paragraphs from *Wissenschaftslehre I–II*. Edited by F. Kambartel, with an introduction, complementary summaries, and indexes.

Boole, G., 1951 [1854], *An Investigation of the Laws of Thought, on Which are Founded the Mathematical Theories of Logic and Probabilities*, New York.

Bosanquet, B., 1911 [1888], *Logic, or the Morphology of Knowledge I–II*. Second edition. Oxford.

Bradley, F. H., 1883, *Principles of Logic I*, London.

Brentano, F., 1956, *Die Lehre vom richtigen Urteil* [The Theory of Right Judgement]. Posthumously edited by F. Mayer-Hillebrand from the lectures on logic and other manuscripts in the field of epistemology. Bern.

Bunge, M., 1973, 'A Program for the Semantics of Science'. In: Bunge, M. (ed.) 1973.

Bunge, M. (ed.), 1973, *Exact Philosophy*, Dordrecht.

Burkamp, W., 1932, *Logik* [Logic], Berlin.

Carnap, R., 1929, *Abriss der Logistik, mit besonderer Berücksichtigung der Relationstheorie und ihrer Anwendungen* [An Outline of Logistics, with Special Emphasis on the Theory of Relations and Its Applications], Vienna.

Carnap, R., 1951 [1950], *Logical Foundations of Probability*, Chicago.

Carnap, R., 1960 [1947], *Meaning and Necessity. A Study in Semantics and Modal Logic*. Enlarged edition. Chicago.

Carnap, R., 1963, 'Intellectual Autobiography'. In: *The Philosophy of Rudolf Carnap*. Edited by P. A. Schilpp. London. The Library of Living Philosophers, Volume XI.

Carroll, L., 1970 [1865, 71], *The Annotated Alice. Alice's Adventures in Wonderland and Through the Looking-Glass*. Introduced and annotated by M. Gardner. Harmondsworth.

Castañeda, H. N., 1972, 'Plato's *Phaedo* Theory of Relations', *Journal of Philosophical Logic* 1, 467–480.

Church, A., 1960, 'Suppositio personalis'. In: Runes, C. D. (ed.) 1960, *Dictionary of Philosophy*, Totowa.

Cicero, 1949, *De inventione. De optimo genere oratorum. Topica*. With an English translation by H. M. Hubbell. London.

Cicero, 1954, *Ad C. Herennium libri IV de ratione dicendi. (Rhetorica ad Herennium)*. With an English translation by H. Caplan. London. Authorship controversial.

Clark, J. T., S. J., 1952, *Conventional Logic and Modern Logic. A Prelude to Transition*. With a preface by W. V. Quine. Woodstock, Md.

Cobham, A., 1956, 'Reduction to a Symmetric Predicate', *The Journal of Symbolic Logic* 21, 56–59.

Cohen, Hermann, 1914 [1902], *System der Philosophie I: Logik der reinen Erkenntnis* [The System of Philosophy I. The Logic of Pure Knowledge]. Second revised edition. Berlin.

Copi, I. M. and Gould, J.A. (eds.), 1970 [1967], *Contemporary Readings in Logical Theory*, New York.

Coreth, E., S.J., 1961, *Metaphysik. Eine Methodisch-Systematische Grundlegung*, Innsbruck.

Coreth, E., S.J., 1968, *Metaphysics*. English edition of Coreth, E., S.J. 1961 by J. Donceel with a critique by B. J. F. Lonergan. New York.

Couturat, L., 1961 [1901], *La Logique de Leibniz, d'après des documents inédits* [The Logic of Leibniz, According to the Unedited Documents], Hildesheim.

Cresswell, M. J., 1968, see: Hughes, G. E. and Cresswell, M. J.

Cresswell, M. J., 1973, *Logics and Languages*, London.

Crittenden, C., 1970, 'Ontology and the Theory of Descriptions', *Philosophy and Phenomenological Research* **31**, 85–96.

Dampier, W. C., 1968 [1929], *A History of Science and its Relations with Philosophy and Religion*. Reprint of the fourth edition with a postscript by I. B. Cohen. Cambridge.

Dassonville, M., see: Ramus, Petrus, 1964b [1555].

Davidson, D. and Harman, G. (eds.), 1972, *The Semantics of Natural Language*, Dordrecht.

De Morgan, A., 1966, *On the Syllogism and Other Logical Writings*. Edited and introduced by P. Heath. London.

Dijksterhuis, E. J., 1950, *De mechanisering van het wereldbeeld*, Amsterdam.

Dijksterhuis, E. J., 1969, *The Mechanization of the World Picture*, Oxford. A translation of Dijksterhuis, E. J. 1950.

Dik, S. C., 1968, *Coordination. Its Implications for the Theory of General Linguistics*, Amsterdam.

Dirven, R., 1971, *Some Problems of Attribution and Predication in English Syntax. A Transformational Approach*. Duplicated (Catholic University of Louvain).

Donnellan, K., 1966, 'Reference and Definite Descriptions', *Philosophical Review* **75**, 281–304. Also in: Steinberg, D. D. and Jakobovits, L. A. (eds.) 1971.

Dooren, W. van, 1965, *Het totaliteitsbegrip bij Hegel en zijn voorgangers* [Hegel and His Predecessors on the Concept of Totality], Assen.

Doyle, J. L., 1953, 'John of St. Thomas and Mathematical Logic', *The New Scholasticism* **27**, 3–38.

Dummet, M. A. E., 1967, 'Gottlob Frege'. In: *Enc. Phil. III*, pp. 225–237.

Dummet, M. A. E., 1973, *Frege: Philosophy of Language*, London.

Dürr, K., 1945, 'Moderne Darstellung der platonischen Logik. Ein Beitrag zur Erklärung des Dialoges "Sophistes"' [A Modern Exposition of Platonic Logic. A Contribution to the Analysis of "the Sophist"], *Museum Helveticum* **2**, 166–194.

Dürr, K., 1949a, 'Die Syllogistik des Johannes Hospinianus (1515–1575)' [The Syllogistic of Johannes Hospinianus], *Synthese* **9**, 472–484.

Dürr, K., 1949b, *Leibniz' Forschungen im Gebiet der Syllogistik* [Leibniz's Investigations in the Field of Syllogistic], Berlin. *Leibniz zu seinem 300. Geburtstag 1646–1946 V*.

Edghill, E. M., see: Aristotle 1963b [1928].

Edwards, P., see: *Enc. Phil.* 1967.

Eley, L., 1969, *Metakritik der formalen Logik. Sinnliche Gewissheit als Horizont der Aussagenlogik und elementaren Prädikatenlogik* [Metacriticism of Formal Logic. Perceptual Certainty as the Frame of Reference of Propositional and First-Order Predicate Logic], The Hague.

Elzer, H. M., 1967, *Bildungsgeschichte als Kulturgeschichte. Einführung in die historische Pädagogiek und in den Pädagogischen Aristotelismus*. II: *Von der Renaissance bis zum Ende der Aufklärung* [History of Education as Cultural History. An Introduction to Historical Pedagogics and to Pedagogical Aristotelianism. II: From the Renaissance to the End of the Enlightenment], Ratingen near Düsseldorf.

Emmet, D. M., 1966 [1945], *The Nature of Metaphysical Thinking*, London.

Enc. Phil., 1967, *The Encyclopedia of Philosophy I–VIII*. Editor-in-chief: P. Edwards. New York.

Erdmann, B., 1907 [1892], *Logik I: Logische Elementarlehre* [Logic I: Elementary Logic]. Second revised edition. Halle.

Faris, J. A., 1968, *Plato's Theory of Forms and Cantor's Theory of Sets*. Inaugural lecture. Belfast.

Fearnside, W. W. and Holther, W. B., 1959, *Fallacy, the Counterfeit of Argument*, Englewood Cliffs, N.J.

Fearnside, W. W. and Holther, W. B., 1963, *Drogreden of Argument*. Utrecht. A Dutch translation of Fearnside, W. W. and Holther, W. B. 1959.

Feldman, F., 1970, 'Leibniz and "Leibniz' Law"', *The Philosophical Review* **79**, 510–522.

Fichte, J. G., 1794, *Grundlage der gesammten Wissenschaftslehre. Als Handschrift für seine Zuhörer von —— [The Foundation of the General Theory of Science. As a Manuscript for His Audience of ——]*. Leipzig.

Findlay, J. N., 1964 [1958], *Hegel. A Re-examination*, London.

Findlay, J. N., 1970, *Ascent to the Absolute. Metaphysical Papers and Lectures*, London.

Fitch, F. B., 1952, *Symbolic Logic*, New York.

Føllesdal, D., 1958, *Husserl und Frege. Ein Beitrag zur Beleuchtung der Entstehung der phänomenologischen Philosophie* [Husserl and Frege. A Contribution to the Clarification of the Origin of Phenomenological Philosophy], Oslo. *Avhandlinger utgitt av Det Norske Videnskaps-Akademi i Oslo II. Hist.-Filos. Klasse* 1958, no. 2.

Føllesdal, D., 1968, 'Husserl's Notion of Noema', *The Journal of Philosophy* **65**, 680–687.

Foster, M. B., 1931, 'The Concrete Universal: Cook Wilson and Bosanquet', *Mind N.S.* **40**, 1–22.

Foucault, M., 1972, *The Archæology of Knowledge*, London. A translation from the French by A. M. Sheridan Smith of *L'Archéologie de Savoir*. Paris 1969.

Fraassen, B. van, see: Lambert, K. and van Fraassen, B.

Freericks, M. J. M., see: Angelinus, O. F. M. Cap.

Frege, G., 1884, *Die Grundlagen der Arithmetik. Eine logisch mathematische Untersuchung über den Begriff der Zahl*, Breslau.

Frege, G., 1959 [1884, 1950], *Die Grundlagen der Arithmetik. Eine logisch mathematische Untersuchung über den Begriff der Zahl. The Foundations of Arithmetic. A Logico-Mathematical Enquiry into the Concept of Number*. English translation by J. L. Austin. Second revised edition. Oxford. Bilingual edition of Frege 1884.

Frege, G., 1960 [1952], *Translations from the Philosophical Writings of Gottlob Frege*. Edited by P. Geach and M. Black. Second revised edition. Oxford.

Frege, G., 1962 [1893, 1903], *Grundgesetze der Arithmetik. Begriffsschriftlich abgeleitet I–II* [Fundamental Laws of Arithmetic. Derived with an Ideography]. Hildesheim.

Frege, G., 1964 [1879], *Begriffsschrift und andere Aufsätze* [Ideography and Other Essays]. With the remarks of E. Husserl and H. Scholz. Edited by I. Angelelli. Hildesheim.

Frege, G., 1967, *Kleine Schriften* [A Collection of Short Papers]. Edited by I. Angelelli. Hildesheim.

Frege, G., 1970, 'Begriffsschrift, a Formula Language, Modeled upon that of Arithmetic, for Pure Thought. A translation of Frege 1964'. In: Heijenoort, J. van (ed.) 1970.

Frege, G., 1971, *Schriften zur Logik und Sprachphilosophie* [Essays on Logic and the Philosophy of Language]. Hamburg. Introduced and annotated by G. Gabriel.

Freytag gen. Löringhoff, B. Baron von, 1961 [1955], *Logik. Ihr System und ihr Verhältnis zur Logistik* [Logic. Its System and Its Relation with Logistics]. Third revised edition. Stuttgart.

Freytag gen. Löringhoff, B. Baron von, 1967, *Logik II: Definitionstheorie und Methodologie des Kalkülwechsels* [Logic II: Theory of Definition and the Methodology of Calculus Change], Stuttgart.

Gadamer, H. G., 1960, *Wahrheit und Methode. Grundzüge einer Philosophischen Hermeneutik* [Truth and Method. An Outline of Philosophical Hermeneutics], Tübingen.

Geach, P. T., 1968 [1962], *Reference and Generality. An Examination of Some Medieval and Modern Theories*. Revised edition. Ithaca, N.Y.

Geach, P. T., 1969, Should Traditional Grammar be Ended or Mended? *Educational Review* 22, 23–49.

Geach, P. T., 1972, *Logic Matters*, Oxford.

Geurts, J. P. M., 1971, *Het ervaringsgegeven in de natuurwetenschappen. Kennistheoretische aantekeningen*. Duplicated (Utrecht University).

Geurts, J. P. M., c1974, *Experiential Data in Scientific Research*. A revised english edition of Geurts, J. P. M. 1971. To appear.

Göldel, R., 1935, *Die Lehre von der Identität in der deutschen Logik-Wissenschaft seit Lotze* [The Theory of Identity in German Logic since Lotze], Leipzig.

Gould, J. A., 1970, see: Copi, I. M. and Gould, J. A.

Gray, J. G., 1968 [1941], *Hegel and Greek Thought*, New York.

Gribble, C. E. (ed.), 1968, *Studies Presented to Professor Roman Jakobson by his Students*, Cambridge, Mass.

Groot, A. D. de, 1964 [1961], *Methodologie. Grondslagen van onderzoek en denken in de gedragswetenschappen*. Second edition. The Hague.

Groot, A. D. de, 1969, *Methodology. Foundations of Inference and Research in the Behavioral Sciences*. An English translation of Groot, A. D. de 1964. The Hague.

Gulielmus Occam, see: William of Ockham.

Günther, G., 1959, *Idee und Grundriss einer Nichtaristotelischen Logik* [Idea and Outline of a non-Aristotelian Logic], Hamburg.

Haenssler, E. H., 1927, *Zur Theorie der Analogie und des sogenannten Analogieschlusses* [On the Theory of Analogy and the so-called Argument by Analogy], Basel.

Halmos, P. R., 1960, *Naive Set Theory*, Princeton, N.J.

Halmos, P. R., 1968, *Intuïtieve verzamelingenleer*, Utrecht. Dutch translation of Halmos, P. R. 1960.

Hamblin, C. L., 1970, *Fallacies*, London.

Hampshire, S., 1960, 'Spinoza and the Idea of Freedom', *Proceedings of the British Academy* 1960. Also in Strawson, P. F. (ed.) 1968.

Harich, W., see: Baumgarten, A., Bloch, E., Harich, W., and Schröter, K.

Harman, G., 1972, see: Davidson, D. and Harman, G.

Hartmann, E. von, 1923 [1896], *Kategorienlehre I-III* [Theory of Categories I-III]. Second edition by F. Kern. Leipzig.

Hartmann, N., 1942, 'Neue Wege der Ontologie' [New Ways in Ontology]. In: *Systematische Philosophie*. Edited by N. Hartmann. Stuttgart.

Havemann, R., 1964, *Dialektik ohne Dogma? Naturwissenschaft und Weltanschauung* [Dialectics without Dogma. Science and Weltanschauung], Reinbek near Hamburg.

Hegel, G. W. F., 1843 [1817], *Werke* [Works]. Edited by C. L. Michelet, J. Schuhe and others. *VI, Encyklopädie der Philosophischen Wissenschaften im Grundrisse I: Die Logik* [VI, Encyclopedia of the Philosophical Sciences in Outline I: Logic]. Edited by L. von Henning. Berlin.

Hegel, G. W. F., 1896, *Lectures on the History of Philosophy III*. A translation of Hegel, G. W. F. 1928 by E. S. Haldane. London.

Hegel, G. W. F., 1912, *Hegel's Doctrine of Formal Logic: Being a Translation of the First Section of the Subjective Logic*. With an introduction and notes by H. S. Macrae. Oxford. A partial translation of Hegel, G. W. F. 1967.

Hegel, G. W. F., 1923, *Sämtliche Werke* [Collected Works]. Edited by G. Lasson. *XVIIIa: Jenenser Logik, Metaphysik und Naturphilosophie* [XVIIIa: The Logic, Metaphysics and Philosophy of Nature of Jena]. Edited from the manuscript by G. Lasson. Leipzig.

Hegel, G. W. F., 1928 [1833–36], *Sämtliche Werke* [Collected Works]. Jubilee Edition by H. Glockner. *IX: System der Philosophie II. Die Naturphilosophie* [IX: A System of Philosophy II. Philosophy of Nature]. *XIX: Vorlesungen über Geschichte der Philosophie III* [XIX: Lectures on the History of Philosophy III]. Stuttgart.

Hegel, G. W. F., 1929, *Hegel's Logic of World and Idea: Being a Translation of the Second and Third Parts of the Subjective Logic*. With an introduction by H. S. Macrae. Oxford. A partial translation of Hegel, G. W. F. 1967.

Hegel, G. W. F., 1967 [1812–16], *Wissenschaft der Logik I-II* [The Science of Logic I-II]. Edited by G. Lasson. Hamburg. Reproduction of the edition as *Sämtliche Werke III, IV* [Collected Works III, IV]. Leipzig 1934.

Hegel, G. W. F., 1969 *Hegel's Science of Logic*. Translated by A. V. Miller. Foreword by J. N. Findlay. London. A translation of Hegel, G. W. F. 1843.

Heidegger, M., 1914, *Die Lehre vom Urteil im Psychologismus. Ein kritisch-positiver Beitrag zur Logik* [Theory of Judgment in Psychologism: a Critical-Positive Essay on Logic]. Doctoral Dissertation. Leipzig.

Heidrich, C. H. (ed.), 1974, *Semantics and Communication*. Proceedings of the 3rd Colloquium of the Institute for Communications Research and Phonetics, University of Bonn February, 17th – 19th, 1972. Amsterdam.

Heijenoort, J. van (ed.), 1967, *From Frege to Gödel. A Source Book in Mathematical Logic, 1879–1931*, Cambridge, Mass.

Hempel, C. G., 1969 [1952], *Fundamentals of Concept Formation in Empirical Science*, Chicago. *International Encyclopedia of Unified Science II, 7*.

Hempel, C. G. and Oppenheim, P., 1936, *Der Typusbegriff im Lichte der neuen Logik. Wissenschaftstheoretische Untersuchungen zur Konstitutionsforschung und Psychologie* [The Concept of Type in the Light of the New Logic. Methodological Investigations into the Study of the Human Constitution and into Psychology], Leiden.

Herbart, J. F., 1834 [1813], *Lehrbuch zur Einleitung in die Philosophie* [Introduction to Philosophy. A Textbook]. Third revised edition. Konigsberg.

Hering, J., 1930 [1921], 'Bemerkungen über das Wesen, die Wesenheit und die Idee. [Remarks on Being, Essence and Idea]. In: *Jahrbuch für Philosophie und phänomenologische Forschung IV*. Halle.

Hesse, M. B., 1965 [1961], *Forces and Fields. The Concept of Action at a Distance in the History of Physics*, Totowa, N.J.

Hessing, J., 1941, *Logica als leer van zuivere rede I: Inleiding* [Logic as a Theory of Pure Reason I: Introduction], Bussum.

Hillebrand, F., 1891, *Die neuen Theorien der kategorischen Schlüsse. Eine logische Untersuchung* [The new Theories of Categorical Arguments. A Logical Investigation], Vienna.

Hintikka, J., 1958, 'Towards a Theory of Definite Descriptions', *Analysis* **19**, 79–85

Hintikka, J., 1969, *Models for Modalities. Selected Essays*, Dordrecht.

Hintikka, J. (ed.), 1969a, *The Philosophy of Mathematics*, Oxford.

Hintikka, J., 1970, 'Existential Presuppositions and Uniqueness Presuppositions'. In Lambert, K. (ed.) 1970.

Hintikka, J., 1973, *Logic, Language Games and Information*, Oxford.

Hintikka, J., Moravcsik, J. M., and Suppes, P. (eds.), 1973, *Approaches to Natural Language*. Proceedings of the 1970 Stanford Workshop on Grammar and Semantics. Dordrecht.

Hispanus, Petrus, see: Peter of Spain.

Hiż, H., 1951, 'On the Inferential Sense of Contrary-to-Fact Conditionals', *Journal of Philosophy* **48**.

Hobbes, Th., 1967 [1909], Hobbes's Leviathan. Reprinted from the Edition of 1651. With an essay by the late W. G. Pogson Smith. Oxford.

Holther, W. B., see: Fearnside, W. W. and Holther, W. B.

Honecker, M., 1927, *Logik. Eine Systematik der logischen Probleme*. [Logic. A Systematic Exposition of Logical Problems], Berlin.

Hubbell, H. M., see: Cicero 1949.

Hughes, G. E. and Cresswell, M. J., 1968, *An Introduction to Modal Logic*, London.

Husserl, E., 1922 [1913], 'Ideen zu einer reinen Phänomenologie und phänomenologischen Philosophie' [Ideas for a Pure Phenomenology and Phenomenological Philosophy]. In: *Jahrbuch für Philosophie und phänomenologische Forschung I*, 1. Halle.

Husserl, E., 1929, 'Formale und transzendentale Logik. Versuch einer Kritik der logischen Vernunft' [Formal and Transcendental Logic. A Critical Essay on Logical Reason]. In: *Jahrbuch für Philosophie und phänomenologische Forschung X*.

Husserl, E., 1948 [1939], *Erfahrung und Urteil. Untersuchungen zur Genealogie der Logik* [Experience and Judgement. Investigations into the Genealogy of Logic]. Edited by L. Landgrebe. Hamburg.

Husserl, E., 1968 [1901], *Logische Untersuchungen II*, 1: *Untersuchungen zur Phänomenologie und Theorie der Erkenntnis* [Logical Investigations II, 1: Investigations into Phenomenology and the Theory of Knowledge]. Fifth edition. II,2: *Elemente einer phänomenologischen Aufklärung der Erkenntnis* [II,2: Elements of a Phenomenological Clarification of Knowledge]. Fourth edition. Halle.

Husserl, E., 1969, *Formal and Transcendental Logic*. Translated by D. Cairns. The Hague. A translation of Husserl, E. 1929.

Husserl, E., 1969, *Ideas: General Introduction to Pure Phenomenology*. Translated by W. R. Boyce Gibson. London, New York. A translation of Husserl, E. 1922.

Husserl, E., 1970, *Logical Investigations*. Translated by J. N. Findlay. London, New York. A translation of Husserl 1968.

Ioannes a Sancto Thoma, see: John of St. Thomas.

Jacoby, G., 1955, *Allgemeine Ontologie der Wirklichkeit II*, pp. 609–1014 [General Ontology of Reality], Halle.

Jacoby, G., 1962, *Die Ansprüche der Logistiker auf die Logik und ihre Geschichtschreibung. Ein Diskussionsbeitrag.* [The Claims of Logisticians with regard to Logic and its Historiography. A Contribution to the Discussion], Stuttgart.

Jakobovits, L. A., 1971, see: Steinberg, D. D. and Jakobovits, L. A.

Jenkinson, A. J., see: Aristotle 1963b [1928].

Jerusalem, W., 1905, *Der kritische Idealismus und die reine Logik. Ein Ruf im Streite* [Critical Idealism and Pure Logic. A Battle-Cry], Vienna.

Jespersen, O., 1924, *The Philosophy of Grammar*, London.

Jevons, W. S., 1958 [1877], *The Principles of Science. A Treatise on Logic and Scientific Method.* Reprint of the second revised edition, with a foreword by E. Nagel. New York. First edition 1874.

Jevons, W. S., 1890, *Pure Logic & Other Minor Works*, London and New York.

John of St. Thomas, 1962 [1955], *Outlines of Formal Logic.* Introduced and translated from the Latin by F. C. Wade, S.J., A.M., S.T.L. Milwaukee, Wisc.

John of St. Thomas, 1965 [1955], *The Material Logic of John of St. Thomas.* Basic Treatises. Translated from the Latin by Y. R. Simon, J. J. Glanville and G. D. Hollenhorst. Prefaces by J. Maritain and Y. R. Simon. Chicago. Translation of the second part of *Ars logica.*

Kalinowski, G., 1972, *La logique des normes* [The Logic of Norms], Paris.

Kamlah, W. and Lorenzen, P., c.1970 [1967], *Logische Propädeutik oder Vorschule des vernunftigen Redens* [Logical Propaedeutics or a Primer to Reasonable Discourse]. Revised edition. Mannheim.

Kauppi, R., 1967, *Einführung in die Theorie der Begriffssysteme* [Introduction to the Theory of Concept-Systems]. Tampere. *Acta Universitatis Tamperentis ser A.* vol. 15.

Kant, I., 1801 [1800], *Logik. Ein Handbuch zu Vorlesungen* [Logic. A Lecture Manual]. Edited by G. B. Jäsche. Reutlingen.

Kant, I., 1905 [1762], 'Die falsche Spitzfindigkeit der vier syllogistischen Figuren erwiesen' [A Demonstration of the Mistaken Subtility of the Four Syllogistic Figures]. In: *Werke II: Vorkritische Schriften II* (1757–1777) [Works II: Pre-Critical Essays]. Berlin. *Gesammelte Schriften II* [Collected Papers II]. Edited by the "Königliche Preussischen Akademie der Wissenschaften."

Kant, I., 1922, *Sämtliche Werke I: Kritik der reinen Vernunft* [Collected Works I: Critique of Pure Reason]. General editor: K. Vorländer. Volume I edited by Th. Valentiner. Leipzig.

Kant, I., 1963 [1885], *Kants Introduction to Logic and His Essay on the Mistaken Subtility of the Four Figures.* Translated by T. K. Abbot. London. A translation of Kant, I. 1905 and Kant, I. 1801.

Kant, I., 1963, *Immanuel Kant's Critique of Pure Reason.* Translated by Norman Kemp Smith. London.

Kant, I., 1968 [1781], *Kritik der reinen Vernunft* [Critique of Pure Reason]. Reprint of the second edition under the editorship of I. Heidemann. Stuttgart.

Kauppi, R., 1967, *Einführung in die Theorie der Begriffssysteme* [Introduction to the Theory of Concept-Systems]. Tampere. *Acta Universitatis Tamperensis ser. A* vol. 15.

Keckermann, B., 1614, *Opera* [Works]. I, column 161: *Systema Logicae Minus* [Minor Logic]. Geneva.

Keckermann, B., 1614a, *Dialectica, dat is: reden-kavelinge ofte Bewijsconst daer een seer corte wech toebereydet, voor de ghene welckers vernuft een corte ende lichte const vereyscht. Met eenighe voor-bereydingen der logischer ofte bewijs-const.* An abridged Dutch translation of Keckermann, B. 1614. Amsterdam.

Keynes, J. N., 1928 [1884], *Studies and Exercises in Formal Logic: Including a Generalization of Logical Processes in Their Application to Complex Inferences*. Reprint of the fourth revised and enlarged edition of 1906. London.

Kierkegaard, S., 1846, *Afsluttende uvidenskabelig Efterskrift til de philosophiske Smuler. Mimisk-pathetisk-dialektisk Sammenskrift, Existentielt Indlæg, af Johannes Climacus*. Copenhagen.

Kierkegaard, S., 1941, *Kierkegaard's Concluding Unscientific Postscript*. A translation of Kierkegaard, S. 1846 by D. F. Swenson and W. Laurie. Oxford.

Klineberg, O., 1954 [1940], *Social Psychology*. Second revised edition. New York.

Klooster, W. G., see: Kraak, A. and Klooster, W. G.

Kneale, W. C. and Kneale, M., 1964 [1962], *The Development of Logic*, Oxford.

Kraak, A., 1966, *Negatieve zinnen. Een methodologische en grammatische analyse* [Negative sentences. A Methodological and Grammatical Analysis], Hilversum.

Kraak, A. and Klooster, W. G., 1968, *Syntaxis* [Syntax], Culemborg.

Kretzmann, N., 1967a, 'History of Semantics', In: *Enc. Phil. VII*, pp. 358–406.

Kretzmann, N., 1967b, 'Peter of Spain', In: *Enc. Phil. VI*, pp. 125–126.

Krings, H., see: Baumgartner, H. M., Krings, H., and Wild, C.

Kruijer, G. J., 1959, *Observeren en redeneren. Een inleiding tot de kennisvorming in de sociologie* [Observing and Reasoning. An Introduction to the Acquisition of Knowledge in Sociology], Meppel.

Kyburg, H. E., Jr., 1970, 'Conjunctivitis', In: Swain, M. (ed.) 1970.

Ladd-Franklin, C., 1890, 'Some Proposed Reforms in Logic', *Mind N.S.* **15**, 79–88.

Lakoff, G., 1970, 'Linguistics and Natural Logic', *Synthese* **22**, pp. 151–271.

Lambert, J. H., 1969 [1764], *Philosophische Schriften* [Philosophical Papers]. Edited by H.-W. Arndt. *I: Neues Organon I* [I: New Organon I]. Hildesheim.

Lambert, K. and van Fraassen, B., 1970, 'Meaning Relations, Possible Objects, and Possible Worlds', In: Lambert, K. (ed.), 1970, pp. 1–19.

Lambert, K. (ed.), 1969, *The Logical Way of Doing Things*, New Haven etc.

Lambert, K. (ed.), 1970, *Philosophical Problems in Logic. Some Recent Developments*, Dordrecht.

Leibniz, G. W., 1882, *Die philosophischen Schriften von* —— [The Philosophical Papers ——]. Edited by C. J. Gerhardt. *V: Leibniz und Locke* [V: Leibniz and Locke]. Berlin. Contains: *Nouveaux essais sur l'entendement* [New Essays on the Understanding].

Leibniz, G. W., 1960, *Fragmente zur Logik* [Fragments on Logic]. Selected, translated and elucidated by F. Schmidt. Berlin.

Leibniz, G. W., 1961, *Die philosophischen Schriften von –* [The Philosophical Papers of –]. Edited by C. J. Gerhardt. Unamended reprint of the Berlin edition from 1890 Hildesheim.

Leibniz, G. W., 1966, *Logical Papers. A selection*. Edited, translated and introduced by G. H. R. Parkinson. Oxford etc.

Leibniz, G. W., 1968 [1934], *Philosophical Writings*. Contains translations of Leibniz 1882 by M. Morris. New York.

Leibniz, G. W., 1969 [1956], *Gottfried Wilhelm Leibniz: Philosophical Papers and Letters*. Edited by L. E. Loemker. Dordrecht.

Leisegang, H., 1928, *Denkformen* [Forms of Thought], Berlin.

Lejewski, C., 1960, 'A Re-examination of the Russellian Theory of Descriptions', *Philosophy* **35**, 14–29.

Lenk, H., 1968 [1966], *Kritik der logischen Konstanten. Philosophische Begründungen der Urteilsformen vom Idealismus bis zur Gegenwart* [A Critique of the Logical Constants. Philosophical Foundations of the Forms of Judgment from Idealism to the Present], Berlin. Revised and enlarged edition of the authors "Habilitationsschrift": *Die Urteilsformen. Darstellung und Kritik ihrer Begründungen in der deutschen Tradition seit Kant* [The Forms of Judgment. Exposition and Critique of Their Foundations in German Tradition since Kant], Berlin 1966.

Lenk, H., 'Logikbegründung und rationalistischer Kritizismus' [The Foundations of Logic and Rationalist Criticism], *Zeitschrift für Philosophische Forschung* **24**, 183–205.

Lewis, C. I., 1918, *A Survey of Symbolic Logic*, Berkeley.

Lewis, D., 1970, 'General Semantics', *Synthese* **22**, 18–67. Also in: Davidson, D. and Harman, G. (eds.) 1972.

Linsky, L., 1967, *Referring*, London.

Linsky, L. (ed.), 1971, *Reference and Modality*, Oxford.

Lloyd, G. E. R., 1966, *Polarity and Analogy. Two Types of Argumentation in Early Greek Thought*, Cambridge.

Locke, J., 1959 [1690], *An Essay Concerning Human Understanding II*. Annoted edition with biographical, critical and historical prolegomena of A. C. Fraser. New York.

Lorenzen, P., 1956, Review of Von Freytag 1955, *Archiv für Philosophie* **6**, 338–339.

Lorenzen, P., 1962, *Metamathematik* [Metamathematics], Mannheim.

Lorenzen, P., 1969, *Normative Logic and Ethics*, Mannheim.

Lorenzen, P., c.1970, see: Kamlah, W. and Lorenzen, P.

Lotze, H., 1879, *System der Philosophie II: Metaphysik* [System of Philosophy II: Metaphysics], Leipzig.

Lotze, H., 1912 [1874], *System der Philosophie I: Logik. Drei Bücher vom Denken, vom Untersuchen und vom Erkennen*. Edited and introduced by G. Misch. Leipzig.

Lotze, H., 1884, *Lotze's System of Philosophy, I: Logic*. English translation of Lotze 1912 edited by B. Bosanquet. Oxford.

Lovejoy, A. O., 1936, *The Great Chain of Being: A Study of the History of an Idea*, Cambridge, Mass.

Łukasiewicz, J., 1958 [1951], *Aristotle's Syllogistic from the Standpoint of Modern Formal Logic*. Second enlarged edition. Oxford.

McTaggart, J. M. E., 1927, *The Nature of Existence*, Vol. II. Edited by C. D. Broad. Cambridge.

Maier, H., 1896, *Die Syllogistik des Aristoteles I: Die logische Theorie des Syllogismus und die Entstehung der aristotelischen Logik. I: Formenlehre und Technik des Syllogismus. 2: Die Entstehung der aristotelischen Logik* [Aristotle's Syllogistic I: The Logical Theory of the Syllogism and the Origins of Aristotelian Logic. I: Theory of Form and the Technique of Syllogism. 2: The Origins of Aristotelian Logic], Tübingen.

Marcus, R. Barcan, 1960, 'Extensionality', *Mind N.S.* **69**, 55–62. Also in Linsky, L. (ed.) 1971.

Maritain, J., 1933 [1923], *Éléments de philosophie II: L'ordre des concepts. I: Petite Logique (logique formelle)*. Eleventh revised edition. Paris.

Maritain, J., 1939 [1920], *Éléments de philosophie I: Introduction générale à la philosophie*. Paris.

Maritain, J., 1930, *An Introduction to Philosophy*. A translation of Maritain, J. 1939 by E. I. Watkin. London.

Maritain, J., 1937, *An Introduction to Logic*. A translation of Maritain, J. 1933. London.

Martin, G., 1960, *Leibniz: Logik und Metafysik*, Cologne.

Martin, G., 1964, *Leibniz: Logic and Metaphysics*. Translated by K. J. Northcott and P. G. Lucas. Manchester.

Martin, R. M., 1963, *Intension and Decision. A Philosophical Study*, Englewood Cliffs, N.J.

Mates, B., 1968, 'Leibniz on Possible Worlds.' In: van Rootselaar, B. and Staal, J. F. (eds.).

Mattens, W. H. M., 1970, *De indifferentialis. Een onderzoek naar het anumerieke gebruik van het substantief in het algemeen bruikbaar Nederlands* [Indifferentialis. An Investigation into the Anumerical Use of the Noun in Common Dutch Usage], Assen.

Matthews, G. B., 1964, 'Ockham's Supposition Theory and Modern Logic', *The Philosophical Review* **78**, 91–99.

McCawley, J., 1971, 'Where Do Noun Phrases Come From?' In: Steinberg, D. D. and Jakobovits, L. A. 1971.

Menne, A., 1954, *Logik und Existenz. Eine logistische Analyse der kategorischen Syllogismus-Funktoren und das Problem der Nullklasse* [Logic and Existence. A Logistic Analysis of the Categorical Functors of the Syllogism and the Problem of the Empty Class], Meisenheim.

Menne, A., 1957, 'Implikation und Syllogistik' [Implication and Syllogistic], *Zeitschrift für philosophische Forschung* **11**, 375–386.

Menne, A., 1966, *Einführung in die Logik* [Introduction to Logic], Bern.

Menne, A., 1973, 'Definition'. In: Baumgartner, H. M., Krings, H., and Wild, C. (eds.) 1973.

Merleau-Ponty, M., 1945, *Phénoménologie de la perception* Fourth edition. Paris.

Merleau-Ponty, M., 1962, *Phenomenology of Perception*. Translated by C. Smith. New York, London.

Meulen, J. A. van der, 1951, *Magdalena Äbi und Kant oder das Unendliche Urteil* [Magdalena Äbi and Kant or the Infinite Judgment], Meisenheim/Glan.

Meulen, J. A. van der, 1958, *Hegel. Die gebrochene Mitte* [Hegel. The Broken Middle], Hamburg.

Mill, J. S., 1965 [1843], *A System of Logic, Ratiocinative and Inductive. Being a Connected View of the Principles of Evidence and the Methods of Scientific Investigation.* Reprint of the eighth edition (1872). London.

Montague, R., 1970, 'Universal Grammar', *Theoria* **36**, 373–398.

Moody, E. A., 1953, *Truth and Consequence in Mediaeval Logic*, Amsterdam.

Moody, E. A., 1965 [1935], *The Logic of William of Ockham*, New York.

Moody, E. A., 1967a, 'Mediaeval Logic'. In: *Enc. Phil. IV*, pp. 528–534.

Moody, E. A., 1967b, 'William of Ockham'. In: *Enc. Phil. VIII*, pp. 306–317.

Moore, G. E., 1959, *Philosophical Papers*, London.

Moravcsik, J. M., see: Hintikka, J., Moravcsik, J. M., and Suppes, P. 1973.

Naess (Næss), A., 1966, *Communication and Argument. Elements of Applied Semantics*, Oslo. Translation by A. Hannay of: *En del elementære logiske emner*.

Naess (Næss), A., 1967, 'Bidrag til analyse av den kognitive struktur i 'fenomenteorien' i L'être et le néant' [A Contribution to the Analysis of the Cognitive Structure of the Theory of the Phenomenon in "L'être et le néant"], *Norsk filosofisk tidsskrift* **2**, 1–17.

Naess (Næss), A., 1969, 'Freedom, Emotion, and Self-subsistence. The Structure of a Small, Central Part of Spinoza's *Ethics*', *Inquiry* **2**, 66–104.

Nagel, E., 1961, *The Structure of Science. Problems in the Logic of Scientific Explanation*, London.

Naville, A., 1909, *La logique de l'identité et celle de la contradiction. Notes critiques* [The Logic of Identity and Contradiction. Critical Notes], Geneva.

Nelson, L., 1962, *Sämtliche Werke VII: Fortschritte und Rückschritte der Philosophie. Von Hume und Kant bis Hegel und Fries.* Posthumous publication edited by J. Kraft. Frankfurt.

Nelson, L., 1970, *Progress and Regress in Philosophy: from Hume and Kant to Hegel and Fries.* Translated from Nelson, L. 1962 by Humphrey Palmer. Oxford.

Nelson, N. E., 1947, *Peter Ramus and the Confusion of Logic, Rhetoric, and Poetry*, Ann Arbor. *The University of Michigan Contributions in Modern Philosophy*, 2.

Nicole, P., see: Arnauld, A. and Nicole, P.

Nidditch, P. H., 1962, *The Development of Mathematical Logic*, London.

Niiniluoto, I., see: Bogdan, R. J. and Niiniluoto, I. (eds.) 1973.

Nuchelmans, G., 1957, 'The Analysis of Counterfactual Conditionals I', *Synthese* 9, 48–63.

Nuchelmans, G., 1969, *Overzicht van de analytische wijsbegeerte* [An Outline of Analytical Philosophy], Utrecht.

Nuchelmans, G., 1970, 'Taaluiting en logische structuur' [Linguistic Utterance and Logical Structure], *Algemeen Nederlands tijdschrift voor wijsbegeerte* 62, 162–174.

Nuchelmans, G., 1972, 'Het Mentaals. De opvatting van het denken als een vorm van spreken in de antieke en middeleeuwse taaltheorie' [Mental. The Conception of Thought as a Form of Speech in Classical and Medieval Theory of Language], *Leuvense Bijdragen. Tijdschrift voor Germaanse Filologie* 61, 295–309.

Nuchelmans, G., 1973, *Theories of the Proposition. Ancient and Mediaeval Conceptions of the Bearer of Truth and Falsity*, Amsterdam.

Olafson, F. A., 1967, 'Jean-Paul Sartre'. In: *Enc. Phil. VII*, pp. 287–293.

Olbrechts-Tyteca, L., *c.* 1964, 'Rencontre avec la rhétorique' [Encounter with Rhetoric]. In: *Argumentation c.* 1964.

Olson, R. G., 1969, *Meaning and Argument. Elements of Logic*, New York.

Ong, S. J., W. J., 1958, *Ramus, Method and the Decay of Dialogue. From the Art of Discourse to the Art of Reason*, Cambridge, Mass.

Oppenheim, P., see: Hempel C. G. and Oppenheim, P.

Owen, G. E. L., 1967 [1965], 'The Platonism of Aristotle', *Proceedings of the British Academy* 1965. Also in Strawson, P. F., (ed.) 1967.

Parkinson, G. H. R., 1965, *Logic and Reality in Leibniz's Metaphysics*, Oxford.

Parkinson, G. H. R., 1966, see: Leibniz, G. W. 1966.

Partee, B. H., 'Comments on Montague's Paper'. In: Hintikka, J., Moravcsik, J. M., and Suppes, P. (eds.) 1973.

Passmore, J. A., 1966 [1957], *A Hundred Years of Philosophy*. 2nd edition. London.

Patzig, G., 1965 [1959], *Die aristotelische Syllogistik Logisch-philosophische Untersuchungen über das Buch A der "Ersten Analytiken"* [Aristotelian Syllogistic. Logico-philosophical Investigations into Book A of the Prior Analytics]. Second revised edition. Göttingen. *Abhandlungen der Akademie der Wissenschaften in Göttingen. Philologisch-historische Klasse* – Dritte Folge, Nr. 42.

Peirce, C. S., 1965–67 [1931–33], *Collected Papers*. Edited by C. Hartshorne and P. Weiss. I: *Principles of Philosophy*. II: *Elements of logic*. III: *Exact logic*. IV: *The Simplest Mathematics*, Cambridge, Mass.

Peter of Spain, 1947, *Petri Hispani: Summulae logicales*. Edited by I. M. Bocheński O.P. Turin.

Peter of Spain, 1964, *Tractatus syncategorematum and Selected Anonymous Treatises*. Translated from the Latin by J. P. Mullaway; introduced by J. P. Mullaway and R. Houde. Milwaukee, Wisc.
Peter Ramus, see: Ramus, Petrus.
Peters, J., 1967, *Metaphysica. Een systematisch overzicht* [Metaphysics. A Systematic Survey], Utrecht.
Petrus Hispanus, see: Peter of Spain.
Peursen, C. A. van, 1967, *Fenomenologie en werkelijkheid* [Phenomenology and Reality], Utrecht.
Pfänder, A., 1963 [1921], *Logik* [Logic]. With a foreword by H. Spiegelberg. Tübingen.
Pivčević, E., 1967, 'Husserl *versus* Frege', *Mind N.S.* **76**, 155–165.
Pivčević, E., 1970, *Husserl and Phenomenology*, London.
Plessner, H. (ed.), 1952, *Symphilosophein. Bericht über den Dritten Deutschen Kongress für Philosophie, Bremen 1950* [Symphilosophein. Proceedings of the Third German Congress for Philosophy, Bremen 1950]. Prepared by I. Pape and W. Stache.
Popma, K. J., 1963, *Wijsbegeerte en anthropologie* [Philosophy and Anthropology]. Papers, collected by J. Stellingwerf. Amsterdam. *Christelijk Perspectief* 3/4.
Potts, T. C., 1973, 'Fregean Categorial Grammar'. In: Bogdan, R. J. and Niiniluoto, I. (eds.) 1973.
Prantl, C., 1955 [1867], *Geschichte der Logik im Abendlande I–IV* [A History of Western Logic], Darmstadt.
Prior, A. N., 1960, 'The Runabout Inference Ticket', *Analysis* **21** (1960–61), 38–39. Also in Strawson, P. F. (ed.) 1967.
Prior, A. N., 1962 [1955], *Formal Logic*. Second revised and enlarged edition. Oxford.
Prior, A. N., 1967, 'The Heritage of Kant and Mill'. In: *Enc. Phil. IV*, p. 549.
Prior, A. N., 1969, Review of von Wright 1969, *The British Journal for the Philosophy of Science* **20**, 372–374.
Protokoll, 1953, *Protokoll der philosophischen Konferenz über Fragen der Logik am 17. und 18. November 1951 in Jena* [Proceedings of the Philosophical Conference on Problems of Logic held in Jena November 17–18, 1951]. Edited by A. Baumgarten, E. Bloch, W. Harich and K. Schröter. Berlin. I. *Beiheft zur Deutschen Zeitschrift für Philosophie*.
Purtill, R. L., 1971, *Logic for Philosophers*, New York.
Quine, W. V. O., 1950, *Methods of Logic*, New York.
Quine, W. V. O., 1962 [1940], *Mathematical Logic*. Revised edition. New York.
Quine, W. V. O., 1963 [1953], *From a Logical Point of View. Nine Logico-Philosophical Essays*. Second revised edition. New York.
Quine, W. V. O., 1966, *The Ways of Paradox and Other Essays*, New York.
Ramus, Petrus (Pierre de la Ramée), 1964a [1543], *Dialecticae institutiones, Aristotelicae animadversiones* [Lessons in Dialectics. Critical Remarks on Aristotelian Logic]. Facsimile of the Parisian edition of 1543, with an introduction of W. Risse. Stuttgart-Bad Cannstatt.
Ramus, Petrus (Pierre de la Ramée), 1964b [1555], *Dialectica* [Dialectics]. Critical edition with an introduction, notes and commentary of M. Dassonville. Geneva.
Ramus, Petrus (Pierre de la Ramée), 1966 [1574], *The Logike 1574*. Leeds. Reprint in facsimile of: *The Logike of the Moste Excellent Philosopher P. Ramus Martyr. Newly Translated, and in Divers Places Corrected, after the Mynde of the Author*. In 1574 translated by R. MacIlmaine from: *Dialecticae libri duo* (1556).
Rescher, N., 1964, *Introduction to Logic*, New York.

Rescher, N., 1966, *The Logic of Commands*, London.
Rescher, N., 1968, *Topics in Philosophical Logic*, Dordrecht.
Risse, W., 1964, 70, *Die Logik der Neuzeit I: 1500–1600* (1964). *II 1640–1780* (1970) [Post-Mediaeval Logic I: 1500–1640. II: 1640–1780]. Stuttgart-Bad Cannstatt.
Risse, W., 1965, Review of de Rijk 1962, *Philosophische Rundschau* **13**, 111–115.
Rijk, L. M. de, 1962, 67, *Logica Modernorum. A Contribution to the History of Early Terministic Logic I: On the Twelfth Century Theories of Fallacy* (1962). *II, I The Origin and Early Development of the Theory of Supposition*. (1967). Assen.
Rijk, L. M. de, 1970, 'On the genuine text of Peter of Spain's *Summule logicales Vivarium* **8**, 10–55.
Rijk, L. M. de, 1971, 1973, 'The Development of Suppositio naturalis in Mediaeval Logic'. Part I: 'Natural Supposition as Non-Contextual Supposition', *Vivarium* q, 71–107. Part II: 'Fourteenth Century Natural Supposition as Atemporal (Omnitemporal) Supposition', *Vivarium* **11**, 43–79.
Rijk, R. P. G. de, 1968, 'St. Augustine on Language'. In: Gribble, C. E. (ed.) 1968.
Robbins, B. L., 1968, *The Definite Article in English Transformations*, The Hague.
Robinson, R., c. 1950, *Definition*, Oxford.
Rolfes, E. see: Aristotle 1920, 1922.
Rootselaar, B. and Staal, J. F. (eds.), 1968, *Logic, Methodology and Philosophy of Science III*. Proceedings of the Third International Congress for Logic, Methodology and Philosophy of Science, Amsterdam 1967. Amsterdam.
Ross, W. D., 1949, see: Aristotle 1949.
Runes, C. D. (ed.), 1962, *Dictionary of Philosophy*, Totowa.
Russell, B., 1905, 'On Denoting', *Mind N.S.* **14**, 479–493.
Russell, B., 1947 [1946], *History of Western Philosophy and Its Connection with Political and Social Circumstances from the Earliest Times to the Present Day*. London.
Russell, B., 1959, 'Logic and Ontology'. In: *My Philosophical Development*, London.
Russell, B., 1962, see: Whitehead, A. N. and Russell, B.
Russell, B., 1963a [1919], *Introduction to Mathematical Philosophy*, London.
Russell, B., 1963b [1940], *An Inquiry into Meaning and Truth*, Harmondsworth.
Russell, B., 1964 [1903], *The Principles of Mathematics*. Reprint of the second edition (1937). London.
Salmon, W. C. 1973, 'Confirmation', *Scientific American* **228**, No. 5.
Samuel, O., 1957, *Die Ontologie der Logik und der Psychologie. Eine meontologische Untersuchung* [The Ontology of Logic and Psychology. A Meontological Investigation], Cologne. *Kantstudien, Ergänzungshefte*, Nr. 74.
Sartre, J. P., 1957 [1943], *L'être et le néant. Essai d'ontologie phénoménologique*. 49th Edition. Paris.
Sartre, J. P., *Being and Nothingness. An Essay on Phenomenological Ontology*. Translated and with an introduction by H. E. Barnes. New York.
Schilpp, P. A. (ed.), 1963, *The Philosophy of Rudolf Carnap*. The Library of Living Philosophers Volume XI. London.
Schmidt, F., see: Leibniz, G. W., 1960.
Schmitt, R., 1967, 'Phenomenology'. In: *Enc. Phil. VI*, pp. 135–151.
Scholz, H., 1941, *Metaphysik als strenge Wissenschaft* [Metaphysics as a Rigorous discipline], Cologne.
Scholz, H., 1967 [1931], *Abriss der Geschichte der Logik* [An Outline of the History of Logic], Freiburg. Original title: *Geschichte der Logik* [History of Logic].
Schoonbrood, O. F. M., C. A., 1961, 'Zur Logik erfahrungstranszenter Erkenntnis'

[On the Logic of Knowledge which Transcends Experience], *Wissenschaft und Weisheit* **24**, 182–199.

Schoonbrood, O. F. M., C. A., 1968, 'Bewijsvoering in de filosofie' [Argumentation in Philosophy], *Wijsgerig perspectief op maatschappij en wetenschap* **8**, 161–171.

Schröder, Ernst, 1966 [1890], *Vorlesungen über die Algebra der Logik (Exakte Logik) I* [Lectures on the Algebra of Logic (Exact Logic) I.]. Second edition. New York.

Schröter, K., see: Baumgarten, A., Bloch, E., Harich, W., and Schröter, K.

Scott, D., 1970, 'Advice on Modal Logic'. In Lambert, K. (ed.) 1970.

Scott Jr., T. K., 1965, 'John Buridan on the Objects of Demonstrative Science', *Speculum* **40**, 654–673.

Seuren, P. A. M., 1969, *Operators and Nucleus. A Contribution to the Theory of Grammar.* Cambridge.

Sigwart, C., 1904 [1873, 78], *Logik*. Third revised edition *I: Die Lehre vom Urteil, vom Begriff und vom Schluss. II: Die Methodenlehre* [I: The Theory of Judgement, Concept and Inference. II Methodology]. Tübingen.

Sigwart, C., 1895, *Logic*. A translation of the second edition of Sigwart, C. 1904 by H. Dendy. London.

Simon, Y. R., see: John of St. Thomas 1965.

Skjervheim, H., 1964, *Vitskapen om mennesket og den filosofiske refleksjon* [The Science of Man and Philosophical Reflection], Oslo.

Smiley, T., 1962, 'Syllogism and Quantification', *The Journal of Symbolic Logic* **27**, 58–72.

Snyder, A., 1971, 'Rules of Language', *Mind N.S.* **80**, 161–178.

Snow, C. P., 1965, *The Two Cultures: and a Second Look*, Cambridge. An enlarged version of: *The Two Cultures and the Scientific Revolution* (1959).

Specht, E. K., 1967, *Sprache und Sein. Untersuchungen zur Sprachanalytischen Grundlegung der Ontologie* [Language and Being. Investigations into the Foundations of Ontology by Means of Linguistic Analysis], Berlin.

Spencer, M., 1971, 'Why the "S" in "Intension"?, *Mind N.S.* **80**, 114–115.

Spiegelberg, H., see: Pfänder, A. 1963 [1921].

Staal, J. F., see: van Rootselaar, B. and Staal, J. F.

Stace, W. T., 1961 [1960], *Mysticism and Philosophy*, London.

Stegmüller, W., 1960, 'Das Problem der Kausalität' [The Problem of Causality]. In: Topitsch, E. (ed.).

Stegmüller, W., 1965a [1952], *Hauptströmungen der Gegenwartsphilosophie. Eine kritische Einführung.* Third enlarged edition. Stuttgart.

Stegmüller, W., 1965b [1956, 57], 'Das Universalienproblem einst und jetzt' [The Problem of Universals Then and Now]. In: *Glauben, Wissen und Erkennen. Das Universalienproblem einst und jetzt* [Knowledge and Belief. The Problem of Universals Then and Now], Darmstadt. Originally appeared in *Archiv für Philosophie* **6**, 192–225 and **7**, 45–81.

Stegmüller, W., *Main Currents in Contemporary German, British and American Philosophy*. A translation of Stegmüller, W. 1965a by A. E. Bloomberg. Dordrecht.

Steinberg, D. D. and Jakobovits, (eds.) 1971, *Semantics. An Interdisciplinary Reader in Philosophy, Linguistics and Psychology*, Cambridge.

Stenius, E., 1969, 'Satsen som funktion hos Frege och Wittgenstein' [Frege and Wittgenstein on Propositions as Functions], *Norsk Filosofisk Tidsskrift* **4**, 9–23.

Strawson, P. F., 1950, 'On Referring', *Mind N.S.* **59**, 320–344.

Strawson, P. F., 1964 [1952], *Introduction to Logical Theory*, London.

Strawson, P. F., (ed.), 1967, *Philosophical Logic*.

Strawson, P. F., 1970, 'Grammar and Philosophy', *Proceedings of the Aristotelian Society N.S.* **52**, 1–20.

Suppes, P., see: Hintikka, J., Moravcsik, J. M., and Suppes, P. (eds.) 1973.

Swing, Th. K., 1969, *Kant's Transcendental Logic*, New Haven.

Swain, M. (ed.), 1970, *Induction, Acceptance and Rational Belief*, Dordrecht.

Symphilosophein, 1952, *Symphilosophein. Bericht über den Dritten Deutschen Kongress für Philosophie, Bremen* 1950 [Symphilosophein, Proceedings of the Third German Congress for Philosophy, Bremen 1950]. Prepared by I. Pape and W. Stache and edited by H. Plessner, München.

Tarski, A., 1941, *Introduction to Logic and to the Methodology of the Deductive Sciences*, New York. Appeared originally in Polish (1936).

Tarski, A., 1964 [1953], *Inleiding tot de logica en tot de methodeleer der deductieve wetenschappen*. A Dutch edition of Tarski 1941 by E. W. Beth. Second revised edition. Amsterdam.

Thomas, I., 1967, 'Interregnum'. In: *Enc. Phil. IV*, pp. 534–537.

Thomason, R. H., 1972, 'A Semantic Theory of Sortal Incorrectness', *Journal of Philosophical Logic* **1**, 209–258.

Topitsch, E. (ed.), 1960, *Probleme der Wissenschaftstheorie. Festschrift für Victor Kraft* [Problems in the Theory of Science. Festschrift for Victor Kraft], Vienna.

Toulmin, S. E., 1964 [1958], *The Uses of Argument*, Cambridge.

Toulmin, S. E., 1972, *Human Understanding*. Vol. I: General Introduction and Part I. Oxford.

Tranøy, K. E., 1954, 'Filosofien i Norge' [Philosophy in Norway], *Syn og Segn* **60**, 260–268.

Trendelenburg, A., 1862 [1840], *Logische Untersuchungen I–II* [Logical Investigations]. Second enlarged edition. Leipzig.

Tugendhat, E., 1970, 'The Meaning of "Bedeutung" in Frege', *Analysis* **30**.

Tymieniecka, A. (ed.), 1965, *Contributions to Logic and Methodology in Honor of J. M. Bocheński*. Edited by A. Tymieniecka in collaboration with C. Parsons. Amsterdam.

Ubbink, J. B., 1962, *Plato's paradox en Bohr's idee* [Plato's Paradox and Bohr's Idea], Arnhem.

Überweg, F., 1882 [1857], *System der Logik und Geschichte der logischen Lehren*. Fifth revised and enlarged edition. Prepared and edited by J. B. Meyer. Bonn.

Überweg, F., 1871, *System of Logic and History of Logical Doctrines*. Translated by T. M. Lindsay. London.

Ujomov, A. I., 1965, *Dinge, Eigenschaften und Relationen* [Things, Properties and Relations]. Translated from the Russian by G. Kröber assisted by P. Bolhagen. Berlin. First edition in Russian 1963.

Vater, H., 1963, *Das System der Artikelformen im gegenwärtigen Deutsch* [The System of Articles in Contemporary German], Tübingen.

Veatch, H. B., 1950, 'Aristotelian and Mathematical Logic', *The Thomist* **13**, 50–96.

Veatch, H. B., 1952a, *Intentional Logic: a Logic Based on Philosophical Realism*, New Haven.

Veatch, H. B., 1952b, 'The Significance of the Current Criticism of the Syllogism', *The Thomist* **15**, 624–641.

Veatch, H.B., 1953, 'For a Realistic Logic'. In: Wild, J. (ed.). 1953.

Veatch, H. B.,1969, *Two Logics. The Conflict Between Classical and Neo-Analytic Philosophy*, Evanston, Ill.

Vendler, Z., 1962, 'Each and Every, Any and All', *Mind N.S.* **71**, 145–160.
Vendler, Z., 1967, *Linguistics in Philosophy*, Ithaca, N.Y.
Venn, J., 1962 [1866], *The Logic of Chance*. Fourth edition. New York.
Vieru, S., 1971, *Embedding of Assertoric Syllogistic into the Predicate Calculus*. Submitted at the IVth International Congress for Logic, Methodology and Philosophy of Science, Bucharest 1971.
Walters, R. S., 1967, 'Contrary-to-Fact Conditional'. In: *Enc. Phil. II*, 212–216.
Weinberg, J. R., 1965, *Abstraction, Relation and Induction. Three Essays in the History of Thought*, Madison.
Weinberger, O., 1964, *Philosophische Studien zur Logik I: Einige Bemerkungen zum Begriff der Identität. II: Philosophische Bemerkungen zur Sollsatzlogik* [Philosophical Studies in Logic I: Remarks on the Concept of Identity. II: Philosophical Remarks on the Logic of Imperatives.]. Prague. *Rozpravy Československé Akademie Věd. Řada společenských věd, Ročník 74 – Sešit 5.*
Weinberger, O., 1965, *Der Relativierungsgrundsatz und der Reduktionsgrundsatz – zwei Prinzipien des dialektischen Denkens. Eine logisch-methodologische Studie* [Two Principles of Dialectic Thought: Relativization and Reduction. A Logico-Methodological Study].
Prague. *Rozpravy Československé Akademie Věd. Řada společenských věd. Ročník 75 – Sešit 5.*
Wellek, A., 1967, 'Wilhelm Wundt'. In: *Enc. Phil. VIII*, pp. 349–351.
Wesly, P., 1970, Review of Specht 1967, *Foundations of Language* **8**, 140–144.
Whately, R., 1859 [1826], *Elements of Logic*. Reprint of the ninth edition. London.
Whitehead, A. N. and Russell, B., 1962 [1910], *Principia Mathematica to *56*. Cambridge.
Wick, W. A., 1942, *Metaphysics and the New Logic*, Chicago.
Wieland, W., 1967, 'Zur Deutung der Aristotelischen Logik' [On Interpreting Aristotelian Logic]. Review of Patzig, G. 1965. *Philosophische Rundschau* **14**, 1–27.
Wilcox, W. C., 1971, 'Another Look at Distribution', *Mind N.S.* **80**, 133–135.
Wild, C., see: Baumgartner, H. M., Krings, H., and Wild, C.
Wild, J. (ed.), 1953, *The Return to Reason: Essays in Realistic Philosophy*, Chicago.
William of Ockham, 1951, 1954, *Summa logicae I* (1951). *II* and *III* (1954). Edited by P. Boehner. New York.
Wilson, Curtis, 1960 [1956], *William Heytesbury. Medieval Logic and the Rise of Mathematical Physics*, Madison, Wisc.
Windelband, W., 1910, *Über Gleichheit und Identität* [On Equality and Identity], Heidelberg. *Sitzungsberichte der Heidelberger Akademie der Wissenschaften, Philosophisch-historische Klasse* 1910, 14. *Abhandlung.*
Wolff, Chr., 1728, *Philosophia rationalis sive logica, methodo scientifica pertractata et ad usum scientiarum atque vitae aptata. Praemittitur discursus praeliminaris de philosophia in genere*. Frankfurt.
Wolff, Chr., 1770, *Logic or Rational Thoughts on the Power of the Human Understanding with their Use and Application in the Knowledge and Search For Truth*. Translated from the German. London.
Wolter O.F.M., A. B., 1946, *The Transcendentals and Their Function in the Metaphysics of Duns Scotus*, St. Bonaventure, N.Y.
Woodworth, R. S., 1952 [1948], *Contemporary Schools of Psychology*, London.
Wright, G. H. von, 1969, *Time, Change and Contradiction*, Cambridge.
Wundt, W., 1893 [1880], *Logik. Eine Untersuchung der Principien der Erkenntniss und*

der Methoden wissenschaftlicher Forschung I: Erkenntnisslehre [Logic. An Investigation of the Principles of Knowledge and of the Methods of Scientific Inquiry I: Theory of Knowledge]. Second revised edition. Stuttgart.

Yotsukura, S., 1970, *The Articles in English. A Structural Analysis of Usage.* The Hague.

Zelený, J., 1970, *Die Wissenschaftslogik bei Marx und "Das Kapital"* [Marx's Logic of Science and "Das Kapital"]. Frankfurt, Vienna.

Ziehen, Th., 1920, *Lehrbuch der Logik auf positivistischer Grundlage mit Berücksichtigung der Geschichte der Logik* [A Textbook of Logic on a Positivistic Basis, with a Survey of the History of Logic], Bonn.

INDEX OF PROPER NAMES

INDEX OF SUBJECTS

as 61 see also: qua
as far as, see: in so far as
as such 14, 133–137, 140, 157, 182, 195, 221, 243, 266, 334f, 341, 345, 444, 460 see also: *als solches, comme tel*
ascensus 360, 473
ascent, ascending 320, 371f, 393, 445, 456, 475
aspect 278, 283, 364, 429f, 455, 460, 465 see also: respect; side; logical space
assertoric judgment 192f, 432
– premiss 277
astrology 19
astronomy 255f
asymmetric(-al) 237, 476 see also: anti-symmetrical
– predicate 96, 226f, 248, 285
– relation 227, 235, 264, 319, 330, 333, 369, 456
asymmetry 236, 284, 333f, 369
atomic 176f, 401f, 404
– predicate 245f, 251
– proposition 10f, 142
attribution theory 205, 207, 213 see also: inherence theory of the copula
Aufhebung 314, 316f, 320, 322ff, 333, 361 see also: degree; *Erhebung*
äusserlich 170
aut-junctive, *aut*-junction 271, 322f, 327, 418, 420, 423, 426, 430, 432, 441, 450 see also: disjunction (exclusive)
auxiliary rule of concept comparison 338, 340f, 346f, 356
average case 117, 197
axioma particulare 102f
– *proprium* 102f, 141

bad (specimen) 325f, 328, 335 see also: hybrid
Bamalip 19ff, 391f, 407
Barbara 150, 158, 174, 181f, 363, 375f
beard; see: Meinong's –; Wundt's –
Begriff 131, 214 see also: concept
Behauptungsgewicht 193
being 129, 197, 229, 410f, 415, 470, 479 see also: existence
Beispiel 153, 167, 349f see also: example; *exemplum*

beliebig 25, 190, 421 see also: arbitrary
belief in angels, see: scope of the –
besondere Urteile 106
bestimmte Grössen 110 see also: definite magnitude
beweisdefinit 68
Beziehung, see: relation
biology 45 see also: arbitrary lion, -rose; botany; hybrid; general dog; morphology; signature; taxonomic; *Urpflanze*; zoology
Bramantip 19 see also: Bamalip
botany 330, 372 see also: biology
Burburu 158, 307, 375, 399, 466

calculus alternativus 418
can be (*kann sein*) 432, 434, 437f see also: indefinite proposition; may be; particular proposition; possible being; possible existence
cardinal numbers 318f; – scale 227
categorial displacement 339ff
categorical argument, see: argument (categorical)
– form 21, 33
– judgment 115f, 189, 191ff, 279, 320f, 461f
– proposition 11, 18, 79, 97, 135, 145, 243, 331, 408f, 426
category 367, 375, 384, 426f, 467
cause 136
CE (complete equality) 247, 249–254, 258–261, 270, 345 see also: degree of CE
chain of being 361, 433
change 240, 284, 369 see also: logical stability
circulus vitiosus in definiendo 326
class 16f, 19, 36, 39, 41, 47f, 60, 92, 139, 206f, 212, 215, 274, 353, 383, 388f, 393f, 409, 413f, 418, 432, 454 see also: logic (class)
– abstraction (operator of) 89
– name 151, 467
– product 431
comme tel 134f, 331 see also: *als solches*; as such
commonplace 472 see also: *locus*; *sedes argumenti*; τόπος

logically true 458
- valid, see: valid(ity)
logician 26f
- (essentialist) 354
- (fifteenth century) 146, 154
- (fourteenth century) 99, 157
- (id-) 278, 315, 318, 445
- (idealist) 51
- (medieval) 97
- (modern, contemporary) 9, 13, 29, 32, 36, 49, 52, 63, 96, 150, 171, 205f, 221, 223, 236f, 246, 281f, 334, 370, 373, 389, 395, 398, 410f, 415, 430, 447
- (neo-Platonic) 377
- (NPF) 233
- (phenomenological) 382
- (pure) 127
- (Renaissance) 101, 105, 154, 158, 231, 360, 427, 478
- (seventeenth century) 151
- (sixteenth century) 151, 154
- (thirteenth century) 83, 97, 99
- (traditional) 9, 17, 45, 61, 69, 75, 116, 121, 128, 150, 177, 185, 205, 209, 226, 230, 238, 247, 252, 266, 269, 270f, 273, 276, 278, 293, 305, 317, 320, 345f, 370, 383, 394, 396f, 408, 418, 428, 431, 450f, 460ff, 465f, 470, 472
- (what-) 241
logicist 123
Logik 7f, 10f, 19, 31, 43, 60, 114, 308
logistic (*Logistik*) 8ff, 13, 19, 31, 224, 234, 369, 453
logique de l'identité 225
- *de l'inhérence* 15, 208, 467 see also: logic (what-)
- *de la rélation* 15, 208 see also: logic (relating)
logophore 35f, 59, 138, 209, 241, 243, 328, 330, 446, 451
logophoric 41, 48, 58, 185, 276, 313, 354, 392, 399, 428, 453, 479
- article, see: article (logophoric)
- concept 37, 303, 421
- entity 47, 364
- expression 39
- indefinite major premiss 294

- – judgment 468
- – proposition 199, 323, 458
- judgment 38, 40, 45f, 59–63, 67f, 110, 186, 188, 224, 324, 373, 385, 399, 443, 458, 460, 467f, 477
- language 50, 285
- logic, see: logic (logophoric)
- meaning 40, 184, 331
- operator 59, 68
- proposition 38ff, 199, 263, 302, 325, 458, 460, 467, 480
- statement 429
- term 40, 44, 48, 50, 65, 180, 213, 243, 266, 314, 324, 329, 424, 451
- tautology 44
- use 38f, 41, 44f, 50, 58, 61, 95, 198, 202, 273, 371, 458f
- value judgment 94, 131
logos 27, 36, 323, 431f
- -level 271

major (premiss) 102, 104f, 143, 163, 174, 183f, 186, 191f, 263, 294, 297f, 300, 303, 313f, 316ff, 320, 322ff, 331, 335, 351, 354, 356ff, 364, 367, 458, 460
many-place predicates 22, 198, 237
Marxist 230, 236, 286, 316
mass propositions 185
material logic, see logic (material)
materialiter loquendo 135
mathematic/-al, -s 9, 23, 26, 45, 49, 55, 68, 108, 110, 123, 126, 168, 227, 411, 413, 419, 469, 475
matière 213
- *contingente* 109, 120f
- *de doctrine* 121
- *morale* 123
- *nécessaire* 109, 120f
may be 432–436, 438, 441, 443f see also: can be (*kann sein*); indefinite proposition; particular proposition; possible being
meaning 17, 43, 45, 57, 65f, 71, 80, 114, 116, 119, 193, 200, 205, 207, 212f, 220, 233, 244, 263, 269, 304, 340, 350, 357, 368, 381, 384, 428, 463 see also: reference; sense
Meinong's beard 4, 377

- premiss 20, 53, 143, 147, 150, 152, 174, 350, 355, 361, 409, 461
- proposition 23, 32, 52, 56, 70, 76, 103–106, 113, 120–125, 127ff, 131, 144f, 150, 156, 161, 164, 166, 170f, 173ff, 178, 198f, 205ff, 212f, 306, 350, 353, 360, 365–368, 378, 382, 392, 399, 406, 408, 413, 423, 455, 458–462, 471, 477
- quantifier 50, 52, 67f, 70, 398, 411, 415, 434
- statement 391, 434f
- subject 77
- term 48, 107, 174
universalité métaphysique 119, 122f; – *morale* 122f
Universalurteil 115f, 139, 189
universe of discourse 21, 45, 51f, 63–67, 108, 389, 409ff, 415
see also: domain
unordered couple 237
Urpflanze 442f, 456
Urteilsrelation 205 see also: judgment (relation of)
ut 133

valid/-ity (in-) 21, 32f, 56ff, 60, 67, 85, 103, 105, 136, 143, 146, 148ff, 152, 154, 156f, 164, 166, 168, 170, 174f, 181–187, 192, 207, 223f, 229, 254, 261, 263, 292, 294, 299f, 309, 324f, 331, 350, 353f, 361, 363, 393, 396, 406f, 409f, 414, 437f, 446, 457, 460, 462, 465, 468, 474
value 5, 96
- of the copula 309
- -jelly 241, 285
- judgment 94, 332, 466f
variable 42, 53, 57, 145, 151, 176, 179,

182, 190, 200, 247f, 254, 265, 318, 344, 357, 378, 383, 387, 407, 420, 428, 433, 438, 461, 463, 467, 470, 476
- copula 192–196, 199, 459
- number 47, 53
Variation (method of) 196f
vector 251
vel-junct/-ion, -ive 415, 418 see also: disjunction (inclusive)
Vergessbarkeitsprinzip, see: principle of forgettability
verneinende Urteile 107, 121, 477
Vernunftlehre 308
vis universalis 161, 182, 267, 285, 306, 329, 351, 356f, 399 see also: logical force
Voraussetzung 356
Vorstellung 214, 387
Vorurteil 315f, 350 see also: presupposition

warheitsdefinit 479
Way Out 221 see also: escape clause
wertdefinit 68, 479
Wesen 34, 82, 196f, 210, 260, 324 see also: essence
Wesensforschung 197
Wesensinduktion 372
Wesenskern 196, 472 see also: logical gene
What 13, 27, 31, 37, 140, 241, 330, 415 see also: essence; logic (what-)
- statement 332, 467
woof 325
Wundt's Beard 340

z 431; -operator 441
zoolog/-ical, -y 20, 452, 456
zoo-method 456

SYNTHESE HISTORICAL LIBRARY

Texts and Studies
in the History of Logic and Philosophy

Editors:

N. KRETZMANN (Cornell University)
G. NUCHELMANS (University of Leyden)
L. M. DE RIJK (University of Leyden)

1. M. T. BEONIO-BROCCHIERI FUMAGALLI, *The Logic of Abelard.* Translated from the Italian. 1969, IX + 101 pp.

2. GOTTFRIED WILHELM LEIBNITZ, *Philosophical Papers and Letters.* A selection translated and edited, with an introduction, by Leroy E. Loemker. 1969, XII + 736 pp.

3. ERNST MALLY, *Logische Schriften,* ed. by Karl Wolf and Paul Weingartner. 1971, X + 340 pp.

4. LEWIS WHITE BECK (ed.), *Proceedings of the Third International Kant Congress.* 1972, XI + 718 pp.

5. BERNARD BOLZANO, *Theory of Science,* ed. by Jan Berg. 1973, XV + 398 pp.

6. J. M. E. MORAVCSIK (ed.), *Patterns in Plato's Thought. Papers arising out of the 1971 West Coast Greek Philosophy Conference.* 1973, VIII + 212 pp.

7. NABIL SHEHABY, *The Propositional Logic of Avicenna: A Translation from al-Shifā': al-Qiyās,* with Introduction, Commentary and Glossary. 1973, XIII + 296 pp.

8. DESMOND PAUL HENRY, *Commentary on De Grammatico: The Historical-Logical Dimensions of a Dialogue of St. Anselm's.* 1974, IX + 345 pp.

9. JOHN CORCORAN, *Ancient Logic and Its Modern Interpretations.* 1974, X + 208 pp.

SYNTHESE LIBRARY

Monographs on Epistemology, Logic, Methodology,
Philosophy of Science, Sociology of Science and of Knowledge, and on the
Mathematical Methods of Social and Behavioral Sciences

Editors:

DONALD DAVIDSON (The Rockefeller University and Princeton University)
JAAKKO HINTIKKA (Academy of Finland and Stanford University)
GABRIËL NUCHELMANS (University of Leyden)
WESLEY C. SALMON (University of Arizona)

17. NICHOLAS RESCHER, *Topics in Philosophical Logic*. 1968, XIV + 347 pp.
18. ROBERT S. COHEN and MARX W. WARTOFSKY (eds.), *Proceedings of the Boston Colloquium for the Philosophy of Science 1966–1968*, Boston Studies in the Philosophy of Science (ed. by Robert S. Cohen and Marx W. Wartofsky), Volume IV. 1969, VIII + 537 pp.
19. ROBERT S. COHEN and MARX W. WARTOFSKY (eds.), *Proceedings of the Boston Colloquium for the Philosophy of Science 1966–1968*, Boston Studies in the Philosophy of Science (ed. by Robert S. Cohen and Marx W. Wartofsky), Volume V. 1969, VIII + 482 pp.
20. J. W. DAVIS, D. J. HOCKNEY, and W. K. WILSON (eds.), *Philosophical Logic*, 1969, VIII + 277 pp.
21. D. DAVIDSON and J. HINTIKKA (eds.), *Words and Objections: Essays on the Work of W. V. Quine*, 1969. VIII + 366 pp.
22. PATRICK SUPPES, *Studies in the Methodology and Foundations of Science. Selected Papers from 1911 to 1969.* 1969, XII + 473 pp.
23. JAAKKO HINTIKKA, *Models for Modalities. Selected Essays.* 1969, IX + 220 pp.
24. NICHOLAS RESCHER *et al.* (eds.), *Essay in Honor of Carl G. Hempel. A Tribute on the Occasion of his Sixty-Fifth Birthday.* 1969, VII + 272 pp.
25. P.V. TAVANEC (ed.), *Problems of the Logic of Scientific Knowledge.* 1969, XII + 429 pp.
26. MARSHALL SWAIN (ed.), *Induction, Acceptance, and Rational Belief.* 1970, VII + 232 pp.
27. ROBERT S. COHEN and RAYMOND J. SEEGER (eds.), *Ernst Mach: Physicist and Philosopher*, Boston Studies in the Philosophy of Science (ed. by Robert S. Cohen and Marx W. Wartofsky), Volume VI. 1970, VIII + 295 pp.
28. JAAKKO HINTIKKA and PATRICK SUPPES, *Information and Inference.* 1970, X + 336 pp.
29. KAREL LAMBERT, *Philosophical Problems in Logic. Some Recent Developments.* 1970, VII + 176 pp.
30. ROLF A. EBERLE, *Nominalistic Systems.* 1970, IX + 217 pp.
31. PAUL WEINGARTNER and GERHARD ZECHA (eds.), *Induction, Physics, and Ethics, Proceedings and Discussions of the 1968 Salzburg Colloquium in the Philosophy of Science.* 1970, X + 382 pp.
32. EVERT W. BETH, *Aspects of Modern Logic.* 1970, XI + 176 pp.
33. RISTO HILPINEN (ed.), *Deontic Logic: Introductory and Systematic Readings.* 1971, VII + 182 pp.
34. JEAN-LOUIS KRIVINE, *Introduction to Axiomatic Set Theory.* 1971, VII + 98 pp.
35. JOSEPH D. SNEED, *The Logical Structure of Mathematical Physics.* 1971, XV + 311 pp.
36. CARL R. KORDIG, *The Justification of Scientific Change.* 1971, XIV + 119 pp.
37. MILIČ ČAPEK, *Bergson and Modern Physics*, Boston Studies in the Philosophy of Science (ed. by Robert S. Cohen and Marx W. Wartofsky), Volume VII. 1971, XV + 414 pp.
38. NORWOOD RUSSELL HANSON, *What I do not Believe, and other Essays* (ed. by Stephen Toulmin and Harry Woolf). 1971, XII + 390 pp.
39. ROGER C. BUCK and ROBERT S. COHEN (eds.), *PSA 1970. In Memory of Rudolf Carnap*, Boston Studies in the Philosophy of Science (ed. by Robert S. Cohen and Marx W. Wartofsky), Volume VIII. 1971, LXVI + 615 pp. Also available as a paperback.
40. DONALD DAVIDSON and GILBERT HARMAN (eds.), *Semantics of Natural Language.* 1972, X + 769 pp. Also available as a paperback.
41. YEHOSUA BAR-HILLEL (ed.), *Pragmatics of Natural Languages.* 1971, VII + 231 pp.
42. SÖREN STENLUND, *Combinators, λ-Terms and Proof Theory.* 1972, 184 pp.

43. MARTIN STRAUSS, *Modern Physics and Its Philosophy. Selected Papers in the Logic, History, and Philosophy of Science.* 1972, X + 297 pp.
44. MARIO BUNGE, *Method, Model and Matter.* 1973, VII + 196 pp.
45. MARIO BUNGE, *Philosophy of Physics.* 1973, IX + 248 pp.
46. A. A. ZINOV'EV, *Foundations of the Logical Theory of Scientific Knowledge (Complex Logic)*, Boston Studies in the Philosophy of Science (ed. by Robert S. Cohen and Marx W. Wartofsky), Volume IX. Revised and enlarged English edition with an appendix, by G. A. Smirnov, E. A. Sidorenka, A. M. Fedina, and L. A. Bobrova. 1973, XXII + 301 pp. Also available as a paperback.
47. LADISLAV TONDL, *Scientific Procedures*, Boston Studies in the Philosophy of Science (ed. by Robert S. Cohen and Marx W. Wartofsky), Volume X. 1973, XII + 268 pp. Also available as a paperback.
48. NORWOOD RUSSELL HANSON, *Constellations and Conjectures* (ed. by Willard C. Humphreys, Jr.), 1973, X + 282 pp.
49. K. J. J. HINTIKKA, J. M. E. MORAVCSIK, and P. SUPPES (eds.), *Approaches to Natural Language. Proceedings of the 1970 Stanford Workshop on Grammar and Semantics.* 1973, VIII + 526 pp. Also available as a paperback.
50. MARIO BUNGE (ed.), *Exact Philosophy – Problems, Tools, and Goals.* 1973, X + 214 pp.
51. RADU J. BOGDAN and ILKKA NIINILUOTO (eds.), *Logic, Language, and Probability.* A selection of papers contributed to Sections IV, VI, and XI of the Fourth International Congress for Logic, Methodology, and Philosophy of Science, Bucharest, September 1971. 1973, X + 323 pp.
52. GLENN, PEARCE and PATRICK MAYNARD (eds.), *Conceptual Change.* 1973, XII + 282 pp.
53. ILKKA NIINILUOTO and RAIMO TUOMELA, *Theoretical Concepts and Hypothetico-Inductive Inference.* 1973, VII + 264 pp.
54. ROLAND FRAÏSSÉ, *Course of Mathematical Logic – Volume I: Relation and Logical Formula.* 1973, XVI + 186 pp. Also available as a paperback.
55. ADOLF GRÜNBAUM, *Philosophical Problems of Space and Time.* Second, enlarged edition, Boston Studies in the Philosophy of Science (ed. by Robert S. Cohen and Marx W. Wartofsky), Volume XII. 1973, XXIII + 884 pp. Also available as a paperback.
56. PATRICK SUPPES (ed.), *Space, Time, and Geometry.* 1973, XI + 424 pp.
57. HANS KELSEN, *Essays in Legal and Moral Philsosphy*, selected and introduced by Ota Weinberger, 1973, XXVIII + 300 pp.
58. R. J. SEEGER and ROBERT S. COHEN (eds.), *Philosophical Foundations of Science. Proceedings of an AAAS Program, 1969.* Boston Studies in the Philosophy of Science (ed. by Robert S. Cohen and Marx W. Wartofsky), Volume XI. 1974, IX + 545 pp. Also available as paperback.
59. ROBERT S. COHEN and MARX W. WARTOFSKY (eds.), *Logical and Epistemological Studies in Contemporary Physics*, Boston Studies in the Philosophy of Science (ed. by Robert S. Cohen and Marx W. Wartofsky), Volume XIII. 1973, VIII + 462 pp. Also available as paperback.
60. ROBERT S. COHEN and MARX W. WARTOFSKY (eds.), *Methodological and Historical Essays in the Natural and Social Sciences. Proceedings of the Boston Colloquium for the Philosophy of Science, 1969–1972*, Boston Studies in the Philosophy of Science (ed. by Robert S. Cohen and Marx W. Wartofsky), Volume XIV. 1974, VIII + 405 pp. Also available as paperback.

63. SÖREN STENLUND (ed.), *Logical Theory and Semantic Analysis. Essays Dedicated to Stig Kanger on His Fiftieth Birthday.* 1974, V + 217 pp.
64. KENNETH SCHAFFNER and ROBERT S. COHEN (eds.), *Proceedings of the 1972 Biennial Meeting, Philosophy of Science Association,* Boston Studies in the Philosophy of Science (ed. by Robert S. Cohen and Marx W. Wartofsky), Volume XX. Also available as paperback.
65. HENRY E. KYBURG, JR., *The Logical Foundations of Statistical Inference.* 1974, IX + 421 pp.
66. MARJORIE GRENE, *The Understanding of Nature: Essays in the Philosophy of Biology,* Boston Studies in the Philosophy of Science (ed. by Robert S. Cohen and Marx W. Wartofsky), Volume XXIII. 1974, XII + 360 pp. Also available as paperback.

In Preparation

61. ROBERT S. COHEN and MARX W. WARTOFSKY (eds.), *For Dirk Struik. Scientific, Historical Essays in Honor of Dirk J. Struik,* Boston Studies in the Philosophy of Science (ed. by Robert S. Cohen and Marx W. Wartofsky, Volume XV. Also available as paperback.
62. KAZIMIERZ AJDUKIEWICZ, *Pragmatic Logic,* transl. from the Polish by Olgiero Wojtasiewicz.
67. JAN M. BROEKMAN, *Structuralism: Moscow, Prague, Paris.*
68. NORMAN GESCHWIND, *Selected Papers on Language and the Brain,* Boston Studies in the Philosophy of Science (ed. by Robert S. Cohen and Marx W. Wartofsky), Volume XVI. Also available as paperback.
69. ROLAND FRAÏSSÉ. *Course of Mathematical Logic* – Volume II: *Model Theory.*